EMBEDDED IMAGE PROCESSING ON THE TMS320C6000™ DSP

Examples in Code Composer Studio™ and MATLAB

EMBEDDED IMAGE PROCESSING ON THE TMS320C6000™ DSP

Examples in Code Composer Studio™ and MATLAB

Shehrzad Qureshi

Shehrzad Qureshi
Labcyte Inc., Palo Alto, CA
USA

Embedded Image Processing on the TMS320C6000™ DSP
Examples in Code Composer Studio™ and MATLAB

Library of Congress Cataloging-in-Publication Data

A C.I.P. Catalogue record for this book is available
from the Library of Congress.

ISBN 0-387-25280-0   e-ISBN 0-387-25281-9   Printed on acid-free paper.
ISBN 978-0387-25280-3

Printed in the United States of America.

9 8 7 6 5 4 3 2 1   SPIN 11055570

springeronline.com

TRADEMARKS

The following list includes commercial and intellectual trademarks belonging to holders whose products are mentioned in this book. Omissions from this list are inadvertent.

Texas Instruments, TI, Code Composer Studio, RTDX, TMS320C6000, C6000, C62x, C64x, C67x, DSP/BIOS and VelociTI are registered trademarks of Texas Instruments Incorporated.
MATLAB is a registered trademark of The MathWorks, Inc.
ActiveX, DirectDraw, DirectX, MSDN, Visual Basic, Win32, Windows, and Visual Studio are trademarks of Microsoft.
Intel, MMX, Pentium, and VTune are trademarks or registered trademarks of Intel Corporation or its subsidiaries in the United States and other countries.

DISCLAIMER

Product information contained in this book is primarily based on technical reports and documentation and publicly available information received from sources believed to be reliable. However, neither the author nor the publisher guarantees the accuracy and completeness of information published herein. Neither the publisher nor the author shall be responsible for any errors, omissions, or damages arising out of use of this information. No information provided in this book is intended to be or shall be construed to be an endorsement, certification, approval, recommendation, or rejection of any particular supplier, product, application, or service.

CD-ROM DISCLAIMER

Dedication

*To my family, for instilling in
me the work ethic that made
this book possible, and Lubna,
for her continued support.*

Contents

Preface

The question might reasonably be asked when first picking up this book – why yet another image processing text when there are already so many of them? While most image processing books focus on either theory or implementation, this book is different because it is geared towards *embedded* image processing, and more specifically the development of efficient image processing algorithms running on the Texas Instruments (TI) TMS3206000™ Digital Signal Processor (DSP) platform. The reason why I wrote this book is that when I first started to search for material covering the TMS3206000 platform, I noticed that there was little focus on image processing, even though imaging is an important market base that TI targets with this particular DSP family. To be sure, there are plenty of books that explain how to implement one-dimensional signal processing algorithms, like the type found in digital communications and audio processing applications. And while I found a chapter here or there, or a lingering section that mentioned some of the techniques germane to image processing, I felt that a significant portion of the market was not being sufficiently addressed. For reasons that will hopefully become apparent as you read this book, image processing presents its own unique challenges and it is my sincere hope that you find this book helpful in your embedded image processing endeavors.

For a myriad of reasons, implementing data intensive processing routines, such as the kind typified by image processing algorithms, on embedded platforms presents numerous issues and challenges that developers who travel in the "workstation" or "desktop" realm (i.e. UNIX or Wintel platforms) typically do not need to concern themselves with. To illustrate just a few of these issues, consider the task of implementing an efficient two-dimensional Discrete Wavelet Transform (DWT) on a DSP, a topic covered extensively in Chapter 6 of this book. Such an algorithm might be needed in a digital camera to save images in the JPEG-2000 format or in a software-based MPEG4 system – wavelet coding of images has in the past few years supplanted the discrete cosine transform as the transform of choice for state-of-the-art image compression.

The DWT, like many other similar transforms commonly encountered in signal and image processing, is a *separable* transform. This means that the two-dimensional (2D) form of the transform is computed by generalizing the one-dimensional (1D) form of it – in other words, by first performing the 1D transform along one dimension (for example, each of the individual rows in the image), and then rerunning the transform on the output of the first transform in the orthogonal direction (i.e., the columns of the transformed rows). A software implementation of a 2D separable transform is relatively straightforward, if a routine exists that performs the 1D transform on a vector of data. Such code would most likely march down each row, invoking the aforementioned routine for each row of image data, thus resulting in a matrix of row-transformed coefficients. This temporary image result could then be *transposed* (the rows become the columns and vice-versa), and the same process run again, except this time as the algorithm marches down the row dimension, the columns are now transformed. Lastly, a second transposition reshuffles the image pixels back to their original orientation, and the end result is the 2D transform. Both the 2D DWT and the 2D Fast Fourier Transform (FFT) can be computed in such a manner.

This implementation strategy has the benefit of being fairly easy to understand. The problem is that it is also terribly inefficient. The two transposition operations consume processor cycles, and moreover lead to increased memory consumption, because matrix transposition requires a scratch array. On typical desktop platforms, with their comparatively huge memory footprint and multi-gigahertz clock frequencies, who cares? On these platforms, it is oftentimes possible to get away with such first-cut implementations, and still obtain more than acceptable performance. Of course, this is not always going to be the case but the name of the game in developing efficient code is to optimize only if needed, and only where needed. Thus if the initial implementation is fast enough to meet the desired specifications, there is little to gain from making algorithmic and subsequent low-level optimizations.

The contrast is quite stark in the embedded DSP world however; embedded DSP cores increasingly find their way into real-time systems, with hard deadlines that must be met. And if the system happens to not be real-time, they are quite often memory and/or resource constrained, perhaps to keep costs and power consumption down to acceptable levels (consider, for example, a camera cell-phone). As a consequence, the situation here is that there is typically far less leeway – the clock speeds are somewhat slower, and memory inefficiencies tend to have an amplified effect on performance. Hence memory, and especially fast on-chip memory (which comes at a premium) absolutely must be managed correctly. With respect to the 2D DWT example, the "canonical" form of the algorithm as just described can be altered so that it produces identical output, but does not require the use of a matrix transposition. Such an optimization is described in Chapter 6. As one would expect, as these types of optimizations are incorporated into an algorithm, its essential nature tends to become clouded with the details of the tricks played to coax more performance out of an implementation. But that is the price one pays for speed – clarity suffers, at least with respect to a "reference" implementation.

And it only gets more involved as the design process continues. Images of any reasonable size will not fit in on-chip RAM, which has just been identified as a rate-limiting factor in algorithm performance. Because the latencies involved with accessing off-chip RAM are so severe, an optimal 2D DWT transform should incorporate strategies to circumvent this problem. This optimization exploits spatial locality by shuttling blocks of data between internal and external RAM. For example, as individual rows or a contiguous block of an image is needed, they should be copied into internal RAM, transformed, and then sent back out to external RAM. This process would then continue to the next block, until the entire image has been transformed. A likely response from seasoned C/C++ developers would probably be to use the venerable memcpy function to perform these block memory copies. As it turns out, in the fully optimized case one should use Direct Memory Access (DMA) to increase the speed of the block memory copy, which has the added benefit of freeing the processor for other duties. Taking matters to the extreme, yet another optimization may very well entail interleaving the data transfer and data processing. First, one might set up the DMA transfer, and then process a block of data while the DMA transfer is occurring in the background. Then, when the background DMA transfer completes, process this new set of data while simultaneously sending the just processed block of data back out to off-chip memory, again via DMA. This procedure would continue until all image blocks have been processed.

As is evident from this description, what was once a reasonably straightforward algorithm implementation has quickly become obfuscated with a myriad of memory usage concerns. And memory is not the end of this story! Many DSPs are "fixed-point" devices, processors where floating-point calculations are to be avoided because they must be implemented in software. Developing fixed-point algorithm implementations, where any floating-point calculations are carried out using integer representations of numbers and the decimal point is managed by the programmer, opens up a whole new can of worms. The algorithm developer must now contend with a slew of new issues, such as proper data scaling, quantization effects of filter coefficients, saturation and overflow, to name just a few. So not only has our optimized 2D DWT algorithm been tweaked so that it no longer requires matrix transpositions and is memory efficient through the usage of DMA block memory copies, but any floating-point data is now treated as bit-shifted integer numbers that must be managed by the programmer.

And even still, there are many other issues that the preceding discussion omits, for example the inclusion of assembly language, vendor-specific compiler intrinsics, and a real-time operating system, but the general idea should now be clear. Even just a few of these issues can turn a straightforward image processing algorithm into a fairly complicated implementation. Taken together and all at once, these concerns may seem overwhelming to the uninitiated. Bridging this gap, between the desktop or workstation arena and the embedded world, is this book's raison d'etre. Developers must fully understand not only the strengths and weaknesses of the underlying technology, but also the algorithms and applications to the fullest extent in order to implement them on a DSP architecture in a highly-optimized form. I come from an image processing background, having developed numerous production-quality algorithms and software spanning the gamut of environments – from embedded DSPs to dual Pentium and SGI workstations – and my primary goal is to ease the transition to the embedded DSP world, which as evidenced by this case study presents itself with a set of very unique challenges.

Another motivation for my writing this book, aside from the fact that there is not currently a book on the market covering embedded image processing, is to shed some light on the "black magic" that seems to accompany embedded DSP development. In comparison to developing non-embedded software, during the development of the software that accompanies this book I was continually stymied by one annoyance or another. Certain operations that would work with one DSP development platform seemingly would not work on another, and vice-versa. Or some simple examples from the TI online help tutorial would not work without certain modifications made to linker files, build settings, or the source code. There were so many issues that to be truthfully honest I have lost track of many of them. I hope that reading this book and using the code and projects that are on the CD-ROM will help you in that ever important, yet elusive, quest for "time to market."

Shehrzad Qureshi
shehrzad_q@hotmail.com

Acknowledgments

There are many fine people who had a hand in this work, and I would like to acknowledge their effort and support. First and foremost, my soon-to-be wife, Lubna, provided invaluable feedback with regards to the composition of the book and spent countless hours reviewing drafts of the material. Without her constant encouragement and total understanding, it is doubtful this book would have ever seen the light of day and for that, I am forever grateful. I have also been blessed to have had the pleasure of working with many brilliant engineers and scientists at various employment stops along the way, and without them to learn from I could never have developed my love of algorithms and programming. I thank Steve Ling and others at Texas Instruments for their generosity in lending DSP development boards that made this book possible, as well as Cathy Wicks for her help along the way. I sincerely appreciate the support extended to me by the folks at The MathWorks, in particular Courtney Esposito who disseminated early drafts of the manuscript for technical review. Finally, I would like to thank Springer for their continued support in this endeavor, especially Melissa Guasch, Deborah Doherty, and Alex Greene for their help in the final preparation of the manuscript.

Chapter 1

INTRODUCTION

When engineers or scientists refer to an image, they are typically speaking of an optical representation of a scene acquired using a device consisting of elements excited by some light source. When this scene is illuminated by a light source, these elements subsequently emit electrical signals that are digitized to form a set of "picture elements" or *pixels*. Together these pixels make up a *digital* image. Many of these devices are now driven by Digital Signal Processors, for reasons explained in 1.6. The imaging device may take the form of a camera, where a photographic image is created when the objects in the scene are illuminated by a natural light source such as the sun, or an artificial light source, such as a flash. Another example is an x-ray camera, in which case the "scene" is typically some portion of the human body or a dense object (e.g., luggage in an airport security system), with the light source consisting of x-ray beams. There are many other examples, some of which do not typically correspond to what most people think of as an image. For example, in the life sciences and biomedical fields there are numerous devices and instruments that can be thought of as cameras in some sense, where the acquisition detectors are photodiodes excited by some type of infrared light. This book describes image processing algorithms that operate on all sorts of images, and provides numerous implementations of such algorithms targeting the Texas Instruments (TI) TMS320C6000™ DSP platform. Prior to embarking on this journey, this first chapter introduces the structure of the book and the representation of digital images, and the second chapter provides background information on the tools used to develop image processing algorithms.

1.1 STRUCTURE AND ORGANIZATION OF THE BOOK

Because the whole thrust of this book are *efficient* implementations of image processing algorithms running on *embedded* DSP systems, it is not sufficient to simply present an algorithm and describe a first-cut implementation that merely produces the correct output. The primary goal is efficient algorithm implementations, while a secondary goal is to learn how to utilize the appropriate TI technologies that aid in the development and debugging of code that can be used in real-world applications. Achieving these goals takes time, and as such we are somewhat constrained by space. As a result, this book is *not* intended to be a complete coverage of image processing theory and first principles, for that the reader is referred to [1] or [2]. However, what you will find is that while such books may give you the mathematical and background knowledge for understanding how various image processing algorithms work in an abstract sense, the transition from theory to implementation is a jump that deserves more attention than is typically given. In particular, taking a description of an image processing algorithm and coding an efficient implementation in the C language on an embedded fixed-point and resource-constrained DSP is not for the faint of heart. Nowadays, given the proliferation of a variety of excellent "rapid-generation" high-level technical computing environments like MATLAB® (especially when coupled with the Image Processing Toolbox) and various software libraries like the Intel® Integrated Performance Primitives, it is not overtly difficult to put together a working image processing prototype in fairly short order. We will use both of the aforementioned software packages in our quest for embedded DSP image processing, but bear in mind that it is a windy road.

The meat of this book is split amongst Chapters 3-6, with some ancillary material appearing in the two appendices. This chapter contains introductory material and Chapter 2 is important background information on the various tools employed throughout the rest of the book. Chapters 3-6 roughly cover four general categories of image processing algorithms:

1. **Chapter 3**: image enhancement via spatial processing techniques (point-processing).
2. **Chapter 4**: image filtering (linear, non-linear, and adaptive).
3. **Chapter 5**: image analysis (edge-detection and segmentation).
4. **Chapter 6**: wavelets (with applications to edge detection and image enhancement).

Due to the challenging nature of embedded development, the strategy is to start off simple and then progressively delve deeper and deeper into the intricate implementation details, with the end game always being an efficient algorithm running on the DSP. The book follows a "cookbook" style, where an image processing algorithm is first introduced in its theoretical context. While this is first and foremost a practitioner's book, it goes without saying that a solid understanding of the theoretical underpinnings of any algorithm is critical to achieving a good implementation on a DSP. After the theoretical groundwork has been laid, examples in MATLAB are used to drive home the core concepts behind the algorithm in question, without encumbering the reader with the details that inevitably will follow. Depending on the situation, it may be the case that the MATLAB code is ported to C/C++ using Visual Studio .NET 2003. These Windows applications allow for interactive visualization and prove invaluable as parallel debugging aids, helping to answer the often posed question "Why doesn't my C code work right on my DSP, when it works just fine on the PC?" And of course, because this book is primarily targeted at those who are implementing embedded image processing algorithms, each algorithm is accompanied by an implementation tested and debugged on either the C6701 Evaluation Module (EVM) or C6416 DSP Starter Kit (DSK). Both of these DSP platforms are introduced in 2.1.3. It should be noted that majority of the TI image-processing implementations that accompany this book use the C6416 DSK, as it contains a more recent DSP and TI appears to be phasing out the EVM platform.

1.2. PREREQUISITES

In order to get the most out of this book, it is expected that the reader is reasonably fluent in the C language and has had some exposure to MATLAB, C++, and the TI development environment. If the reader is completely new to embedded development on the TI DSP platform, but does have some experience using Microsoft Visual Studio or a similar integrated development environment (IDE), it should not be too difficult to pick up Code Composer Studio™ (CCStudio). CCStudio is heavily featured in this book and is TI's answer to Visual Studio. While one can always fall back to command-line compilation and makefiles, CCStudio incorporates many advanced features that make programming and debugging such a joy, as compared to the days of gdb, gcc, vi, and emacs.

One does not need to be an expert C++ programmer in order to make sense of the Visual Studio projects discussed in the book. The use of C++ is purposely avoided on the DSP, however it is the language of choice for many programmers building scientific and engineering applications on Windows and UNIX workstations. For high-performance, non-embedded image processing, I would hasten to add that it is the only choice, perhaps leaving room for some assembly if need be. Nevertheless, I have eschewed many cutting-edge C++ features – about the most exotic C++ code one will encounter is the use of namespaces and perhaps a sprinkling of the Standard Template Library (STL). An understanding of what a class and method are, along with some C++ basics such as exception handling and the standard C++ library is all that is required, C++-wise.

The source code for all Visual Studio projects utilize Microsoft Foundation Classes (MFC) and GDI+ for their GUI components (see 2.4). For the most part, the layout of the code is structured such that one need not be a Windows programming guru in order to understand what is going on in these test applications. Those that are not interested in this aspect of development can ignore MFC and GDI+ and simply treat that portion of the software as a black box. Any in-depth reference information regarding Microsoft technologies can always be found in the Microsoft Developer's Network (MSDN)[3].

Lastly, a few words regarding mathematics, and specifically signal processing. In this book, wherever image processing theory is presented, it is just enough so that the reader is not forced to delve into an algorithm without at least the basics in hand. Mainly this is due to space constraints, as the more theory that is covered, the fewer algorithms that can fit into a book of this size. When it comes right down to it, in many respects image processing is essentially one-dimensional signal processing extended to two dimensions. In fact, it is often treated as an offshoot of one-dimensional signal processing. Unfortunately, there is not enough space to thoroughly cover the basics of one-dimensional DSP applications and so while some signal processing theory is covered, the reader will gain more from this book if they have an understanding of basic signal processing topics such as convolution, sampling theory, and digital filtering. Texts covering such one-dimensional signal processing algorithms with applications to the same TI DSP development environments utilized in this book include [4-6].

1.3 CONVENTIONS AND NOMENCLATURE

Many of the featured image processing algorithms are initially illustrated in pseudo-code form. The "language" used in the pseudo-code to describe the algorithms is not formal by any means, although it does resemble procedural languages like C/C++ or Pascal. For these pseudo-code listings, I have taken the liberty of loosely defining looping constructs and high-level assignment operators whose definitions should be self-explanatory from the context in which they appear. The pseudo-code also assumes zero-based indexing into arrays, a la C/C++.

Any reference to variables, functions, methods, or pathnames appear in a non-proportional `Courier` font, so that they stand out from the surrounding text. In various parts of the book, code listings are given and these listings use a 10-point non-proportional font so they are highlighted from the rest of the text. This same font is used wherever any code snippets are needed. Cascading menu selections are denoted using the pipe symbol and a bold-faced font, for example **File|Save** is the "Save" option under the "File" main menu.

With any engineering discipline, there is unfortunately a propensity of acronyms, abbreviations, and terminology that may seem daunting at first. While in most cases the first instance of an acronym or abbreviation is accompanied by its full name, in lieu of a glossary the following is a list of common lingo and jargon the reader should be familiar with:

- **C6x**: refers to the TMS320C6000 family of DSPs, formally introduced in 2.1. The embedded image processing algorithms implemented in this book target this family of DSPs. C6x is short-hand for the C62x, C67x, and C64x DSPs.
- **IDE**: integrated development environment. The burgeoning popularity of IDEs can be attributed in part to the success of Microsoft Visual Studio and earlier, Borland's Turbo C and Turbo Pascal build systems. IDEs combine advanced source code editors, compilers, linkers, and debugging tools to form a complete build system. In this book we utilize three IDEs (MATLAB, CCStudio, and Visual Studio), although MATLAB is somewhat different in that it is an interpreted language that does not require compilation or a separate linking step.
- **TI**: Texas Instruments, the makers of the C6x DSPs.
- **CCStudio**: abbreviation for the Code Composer Studio IDE, TI's flagship development environment for their DSP products.
- **M-file**: a collection of MATLAB functions, analogous to a C/C++ module or source file.
- **MEX-file**: MATLAB callable C/C++ and FORTRAN programs. The use and development of MEX-files written in C/C++ is discussed in Appendix A.
- **toolbox**: add-on, application-specific solutions for MATLAB that contain a family of related M-files and possibly MEX-files. In this book, the Image Processing Toolbox, Wavelet Toolbox, and Link for Code Composer Studio are all used. For further information on MATLAB toolboxes, see [7].

- **host**: refers to the PC where CCStudio is running. The host PC is connected to a TI DSP development board via USB, PCI, or parallel port.
- **EVM**: an abbreviation for evaluation module, introduced in 2.1.3. The EVM is a PCI board with a TI DSP and associated peripherals used to develop DSP applications. All EVM code in this book was tested and debugged on an EVM containing a single C6701 DSP, and this product is referred to in the text as the C6701 EVM.
- **DSK**: refers to a "DSP starter kit", also introduced in 2.1.3. The DSK is an external board with a TI DSP and associated peripherals, connected to a host PC either via USB or parallel port (see Figure 2-4). All DSK code in this book was tested and debugged on a DSK with a C6416 DSP, and this product is referred to in the text as the C6416 DSK.
- **target**: refers to the DSP development board, either an EVM or DSK.

Finally, a few words regarding the references to TI documentation – the amount of TI documentation is literally enormous, and unfortunately it is currently not located in its entirety in a central repository akin to MSDN. Each TI document has an associated "literature number", which accompanies each reference in this book. The literature number is prefixed with a four-letter acronym – either SPRU for a user manual or SPRA for an application report. Some of these PDFs are included with the stock CCStudio install, but all of them can be downloaded from www.ti.com. For example, a reference to SPRU653.pdf (user manual 653) can be downloaded by entering "SPRU653" in the keyword search field on the TI web-site, if it is not already found within the `docs\pdf` subdirectory underneath the TI install directory (typically `C:\TI`).

1.4 CD-ROM

All the source code and project files described in Chapters 3-6 and the two appendices are included on the accompanying CD-ROM. Furthermore, most of the raw data – images and in one case, video – can also be found on the CD-ROM. The CD is organized according to chapter, and the top-level `README.txt` file and chapter-specific `README.txt` files describe their contents more thoroughly.

The host PC used to build, test, and debug the DSP projects included on the CD-ROM had two versions of CCStudio installed, one for the C6701 EVM and another for the C6416 DSK. As a result, chances are that the projects will not build on your machine without modifications made to the

CCStudio project files. The default install directory for CCStudio is C:\TI, and as the DSP projects reference numerous include files and TI libraries, you will most likely need to point CCStudio to the correct directories on your machine (the actual filenames for the static libraries and header files should remain the same). There are two ways of going about this. One way is to copy the project directory onto the local hard drive, and then open the project in CCStudio. CCStudio will then complain that it cannot find certain entities referenced in the .pjt (CCStudio project) file. CCStudio then prompts for the location of the various files referenced within the project which it is unable to locate. This procedure is tedious and time-consuming and an alternate means of accomplishing the same thing is to directly edit the .pjt file, which is nothing more than an ASCII text file. Listing 1-1 shows the contents of the contrast_stretch.pjt CCStudio (version 2.20) project file, as included on the CD-ROM. The lines in bold are the ones that need to be tailored according to your specific installation.

Listing 1-1: The contents of an example CCStudio project file from Chapter 3, contrast_stretch.pjt. If the default options are chosen during CCStudio installation, the bold-faced directories should be changed to C:\TI – otherwise change them to whatever happens to be on your build machine.

```
CPUFamily=TMS320C67XX
Tool="Compiler"
Tool="DspBiosBuilder"
Tool="Linker"
Config="Debug"
Config="Release"

[Source Files]
Source="C:\TIC6701EVM\c6000\cgtools\lib\rts6701.lib"
Source="C:\TIC6701EVM\c6200\imglib\lib\img62x.lib"
Source="C:\TIC6701EVM\myprojects\evm6x\lib\evm6x.lib"
Source="C:\TIC6701EVM\myprojects\evmc67_lib\Dsp\Lib\devlib\Dev6x.lib"
Source="C:\TIC6701EVM\myprojects\evmc67_lib\Dsp\Lib\drivers\Drv6X.lib"
Source="contrast_stretch.c"
Source="contrast_stretch.cmd"

["Compiler" Settings: "Debug"]
Options=-g -q -fr".\Debug"
-i "C:\TIC6701EVM\myprojects\evm6x\dsp\include"
```

-i "C:\TIC6701EVM\c6200\imglib\include" -d"_DEBUG" -mv6700

["Compiler" Settings: "Release"]
Options=-q -o3 -fr".\Release"
-i "C:\TIC6701EVM\myprojects\evm6x\dsp\include"
-i "C:\TIC6701EVM\c6200\imglib\include" -mv6700

["DspBiosBuilder" Settings: "Debug"]
Options=-v6x

["DspBiosBuilder" Settings: "Release"]
Options=-v6x

["Linker" Settings: "Debug"]
Options=-q -c -m".\Debug\contrast_stretch.map"
-o".\Debug\contrast_stretch.out" -w -x

["Linker" Settings: "Release"]
Options=-q -c -m"."Release"contrast_stretch.map"
-o".\Release\contrast_stretch.out" -w -x

Upon opening one of these projects, if CCStudio displays an error dialog with the message "Build Tools not installed", verify that the project directory is valid. If that does not correct the issue, then the problem is most likely the CPUFamily setting at the beginning of the .pjt file. Chances are that there is a mismatch between the current CCStudio DSP configuration and whatever is defined in the project file. This setting can be changed to whatever is appropriate (i.e. TMS320C64XX), although this modification is fraught with peril. At this point, it would be best to re-create the project from scratch, using the provided project file as a template. See [8] for more details on CCStudio project management and creation.

Some of the Visual Studio .NET 2003 projects on the CD-ROM may also need to be modified in a similar fashion, depending on where the Intel Integrated Performance Primitives Library and CCStudio have been installed. In this case, direct modification of the .vcproj (Visual Studio project) file is not recommended, as the format of these text files is not as simple as .pjt files. Instead, modify the include and library path specifications from within the project properties dialog within Visual Studio, via the **Project|Properties** menu selection.

1.5 THE REPRESENTATION OF DIGITAL IMAGES

Except for a single example in 3.4, this book deals entirely with digital *monochrome* (black-and-white) images, oftentimes referred to as "intensity" images, or "gray-scale" images. A monochrome digital image can be thought of as a discretized two-dimensional function, where each point represents the light intensity at a particular spatial coordinate. These spatial coordinates are usually represented in a Cartesian system as a pair of positive integer values, typically denotes in this book as *(i,j)* or *(x,y)*. The spatial coordinate system favored in this book is one where the first integer *i* or *x* is the row position and the second integer *j* or *y* is the column position, and the origin is the upper left corner of the image. Taken together, the discrete image function *f(i,j)* returns the pixel at the ith row and j^{th} column. Depending on the language, the tuples *(i,j)* or *(x,y)* may be zero-based (C/C++) or one-based (MATLAB) language. A digital image is usually represented as a *matrix* of values, and for example in MATLAB you have

$$f(i,j) = \begin{bmatrix} f(1,1) & f(1,2) & f(1,3) & \cdots & f(1,N) \\ f(2,1) & f(2,2) & f(2,3) & \cdots & f(2,N) \\ \vdots & \vdots & \vdots & & \vdots \\ f(M,1) & f(M,2) & f(M,3) & \cdots & f(M,N) \end{bmatrix}$$

The number of rows in the above image is *M* and the number of columns is *N* – these variable names show up repeatedly throughout the MATLAB code in this book. In the C code, the preprocessor symbols X_SIZE and Y_SIZE refer to the number of rows and number of columns, respectively.

In the C and C++ languages, we represent an image as an array of two dimensions, and since both C and C++ use zero-based indices and brackets to specify a pointer dereference, the C/C++ equivalent to the above image matrix is

$$f(i,j) = \begin{bmatrix} f[0][0] & f[0][1] & f[0][2] & \cdots & f[0][N-1] \\ f[1][0] & f[1][1] & f[1][2] & \cdots & f[1][N-1] \\ \vdots & \vdots & \vdots & & \vdots \\ f[M-1][0] & f[M-1][1] & f[M-1][2] & \cdots & f[M-1][N-1] \end{bmatrix}$$

As explained in 3.2.2, for performance reasons we usually do not store images in the above fashion when coding image processing algorithms in

C/C++. Very often, the image matrix is "flattened" and stored as a one-dimensional array. There are two prevalent ordering schemes for storing flattened 2D matrices, *row-major* and *column-major*. In row-major ordering, the matrix is stored as an array of rows, and this is the format used in C/C++ when defining 2D arrays. In the column-major format, which MATLAB and FORTRAN use, the matrix is ordered as an array of columns. Table 1-1 illustrates both of these formats, for the above example image matrix.

Table 1-1. Flattening an image matrix (with *M* columns and *N* rows) into a one-dimensional array.

row-major	column-major
$f[0][0]$	$f[0][0]$
$f[0][1]$	$f[1][0]$
$f[0][2]$	$f[2][0]$
…	…
$f[0][N\text{-}1]$	$f[M\text{-}1][0]$
$f[1][0]$	$f[0][1]$
$f[1][1]$	$f[1][1]$
$f[1][2]$	$f[2][1]$
…	…
$f[1][N\text{-}1]$	$f[M\text{-}1][1]$
…	…
$f[M\text{-}1][N\text{-}1]$	$f[M\text{-}1][N\text{-}1]$

The individual pixels in digital monochrome images take on a finite range of values. Typically, the pixel values *f(i,j)* are such that

$$0 \leq f(i,j) < 2^{\text{bpp}}$$

where "bpp" is bits-per-pixel. An individual pixel value *f(i,j)* goes by many commonly used names including: "gray-level intensity", "pixel intensity", or sometimes simply "pixel". In this book, we largely deal with monochrome 8 bpp images, with pixel intensities ranging in value from 0 to 255. These types of images are sometimes referred to as 8-bit images.

1.6 DSP CHIPS AND IMAGE PROCESSING

For the most part, image processing algorithms are characterized by repetitively performing the same operation on a group of pixels. For example, some common arithmetic image operations involve a single image and a single scalar, as in the case of a scalar multiply:

$$g(x,y) = \alpha f(x,y)$$

The above expression translates in code to multiplying each pixel in the image f by the constant α. Bilinear operations are pixel-wise operations that assume images are of the same size, for example:

- $g(x,y) = f_1(x,y) + f_2(x,y)$
- $g(x,y) = f_1(x,y) - f_2(x,y)$
- $g(x,y) = f_1(x,y) * f_2(x,y)$
- $g(x,y) = f_1(x,y) / f_2(x,y)$

Another very common category of image processing algorithms are mask or filter operations, where each pixel $f(x,y)$ is replaced by some function of $f(x,y)$'s neighboring pixels. This digital filtering operation is of critical importance, and this class of algorithms is introduced in Chapter 4. Filtering a signal, either one-dimensional or multi-dimensional, involves repeated *multiply-accumulate*, or MAC, operations. The MAC operation takes three inputs, yields a single output, and is described as

MAC = (a x b) + c

As described in 2.1.2, in fixed-point architectures a, b, and c are integer values whereas with floating-point architectures those three are either single-precision or double-precision quantities. The MAC operation is of such importance that is has even yielded its own benchmark, the M-MAC, or million of multiply-accumulate operations per second (other benchmarks include MIPS, or million of instructions per second and MFLOPS, or million of floating-point instructions per second). Digitally filtering an image involves repeatedly performing the MAC operation on each pixel while sliding a mask across the image, as described in Chapter 4.

All of the above operations have one overriding characteristic that stands out among all others – they involve repetitive numerical computations requiring a high memory bandwidth. In short, image processing is both very compute- *and* data-intensive. In addition, image processing applications are increasingly finding themselves in embedded systems, oftentimes in settings where real-time deadlines must be met. With respect to embedded systems, consider the digital camera or camera cell phone. Such a device requires a computing brain that performs the numerical tasks described above highly efficiently, while at the same time minimizing power, memory use, and in the case of high-volume products, cost. Add to these real-time constraints, like the type one may encounter in surveillance systems or medical devices such as ultrasound or computer-aided surgery, and all of a sudden you now find yourself in a setting where it is very likely that a general purpose processor (GPP) is not the appropriate choice. GPPs are designed to perform a diverse range of computing tasks (many of them not numerically oriented) and typically run heavy-weight operating systems definitely not suited for embedded and especially real-time systems.

Digital Signal Processors arrived on the scene in the early 1980s to address the need to process continuous data streams in real-time. Initially they were largely used in 1D signal processing applications like various telecommunication and audio applications, and today this largely remains the case. However, the rise of multimedia in the 1990s coincided with an increasing need to process images and video data streams, quite often in settings where a GPP was not going to be used. There has been a clear divergence in the evolution of DSPs and GPPs, although every manufacturer of GPPs, from Intel to SUN to AMD, has introduced DSP extensions to their processors. But the fact remains that by and large, DSP applications differ from their GPP counterparts in that they are most always characterized by relatively small programs (especially when compared to behemoths like databases, your typical web browser, or word processor) that entail intensive arithmetic processing, in particular the MAC operation that forms the building block of many a DSP algorithm. There is typically less logic involved in DSP programs, where logic refers to branching and control instructions. Rather, what you find is that DSP applications are dominated by tightly coded critical loops. DSPs are architected such that they maximize the performance of these critical loops, sometimes to the detriment of other more logic-oriented computing tasks. A DSP is thus the best choice for high-performance image processing, where the algorithms largely consist of repetitive numerical computations operating on pixels or groups of pixels, and where such processing must take place in a low-cost, low-power, embedded, and possibly real-time, system.

There are further unique architectural characteristics of DSPs that give them an edge in signal and image processing algorithms, including zero overhead loops, specialized I/O support, unique memory structures characterized by multiple memory banks and buses, saturated arithmetic, and others. In particular, there are certain vector operations that enable a huge boost in computational horsepower. These so-called SIMD (single instruction, multiple data) instructions, designed to exploit instruction level parallelism, crop up throughout this book and are covered in Appendix B. The C64x DSP, discussed in the next chapter, is particularly well suited for image processing applications as it is a high-speed DSP with numerous instructions that map very well to the efficient manipulation of 8-bit or 16-bit pixels. For a more thorough discussion of DSP architectures and the history of the evolution of DSPs, the reader is referred to [9].

1.7 USEFUL INTERNET RESOURCES

There are a number of very useful web-sites pertaining to C6x development, and will leave the appropriate Internet search as an exercise for the reader. That being said, there is one resource above all others that anyone serious about C6x development should be aware of, and that is the Yahoo! group "c6x" (groups.yahoo.com/group/c6x/). This discussion forum is very active, and there are a few expert individuals who actively monitor the postings and are always giving their expert advice on a wide variety of topics pertaining to C6x DSP development.

In addition, the following USENET newsgroups are also important resources that should never be overlooked during an Internet search:

- **comp.dsp**: discussions on signal processing and some image processing applications, as well as general DSP development.
- **comp.soft-sys.matlab**: MATLAB-related programming.
- **sci.image.processing**: image processing algorithms.

The MathWorks also maintains the MATLAB Central File Exchange (www.mathworks.com/matlabcentral/fileexchange/), a very handy repository of user-contributed M-files. This site should be your first stop when searching for a particular MATLAB-based algorithm implementation. Finally, for Windows-related programming issues the following web-sites are highly recommended: www.codeproject.com and www.codeguru.com.

REFERENCES

1. Gonzalez, R., and Woods, R., *Digital Image Processing* (Addison-Wesley, 1992).
2. Russ, J., *The Image Processing Handbook* (CRC Press, 1999).
3. Microsoft Developer Network, http://www.microsoft.msdn.com
4. Chassaing, R., *DSP Applications Using C and the TMS320C6x DSK*, (Wiley, 2002).
5. Dahnoun, N., *Digital Signal Processing Implementation using the TMS320C6000 DSP Platform* (Prentice-Hall, 2000).
6. Tretter, S., *Communication System Design Using DSP Algorithms, With Laboratory Experiments for the TMS320C6701 and TMS320C6711* (Kluwer Academic/Plenum, 2003).
7. http://www.mathworks.com/products/product_listing/index.html?alphadesc
8. Texas Instruments, *Code Composer Studio Getting Started Guide* (SPRU509C).
9. Lapsley, P., Bier, J., Shoham, A., Lee, E., *DSP Processor Fundamentals* (Wiley, 1996).

Chapter 2

TOOLS

Even though this book has a narrow focus, it calls for a wide array of tools, some hardware (DSP development boards) but mostly software. It is the author's strong belief that the development of embedded algorithms should proceed from a high-level vantage point down to the low-level environment, in a series of distinct, clearly defined milestones. The risk in jumping headlong into the embedded environment is getting bogged down in countless details and the inevitable unforeseen engineering challenges that may or may not be directly related to the algorithm. Embedded development is *hard*, and some may claim much harder than coding a web application, GUI application, Java program, or most other software intended to run on a desktop machine. Embedded developers are much closer to the hardware, and usually have fewer computing resources available at their disposal. The saving grace is that the programs typically running on an embedded DSP system are of a much smaller footprint than a desktop or server application. Nevertheless, when you are that much closer to the hardware there are many, many issues that must be taken into account and hence the development strategy put forth in this book – start with the background, prototype whatever operation needs to be implemented, and slowly but surely work your way down to the DSP.

Although this description of embedded image processing development may appear to be characterized by fire and brimstone, in reality the overall situation has gotten much better over the years. Ease of development is proportional to the quality of the tools at your disposal, and in this chapter all of the tools encountered in this book are formally introduced. These include:

- The TMS320C6000 line of DSPs, in particular the C6416 DSP Starter Kit (DSK) and the C6701 Evaluation Module (EVM).
- MATLAB and the various toolboxes used in Chapters 3-6 to prototype image processing algorithms.
- Visual Studio .NET 2003, and the libraries used to build image processing applications that run on the various flavors of Microsoft Windows.

A small amount of background information on the TI line of DSPs is appropriate before jumping into the fray. Thus we first take a slight detour into the land of computer architecture and computer arithmetic, before getting down to business and describing the contents of our tool chest we use throughout the rest of this book.

2.1. THE TMS320C6000 LINE OF DSPS

In 1997, Texas Instruments introduced the C6x generation of DSPs. These chips were unique in that they were the first DSPs to embrace the Very Long Instruction Word (VLIW) architecture[1]. This architectural aspect of the DSP, deemed VelociTI^{TM}, enabled a degree of parallelism previously unheard of in processors of its class and is the key to their high performance. The first members of the C6x family were the fixed-point C62x (C6201, C6211, etc.) and floating-point 67x (C6701, C6711, C6713, etc.) series. These DSPs feature eight functional units (two multipliers and six arithmetic logic units, or ALUs) and are capable of executing up to eight 32-bit instructions per cycle. The C62x/C67x was followed in 2001 by the C64x series of fixed-point DSPs, which represented a large step up in processor speeds (scalable up to 1.1 GHz versus approximately 250-300 MHz for the C62x/C67x at the time of this writing) and also introduced extensions to VelociTI that have important ramifications on high-performance image processing. Figure 1-1 shows block diagrams of both architectures, illustrating their common roots.

What Figure 1-1 does not show are all of the associated peripherals and interconnect structures that are equally important to understanding the C6000 architecture. This relationship is shown in Figure 1-2 for the case of the C62x/C67x DSP. Figure 1-2 shows a processor core surrounded by a variety of peripherals and banks of memory with data shuttling across separate program and data buses. The C6000 has a relatively simple memory architecture, with a flat, byte-addressable 4 GB address space, split into smaller sections mapped to different types of RAM (SDRAM or SBSRAM). As shown in both Figures 1-1 and 1-2, there are actually two data paths in the processor core, each containing four functional units. In the C62x/C67x, each data path has 16 32-bit registers, while in the C64x this register file is augmented with an additional 16 32-bit registers per data path. In addition, the C64x features an enhanced multiplier that doubles the 16-bit multiply rate of the C62x/C67x[2]. A full description of the entire C6x architecture is covered in [1-5], so in this section we instead focus on explaining the rationale and motivation behind VLIW and how it fits into the C6000 architecture, before moving to an important discussion on the difference between fixed-point and floating-point architectures. We conclude the section by introducing the two TI development environments used in this book.

2.1.1 VLIW and VelociTI

In modern Complex Instruction Set Computer (CISC) processors and to a lesser extent, Reduce Instruction Set Computer (RISC) processors, there is an incredible amount of hardware complexity designed to exploit instruction level parallelism (ILP). With sincere apologies for the multitude of acronyms (at least you were warned!), the general concept is that deeply pipelined and superscalar GPPs, with their branch prediction and out-of-order execution, perform significant analysis on the instruction stream at run-time, and then transform this instruction stream wherever possible to keep the processor's pipeline fully utilized. When the processor's pipeline is full, it completes an instruction with every clock cycle. As one might imagine, this analysis is an extraordinarily difficult task, and while the compiler does do some things to alleviate the burden on the CPU, the CPU is very much in the loop with regards to optimizing the flow of instructions through the pipeline.

With VLIW on the other hand, the onus is completely on the compiler, and the processor relies on the compiler to provide it with a group of instructions that are guaranteed to not have any dependencies among them. Hence, the compiler, rather than the hardware, is the driving element behind taking advantage of any ILP. The VLIW concept is an outgrowth of vector processors, like the Cray supercomputers from the 1970s, which were based on this idea of the exact same operation being performed on an array of data. We can illustrate the rationale behind this concept, by considering at a very high-level the steps any processor must take to execute an instruction stream:

1. Fetch the next instruction.
2. Decode this instruction.
3. Execute this instruction.

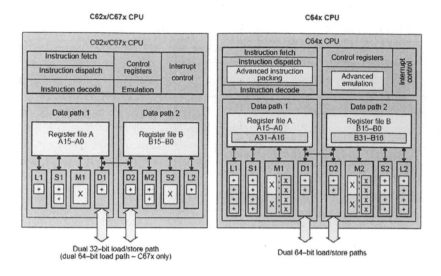

Figure 2-1. C62x/C67x and C64x DSPs. Figure reprinted courtesy of Texas Instruments.

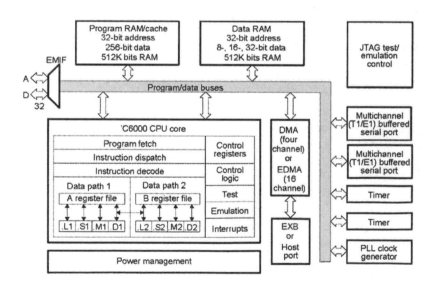

Figure 2-2. TMS320C62x/C67x block diagram, with peripherals and interconnect structure. Figure reprinted courtesy of Texas Instruments.

The classic example in vector processing is the addition of two vectors of length N. Instead of fetching, decoding, and executing the first instruction, fetching, decoding, and executing the second instruction, and so on up to N, why not perform steps 1 and 2 a single time and then execute this same instruction for each vector element? Hence we amortize the fetch and decode overhead over N iterations and the "Single Instruction, Multiple Data," or SIMD paradigm is born. Beginning in the 1980s, this idea was extended with VLIW architectures "as a somewhat liberated SIMD," with the various functional units of a processor "still in lockstep and under centralized control … each performing a different (through carefully preplanned) operation on different data elements."[6] The operative term here is "carefully preplanned," as the compiler is now responsible for instruction scheduling. The compiler groups together multiple instructions that do not have any dependencies into a single VLIW instruction, or *fetch packet*, to borrow TI's nomenclature. The processor, upon decoding this fetch packet, is now free to concurrently dispatch all of the fetch packet's constituent instructions to available functional units for execution because there are no dependencies between any of them. Since the compiler has done the difficult work of creating the VLIW instruction stream, the processor can now be significantly simpler, as the massively complicated hardware blocks designed for run-time extraction of ILP present within CISC/RISC GPPs are no longer needed. This has two important ramifications for embedded DSPs: the clock speed increases while power consumption remains manageable.

The downside to VLIW is what happens if there are not enough independent instructions to completely fill up the fetch packet, or even worse, if the compiler is not intelligent enough perform the requisite code analysis and transformations to remove such dependencies should they appear (which they will). In this case, the compiler will be forced to pad the fetch packet with NOP (no operation) instructions. If the operations the source code lays out have a significant amount of dependencies between closely grouped instructions, the net effect is wasted processor resources. Memory is wasted with fetch packets consisting largely of NOPs, while available functional units lie idle, a veritable double whammy that completely stymies high-performance computing. Thankfully, in the world of DSP this situation is rare. For starters, compiler technology has progressed greatly since the advent of VLIW so compiler limitations are not so much of a concern nowadays, at least on mature architectures. But more significantly, the types of operations frequently encountered in DSP applications – highly repetitive numerical manipulations of large sequences of samples – are more often than not *data parallel*, meaning which there are comparatively fewer dependencies between successive instructions within a loop kernel. As a consequence, you have an ideal situation for the deployment of VLIW technology.

Texas Instruments broke with classical VLIW in an important aspect that led to the VelociTI technology at the heart of the C6x architecture. In conventional VLIW, the processor executes all instructions in the fetch packet with every clock cycle. Hence, if there are NOPs injected in the packet due to data dependencies, the processor's resources are not fully utilized. With VelociTI, TI defined an *execute packet* as a group of instructions within the fetch packet that are in turn guaranteed to have no dependencies with respect to each other. Individual C6x instructions are 32-bits in length, and fetch packets consist of eight instructions, one per functional unit, for a total of 256 bits. Thus, even though all eight instructions are fetched at once, groups out of this eight can be dispatched simultaneously. For example, an eight-instruction fetch packet could be executed in the following manner:

- **Clock 1**: dispatch first three instructions
- **Clock 2**: dispatch next three instructions
- **Clock 3**: dispatch final two instructions

The key breakthrough was in allowing for the length of the execute packet to differ from that of the fetch packet. In the ideal case, we would hope that the execute packet equals the fetch packet, but at least TI has covered their tracks somewhat if this cannot be done. In this fashion, VelociTI mitigates the memory usage problem associated with traditional VLIW (in which the fetch packet must be padded with numerous NOPs for code that is not fully data parallel), while also yielding smaller code size.

TI made another step forward with their second generation VLIW-based architecture, VelociTI.2^{TM}, which was introduced alongside the C64x. This augmented technology included new specialized instructions for "packed data processing" for accelerating many of the key applications targeted by the C6x family. One of these applications is high-performance and low-cost image processing, and these architectural enhancements are utilized throughout the algorithms presented in this book. With the proper utilization of VelociTI.2^{TM} via sensible coding techniques and certain optimizations, the C64x both increases the computational throughput for digital media algorithms and applications, while further reducing code size.

One consequence of the C6x VLIW-based architecture is that there is no MAC instruction, per se. This will probably elicit a strong reaction, especially given the prominence accorded to this fundamental media processing operation (see 1.6). VelociTI largely eschews complex and compound instructions in favor of RISC-like simple instructions, and hence the MAC operation is handled as a multiply followed by an addition. The way the C6x makes up for this lack of an explicit MAC instruction is through the use of "software pipelining". The fixed-point chips are capable of performing two multiplies, four adds, and two address calculations *per cycle*, and so with a loop whose kernel consists of MAC operations (e.g. a dot product of two vectors) the processor *pipelines* the entire operation, that is, it begins successive operations before their predecessors have actually completed. This can only be done with loops where there are no loop-carried dependencies, and by overlapping operations across several iterations of a loop it becomes possible to achieve throughput rates of better than a MAC per cycle. In fact, this aspect of VLIW-based DSP highlights the need for different performance metrics than the standard Millions of Instructions Per Cycle (MIPS), which can be deceiving in light of the parallelism in a pipelined loop when it reaches its steady state. A more suitable metric is Millions of MACs, or MMACs. With VelociTI.2, some additional MAC-like instructions were added to the instruction set, and in Appendix B programming examples are given that illustrate their usage in the context of some of the image processing algorithms studied in Chapters 4-6.

2.1.2 Fixed-Point versus Floating-Point

The terms "fixed-point architecture" and "floating-point architecture" have been bandied about thus far without really having been fully explained. With floating-point arithmetic, the set of real numbers is represented using 32 (the *single-precision* `float` data type) or 64 (the *double-precision* `double` data type) bits by reserving fields for a fraction containing the number of significant digits (the *mantissa*), an exponent, and a sign bit. Here, the term "real number" refers to the mathematical notion of the ordered set of non-imaginary numbers with a decimal point. Floating-point numbers basically represent this set in scientific notation, and they are capable of representing a large range of values. The IEEE Standard 754 defines a format for the floating-point representation of real numbers[7], and is the most widely used representation today.

Floating-point arithmetic is very powerful, and quite convenient for the programmer. Apart from having to worry about some nasty conditions like underflow, overflow, granularity, and some other special cases, for the most part floating-point arithmetic is transparent to the programmer. But this power and convenience comes at a cost. Performing floating-point arithmetic computations in hardware is far more complex than the equivalent integer operations. This may seem like the dark ages, but it was actually the case in the early 1980s that PCs based on the predecessor to the Pentium did not come standard with a "math coprocessor" whose job it was to accelerate floating-point math. The coprocessor was an option and without it floating-point arithmetic was performed in software. Nowadays of course, all Pentium-class processors come standard with the Intel math coprocessor, which in fact uses 80 bits (`long double`) to represent IEEE floating-point numbers. Processors which have the ability to perform floating-point arithmetic in hardware are floating-point architectures. The major advantage of fixed-point processors, where the various functional units only know how to deal with integer quantities and do not have innate support for floating-point numbers, is that the hardware becomes much simpler. The comparative simplicity of the hardware then makes it possible to crank up the clock speed of the processor while maintaining a handle on the overall power consumption

Great, so in fixed-point architectures we are forever stuck only performing integer arithmetic, and there is no concept of a number with a decimal point? Clearly that is severely limiting and not very useful! Of course, floating-point math can still be implemented in software, using "bit banging" routines that perform the bit manipulations in code, but this strategy will have deleterious effects on overall system performance. We must be able to deal with real numbers and on fixed-point devices we use fixed-point arithmetic to manage an implicit decimal point. In essence, to draw an analogy with football (of the American variety), the processor has basically punted the problem of dealing with real numbers to the programmer.

With fixed-point arithmetic, this implicit decimal point, or "radix point" as it is commonly referred to, is placed somewhere in the middle of the digits in a binary encoding. The bits to the right of this "binal point" maintain a fixed number of fraction digits, and it is entirely up to the programmer to manage this binal point during subsequent arithmetic operations. What this entails is a propensity for numerous bit shifts in the code to align the operands correctly. While it makes for less aesthetically pleasing code, once you get the hang of it, the machinations of fixed-point arithmetic are really not that difficult to deal with. What can be difficult to deal with are the reduced accuracy and quantization effects that fixed-point arithmetic can have on numerical computations – these are algorithm-specific issues that must be tackled on a case-by-case basis, as we will do throughout this book. An example fixed-point representation is shown in Figure 2-3, for the case of an 8-bit encoding.

With fixed-point arithmetic, we almost always use more than eight bits in order to gain additional dynamic range and fractional resolution. Programming some of the older non-C6x TI DSPs is actually an odd experience at first for C programmers, as these processors only deal with 32-bit quantities. As a result, `char`, `short`, and `int` variables are all 32 bits

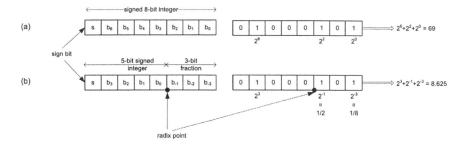

Figure 2-3. Fixed-point representation of numbers. If the sign bit were flipped, then the final result would be negated. (a) A quantity akin to a normal (one byte) signed char, where the integer number is calculated according to the rules of a two's complement format. (b) An example fixed-point representation using the same bits as above, with the radix point placed between the third and fourth least significant bits.

in length. On the C6000 architecture, there are no such shenanigans and the size of the basic C integer data types are mostly what a C/C++ developer might expect, as shown in Table 2-1.

Table 2-1. TMS320C6000 C/C++ Integer Data Types. Note that in Visual C++ a long int is 32 bits in length, and a long long is 64 bits in length.

Type	Size (bits)
signed/unsigned char	8
signed/unsigned short	16
signed/unsigned int	32
signed/unsigned long	40

Fractional Q formats are one way of codifying fixed-point representations. One of the most common representations is the so-called "Q15" format, or more specifically Q0.15, where 16-bit short integers are treated as having their radix point between the most-significant bit (MSB) and the second MSB, leaving a fractional component of 15 bits. As described in [8], in general Qm.n fixed-point numbers split the m+n+1 bits in a binary encoding into two quantities, each of which are stored in two's complement format: the m bits to the left of the radix point constitutes the integer portion of the quantity and the n bits to the right of the radix point is the fractional component of the quantity, a la Figure 2-3b.

The remaining bit is reserved as a sign bit, and thus Figure 2-3b is an example of a Q4.3 number. When interpreting a Q0.15 or more simply Q15 encoding of a floating-point quantity, this number takes on a value between −1.0 and 1.0. With fixed-point arithmetic, the proper scaling of the data is of the utmost importance. For example before conversion to the Q15 format the data must be normalized such that the maximum value is less than or equal to 1.0, and the minimum value is greater than or equal to −1.0. Conversion of a floating-point number into Q15 format is accomplished by first multiplying by 2^{15} − which is equivalent to shifting left by 15 bits in an integer representation − and then truncating down to an integer, by lopping off whatever fractional component remains after the multiplication. In C, this translates in code to

```
short int xQ15 = x*32768; /* 32768 = 2^15 */
```

if x is originally of type `float` or `double`, or alternatively

```
short int xQ15 = x;
xQ15 <<= 15;
```

Now we can perform our arithmetic operations and the hardware is none the wiser. To convert back from Q15 format to a floating-point number we can use the following code:

```
float x = (xQ15>>15);
```

In reality, things are not always so simple. It may be the case that Q15 numbers do not offer sufficient accuracy for the task at hand, and this implies some up front analysis on our part. But just as important, there are other cases that must be taken into account. First, consider the multiplication of two Q0.N numbers:

$$QMULT = Q_1N \times Q_2N$$

If both Q_1N and Q_2N are stored in 16-bit words, then QMULT requires 32 bits of storage. Now suppose Q_1 and Q_2 are still 16-bit words but have different Q formats, i.e.

$$QMULT = Q_1X \times Q_2Y$$

Suppose X=7 and Y=8, then the result of the multiplication QMULT still requires 32 bits but now we have:

$$QMULT = Q_17 \times Q_28$$

The rule of thumb for mixed Q multiplication is that QMULT has its radix point to the left of the (X+Y) most significant bit. So for the above example we now have a Q15 number stored in 32 bits. If QMULT is to be used as part of another computation further down the road, we will need to realign this number by bit shifting to isolate the fractional bits we wish to retain, taking extreme care to guard against overflow. This example assumes there are no integer bits (consider the case if QMULT = $Q_1 1.7$ x $Q_2 2.7$) and then things *really* start to get hairy. In some respects, the use of fixed-point arithmetic can be intimidating and has the potential to open up that veritable can of worms, but generally speaking, in image processing we usually will not encounter such situations. For the most part, the fixed-point programs in the book follow this modus operandi:

1. Convert from a floating-point to a fixed-point representation that offers sufficient dynamic range and accuracy, by left bit-shifting.
2. Perform the numerical operations that constitute the image processing algorithm.
3. Scale back to the original input range (most often 0-255 for 8 bpp images) by shifting to the right however many bits we shifted to the left in step 1.

The majority of the embedded image processing programs in this book are written for the fixed-point C6416 DSP. Those that are not ported to the floating-point C6701 DSP are easily done so. Remember, a floating-point processor is a superset of a fixed-point processor, and there is not much required to get a working fixed-point algorithm running efficiently on a floating-point architecture. The converse is definitely *not* true, as the compiler will be forced to inject slow software floating point routines wherever it encounters arithmetic between floating-point operands.

For further information on the vagaries of floating-point arithmetic, the reader is referred to [9-13].

2.1.3 TI DSP Development Tools (C6701 EVM and C6416 DSK)

The TMS320C6701 Evaluation Module is a PCI board with the 150 MHz floating-point TMS320C6701 DSP at its core[14]. Both the C6701 EVM, and its fixed-point cousin the C6201 EVM have DSPs with 64 KB of on-chip program and 64 KB of on-chip data memories. They also both have 256 KB of 133 MHz Synchronous Burst Static RAM (SBSRAM) and 8 MB of 100 MHz SDRAM in external (off-chip) memory banks. Aside from the fact that the C62x are fixed-point DSPs, another difference between the C67x and

C62x is that the C67x is capable of taking advantage of double word-wide accesses via the LDDW instruction (see B.1.1), which is not available on the C62x.

The EVM boards also feature numerous interfaces for possible expansion, such as extra memory or peripheral daughter cards. However, as of late 2004 TI has discontinued the C6701 EVM product. This strategy is not because the C67x line is a dead-end, in contrast the latest generation of the C6x floating-point family is the C6713 that now reaches clock speeds of 300 MHz. Instead, TI is pushing their "DSP starter kit," or DSK (see Figure 2-4), as the avenue of choice for those wanting a cheap entry-level DSP development platform. The DSK is an external board that sits outside the host PC, and is expandable via daughter cards conforming to the TI TMS320 cross-platform daughtercard specification. This book features the TMS320C6416 DSK[15], which contains a 600 MHz C6416 DSP, with 1 MB of L2 internal memory and 16 MB of external SDRAM. One issue with the older DSKs was that they were somewhat slow, because their connectivity to the host PC was over a parallel port. This has ceased to be a problem because they now come in USB flavors, making them much faster – definitely a welcome development as single-stepping through a debugger over a parallel port on the older DSKs used to be painfully slow!

Most of the DSP implementations in this book, starting with Chapter 4, target the C6416 DSK. The EVM is useful in its own right because with it we can demonstrate usage of the host port interface (HPI), which is not available on the C6416 DSK but is used in other development boards and third-party products (of which there are many). However, it should be stressed that porting C6701 EVM code to another floating-point development platform such as the C6713 DSK is not a difficult task. Most of what will be needed involves changing the linker command file (see 3.2.2) to reflect the proper memory map and replacing some instances of the EVM support library with equivalent Chip Support Library (CSL) or DSP/BIOS API calls. Likewise, C6416 DSK to C67x DSK ports require modifications of the same magnitude.

2.2 TI SOFTWARE DEVELOPMENT TOOLS

Development of all the DSP applications in the book was done using version 2.20 or 2.21 of Code Composer Studio. As this book was going to press, version 3 of Code Composer Studio has been recently released and the source code and project files should be easily ported to this new version of the IDE. The CCStudio manual contains a wealth of information pertaining to the various compiler and linker options, however the manuals do not

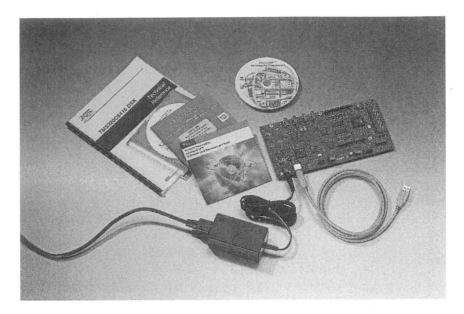

Figure 2-4. C6416 DSK (image courtesy of Texas Instruments). In its nominal form, connecting the DSK to the host is simple: there is a power cord that connects to the board and a USB cable to runs from the board to the USB port on the PC.

address the temperamental nature of both the EVM and DSK. In particular, there are times when one needs to reset the systems, or else CCStudio will not work correctly. With respect to the EVM, if you receive an error message along the lines of "File <some .out file> Does not match the target type, not loaded" then the CCStudio application needs to be closed, and the EVM board reset. The EVM reset is done using a batch file named evm6xrst.bat, which is included in the CCStudio bin directory. After resetting the board, re-open CCStudio and attempt the load operation again. In addition, it has been observed numerous times in Windows XP that the CCStudio application hangs around as a zombie process even after exiting, preventing normal operation the next time it is opened. In this event, the Windows process manager can be used to kill the cc_app.exe process. The DSK, in comparison to the EVM, is actually far more reliable. However, there are some cases when CCStudio and the DSK diagnostics utility claim they are unable to make a connection to the board. In the event that this occurs, the best course of action is to cycle the power on the DSK, making sure to explicitly kill any cc_app.exe processes that may be lingering around via the Windows process manager.

This book features a variety of TI-related software toolkits. These technologies and libraries are briefly introduced in the remainder of this section, along with pointers on where to obtain a few of them that are not included with the stock CCStudio install. In contrast to integrating libraries with applications built using Visual Studio, which very often involves "DLL hell," arcane compilation strategies, and various other annoyances, the DSP software integration process is decidedly simpler. All of the libraries described in this section are available in the form of precompiled static libraries, and incorporating them into CCStudio projects entails two basic steps: including the header file directory into the C preprocessor include path and linking the static library to the project COFF executable. The former can be set from the project settings within CCStudio, and the latter is achieved by either explicitly adding the static library to the project, or adding it to the linker configuration in the project settings.

2.2.1 EVM support libraries

As shown in Figure 2-2, the DSP development platforms feature a variety of peripherals. All of these peripherals, and the board itself, need to be initialized and then preferably accessed using C functions which hide the nastiness of twiddling with memory-mapped registers and the like. For example, prior to accessing any external RAM, the external memory interface (EMIF) peripheral needs to initialized properly, and failure to do so results in undefined behavior. Another prominent example includes Direct Memory Access (DMA), which is used extensively in Chapter 4 and parts of Chapter 6.

Many of the EVM programs in this book use an older support library to perform such duties, as it is the author's experience that this particular library is more reliable on the EVM platform than some of the newer libraries that TI includes with more current versions of CCStudio, such as the Chip Support Library. This C library is the "EVM DSP Support Library," is documented in [16] and can be downloaded from [17]. The use of this library will be explained on a per-need basis, as we proceed to build EVM applications in Chapters 3-6.

2.2.2 Chip Support Library

The Chip Support Library (CSL) is a newer C API that offers similar functionality to the library described in the preceding section, but supports all C6x devices. The CSL can be used with both EVM and DSK platforms, although as previously stated there are some cases, notably involving host/target communication, where we use a different library for the EVM. However, on the DSK platform, any programs that do not use the DSP/BIOS real-time operating system will use the CSL to initialize peripherals like the EMIF, EDMA, and clock (used for profiling).

2.2.3 DSP/BIOS

DSP/BIOS is a scalable real-time operating system that is supported across a range of TI DSPs. The operating system is configured by a GUI that allows one to tailor the operating system to the application at hand, and it contains various "modules" (basically C APIs) that mirror the functionalities in the CSL. DSP/BIOS is used in Chapter 5, and examples of how to incorporate the operating system into both EVM and DSK projects is fully explained in 5.1.2.

2.2.4 FastRTS

The TMS320C62x/C64x Fast Run-Time Support Library, or FastRTS for short, is a C-callable library that contains a set of optimized floating-point functions that can be used in fixed-point C6x DSPs. While this library is not extensively utilized, for performance reasons described in 2.1.2, there are a few instances where it is required. This library is documented in [18] and is not included with the C6416 DSK CCStudio. It can be downloaded from [19].

2.2.5 DSPLIB and IMGLIB

Developing software is a labor-intensive task. Developing highly efficient DSP software is even more labor-intensive, and in this book we attempt to leverage existing TI libraries wherever possible. TI has developed and actively maintains a set of functions implementing a number of common one-dimensional signal processing (DSPLIB) and two-dimensional image processing algorithms (IMGLIB). Both libraries are highly optimized implementations, and come standard with CCStudio. The C62x/C67x DPSLIB directory, which includes the header files, static library and even the source code (in text files with the `.src` extension), is located in `[TI]\c6200\dsplib`, where [TI] should be replaced with whatever the root path of your particular installation is (typically `C:\TI`). The C64x DSPLIB directory is `[TI]\c6400\dsplib`. The C62x/C67x and C64x IMGLIB directories are located in the same directory tree, in sub-directories named `imglib`.

2.3 MATLAB

MATLAB is an IDE designed by The MathWorks and is the lingua franca for rapid-development technical computing, both in academia and industry (a low-cost student edition of MATLAB can be purchased from www.mathworks.com). It is an interpreted language, and the MATLAB code in this book has been developed and tested using Release 14 (service pack 1). In addition, much of it has also been tested on the previous version, Release 13 with service pack 1. Anyone familiar with C should be able to quickly pick up the MATLAB language – readers with absolutely no prior MATLAB experience are encouraged to peruse [20], and more advanced MATLAB idioms and programming techniques are explained throughout the rest of this book.

MATLAB code is structured around the concept of an M-file, which is analogous to a C module. In addition, this book also describes the use of MEX-files, implemented in C/C++ using Visual Studio .NET 2003 as a mechanism for adding new functionality to MATLAB (see Appendix B). One the main selling points of MATLAB is the many application-specific "toolboxes" that contain functions, test data, and GUIs. In this book, there are three such prominently featured toolboxes:

1. **Image Processing Toolbox**: referenced extensively throughout Chapters 3-6, this toolbox contains a wealth of M-files and demos for numerous image processing algorithms.
2. **Link for Code Composer Studio**: introduced in Chapter 5 (see 5.1.2), this toolbox has objects for communicating with CCStudio from MATLAB.
3. **Wavelet Toolbox**: used in Chapter 6 to prototype image processing algorithms based on the two-dimensional discrete wavelet transform.

The Student Edition of MATLAB is fully-featured, and each of these toolboxes can be purchased for academic use at much reduced prices, provided they are not used for commercial purposes. Refer to The MathWorks web-site for further information.

2.4. VISUAL STUDIO .NET 2003

All of the source code for the Windows C++ applications accompanying this book were built using Microsoft Visual Studio .NET 2003. Visual Studio is the standard bearer when it comes to PC IDEs, and has been the most popular Windows compiler since the release of Visual Studio 5 in the mid-90s. Beginning in 1998, Visual Studio version 6 cemented its place as the dominant C/C++ compiler and this compiler remained fairly static until .NET arrived. The next generation of Visual Studio, deemed "Visual Studio .NET" was released in 2002 and came with compilers for various .NET languages, most notably C#, Visual Basic .NET, and of course C/C++. This was followed a year later with Visual Studio .NET 2003, which fixed numerous bugs and added many improvements to the compilers for all of the various languages. As this book was going to press, the next version of Visual Studio, Visual Studio 2005 (code-named "Whidbey") was in beta release.

With the advent of Microsoft's .NET initiative, there are now two distinct flavors of C++: so-called "managed" C++, and "unmanaged" or "native" C++. The difference between the two is that managed C++ is a .NET-compliant language fully aware of the .NET environment and the virtual machine that the environment is predicated on, the common language runtime or CLR. All of the C++ applications in this book use the more traditional unmanaged C++, so that the source code should be portable to other PC compilers like Borland's C++ Builder or Intel's C++ Compiler for Windows. The knock on previous versions of Microsoft's C++ compilers have been their lack of compliance to the ISO C++ standard, and Microsoft's lackadaisical attitude towards rectifying the situation. With the 2003 edition of Visual Studio however, this problem seems to have been solved, and the C++ compiler is now one of the more standard compliant on the market. Furthermore, Microsoft appears to remain committed to maintaining this compliance.

One of the reasons why application development on the PC can sometimes appear daunting is the overwhelming array of class libraries, frameworks, and Application Programming Interfaces (APIs) made available to users to Visual Studio. From the Win32 API to DirectDraw and DirectShow, the number of libraries can be simply dizzying. Visual Studio .NET 2003 applications in this book draw on three major libraries: Microsoft Foundation Classes (MFC), Graphics Device Interface Plus (GDI+), and the Intel Integrated Performance Primitives (Intel IPP) package. Figure 2-5 shows the relationship between these three libraries, and in this book the implementation of a class that facilitates the interactions between the MFC, GDI+, and Intel IPP triumvirate is fully detailed. The remainder of this section briefly introduces each of these three libraries.

2.4.1 Microsoft Foundation Classes (MFC)

Prior to the advent of MFC, building Windows GUIs in Visual Studio or other compilers was done using the raw Win32 API, which consists of literally hundreds, perhaps even thousands, of functions littered all over the Windows programming landscape[21]. As C++ began to attain more popularity in the early 1990s, Microsoft released MFC, which is a collection of classes that constitute a "thin wrapper" around the various C functions and data structures that make up the Win32 API. Essentially, MFC is implemented as a thin object-oriented veneer around Win32, and if you understand the basics of Win32, MFC should be reasonably easy to pick up.

The GUI applications featured in this book are rather simplistic, as our goal is embedded deployment. However, there are cases where we either desire a reference implementation in C/C++ closer to the eventual DSP implementation than MATLAB prototypes, or we need front-end interfaces for DSP-based image processing systems (residing on the EVM or DSK). To be useful, in both instances we require some minimal GUI facilities as well as visualization capabilities. Rendering of images can be done in MFC, but in this book we use the more modern GDI+ (discussed below) to perform these duties. However, MFC is utilized for the rest of the GUI portion of these applications. All of the Windows GUIs in this book are built as so-called "dialog" applications, meaning which they are composed of a single modal dialog whose implementation is entirely encapsulated within a single C++ class that inherits from the MFC base class `CDialog`.

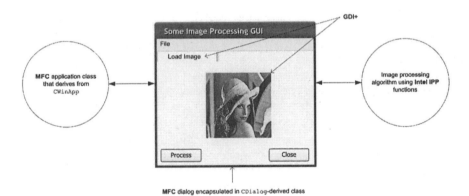

Figure 2-5. Diagram depicting the general software structure of the Windows applications developed in this book. They are MFC applications, with an MFC application class (left) responsible for instantiating a main dialog object. The dialog (middle) typically contains numerous child controls, and the callbacks that deal with reading image files and displaying monochrome 8-bit images use GDI+. The image processing algorithms call on Intel IPP functions (right). The integration of these disparate components and frameworks is managed by the `Image8bpp` class, introduced in Chapter 3.

The behavior of the GUI comes from what happens when the user interacts with the various controls that lie within the main dialog. When the user interacts with such controls, the MFC framework calls the appropriate callback methods in the `CDialog`-derived class. These callback methods are hooked into the application using the Visual Studio .NET 2003 GUI designer (previous versions of Visual Studio used something called the "class wizard" to a similar effect). The Visual Studio GUI designer takes care of inserting all of the infrastructure code that maps button clicks to method invocations. MFC does this through a well-defined command update mechanism via a process known as dynamic data exchange.

2.4.2 GDI+

GDI+ is an object-oriented extension to Microsoft's venerable Graphical Device Interface API, and frankly has been sorely needed for quite some time. GDI is a C library of functions and type definitions used for 2D graphics, providing exactly the type of graphics functionality needed for visualizing the results of image processing algorithms – namely the rendering of bitmapped images. DirectDraw, which is a subset of DirectX, is a COM-based API that can provide similar functionality but is more complicated than GDI and more suited to streaming video and the type of 3D graphics prevalent in games and CAD applications. OpenGL is another option (in the Windows world, there is no dearth of options when it comes to programming libraries) but integration with MFC and Win32 in general is less straightforward than with GDI.

The historic problem with GDI has been its verbosity and vulnerability to resource leaks (the programmer must be diligent in explicitly freeing GDI resources), both of which can be traced back to the library's C heritage. GDI+ corrects many of these problems, by providing convenient wrapper classes around core GDI types like `BITMAP` and `HDC` (a handle to a device context). Of course, there is no such thing as a free lunch, and the cost of increased simplicity is performance – there is a slight performance hit when using GDI+, when compared to coding against the raw GDI API.

For the most part, the image processing applications in this book use GDI+ classes and functionality to accomplish two distinct tasks: parsing image files and displaying bitmapped images. Reading in and parsing a Windows BMP or JPEG file used to be a tedious, mind-numbing chore with GDI. Now with GDI+, an image object can be constructed with nothing more than just the image file pathname, as so:

```
Bitmap myImage(L"some_image.bmp");
```

The 'L' in front of the pathname string denotes a wide-character (16-bit) string literal. Saving an image using GDI+ is equally straightforward:

```
// 1st obtain CLSID descriptor for JPEG format
CLSID jpgClsid;
GetEncoderClsid(L"image/jpeg", &jpgClsid);
// Save the image in JPEG format (last parameter optional)
jpgClsIdmyImage.save(L"some_image.jpg", &jpgClsid, NULL);
```

Likewise, prior to the introduction of GDI, rendering a bitmap onto a Windows GUI was quite a bit more difficult than it should have been. Essentially, efficient display of a bitmap image boils down to an operation known as a "bit-block transfer", for which the legacy `BitBlt` GDI function could be used. This function accepts eight input parameters and moreover, the simple act of rendering an image required a multitude of related GDI functions for initializing the device context, setting up the coordinate system, and so on. Finally, all resource handles must be managed very carefully and explicitly deallocated using appropriate GDI function calls, or else insidious and hard-to-track down memory leaks end up plaguing the application. With the object-oriented GDI+ library, all of the above and more are elegantly encapsulated by the `Bitmap` object. Both GDI and GDI+ come standard with Visual Studio .NET 2003.

2.4.3   Intel Integrated Performance Primitives (IPP)

The Intel IPP library is a collection of optimized C-callable functions targeting multimedia operations such as audio processing, video processing, data compression, and most importantly to us, image processing. You would be hard pressed to find faster implementations of the hundreds of basic operations that the Intel IPP library provides for the IA-32 and Itanium® architectures. The Windows applications in this book use the latest version of the library (4.1), although all of the code has also been tested with the earlier 4.0 version and some with the earlier 3.0 version.

The IPP library can be purchased and downloaded from the Intel web site at http://www.intel.com/software/products/ipp/index.htm. In addition, a trial version of the library can also be downloaded for the purposes of building and using the code that accompanies this book. By default, the IPP installation places the header files, static libraries, and DLLs within `C:\Program Files\Intel\IPP`, and the Visual Studio projects on the CD-ROM assume this location. If you install the IPP library to a different location, you will need to modify the projects to point to the correct location.

REFERENCES

1. Turley, J., Hakkarainen, H., "TI's New 'C6x DSP Screams at 1,600 .MIPS.", *The Microprocessor Report*, 11:14-17, 1997.
2. Texas Instruments, *TMS320C64x Technical Overview* (SPRU395b.pdf).
3. Texas Instruments, *TMS320C6000 Technical Brief* (SPRU197d.pdf).
4. Dahnoun, N., *Digital Signal Processing Implementation using the TMS320C6000 DSP Platform* (Prentice-Hall, 2000), Chapter 2, *The TMS320C6000 Architecture*.
5. Morgan, D., "TMS320C62x/C67x DSPs", *Embedded Systems Programming*, Feb 1999, available online at http://www.embedded.com/1999/9902/9902spectra.htm
6. Almasi, G., Gottlieb, A., *Highly Parallel Computing*, (Benjamin/Cummings, 1994).
7. Hollasch, S., "IEEE Standard 754 Floating Point Numbers", retrieved Dec. 2004 from http://stevehollasch.com/cgindex/coding/ieeefloat.html
8. Texas Instruments, *TMS320C62x DSP Library Programmer's Reference* (SPRU402b.pdf), Appendix A, section A.2, *Fractional Q Formats*.
9. Chassaing, R., *DSP Applications Using C and the TMS320C6x DSK* (Wiley, 2002), Appendix C, *Fixed-Point Considerations*.
10. Labrosse, J., "Fixed-Point Arithmetic for Embedded Systems," C/C++ User's Journal, Feb. 1998; pp. 21-28.
11. Gordon, R., "A Calculated Look at Fixed-Point Arithmetic," *Embedded Systems Programming*, Apr. 1998, available online at http://www.embedded.com/98/9804fe2.htm
12. Lemiex, J., "Fixed-Point Math in C," *Embedded Systems Programming*, retrieved December 2004 from http://www.embedded.com/showArticle.jhtml?articleID=15201575
13. Allen, R., "Converting floating-point applications to fixed-point," *Embedded Systems Programming*, retrieved December 2004 from http://www.embedded.com/showArticle.jhtml?articleID=47903200
14. Texas Instruments, *TMS3206201/6701 Evaluation Module Technical Reference* (SPRU305.pdf).
15. Spectrum Digital, *TMS3206416 DSK Technical Reference* (505945-001 Rev. A).
16. Texas Instruments, *TMS320C6201/6701 Evaluation Module Technical Reference* (SPRU305.pdf).
17. "C6000 EVM Support Files," retrieved December 2004 from http://www.ti.com/sc/docs/tools/dsp/c6000/c62x/evmfiles.htm
18. Texas Instruments, *TMS320C62x/64x FastRTS Library Programmer's Reference* (SPRU653.pdf).
19. "TMS320C62x/TMS320C64x FastRTS Library," retrieved December 2004 from http://focus.ti.com/docs/toolsw/folders/print/sprc122.html
20. The MathWorks, *Getting Started With MATLAB® Version 7* Users Manual (2004).
21. Petzold, C., *Programming Windows Fifth Edition* (Microsoft Press, 1999).

Chapter 3

SPATIAL PROCESSING TECHNIQUES

Broadly speaking, image enhancement algorithms fall into two categories: those that operate in the "spatial domain", and those that operate in the "frequency domain". Spatial processing of images works by operating directly on an image's pixel values. In contrast, frequency domain methods use mathematical tools such as the Discrete Fourier Transform to convert the 2D function an image represents into an alternate formulation consisting of coefficients correlating to spatial frequencies. These frequency coefficients are subsequently manipulated, and then the inverse transform is applied to map the frequency coefficients back to gray-level pixel intensities.

In this chapter, we discuss a variety of common spatial image processing algorithms and then implement these algorithms on the DSP. Image *enhancement* per se sometimes means more than enhancing the subjective quality of a digital image, as viewed from the vantage point of a human observer. The term may also refer to the process of accentuating certain characteristics of an image at the expense of others, for the purpose of further analysis (by either a computer or human being).

3.1 SPATIAL TRANSFORM FUNCTIONS AND THE IMAGE HISTOGRAM

Spatial processing methods are one of the most common forms of image processing, as they are relatively simple but quite effective. The general idea is to enhance the appearance of an image f by applying a gray-level transformation function T that maps the input pixels of f to output pixels in an image g, as shown in Figure 3-1.

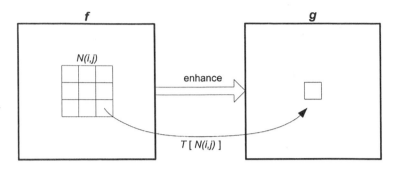

Figure 3-1. Image enhancement via spatial processing.

If the neighborhood *N(i,j)* consists of a single pixel, then this process entails a remapping of pixel intensities. If *r* denotes the input pixel values and *s* denotes the output pixel values, then

$$s = T(r)$$

where *T* is a monotonic mapping function between *r* and *s*, and

$$r = N(i, j) = f(i, j)$$

and thus

$$s = g(i, j) = T[f(i, j)]$$

These types of spatial transform functions, where the transform function *T* is based solely on the input pixel's value and is completely independent of location or neighboring pixel values, are called *point-processing operations*. For example, consider the case of brightening an image by adding a constant *B* to each pixel in the image *f*. Assuming pixel values are normalized to lie within the range [0,1], the gray-level transformation function would look something like Figure 3-2, where the brightening is accomplished by the addition of a constant value *B* to each pixel intensity. Another point-processing operation, which is the focus of this chapter, is histogram modification. We can treat the population of image pixels as a random variable, and with this in mind, the histogram of a digital image is an approximation to its probability density function (PDF). The image histogram, the discrete PDF of the image's pixel values, is defined mathematically as:

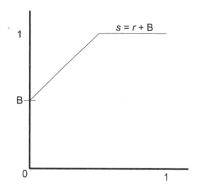

Figure 3-2. Image brightening transformation function.

$$\text{histogram} = p(k) = \frac{n_k}{N}$$

where

- q = bits per pixel, and $k = 0,1,2, ..., 2^q-1$
- n_k = number of pixels in image with intensity k
- N = total number of pixels in image

The histogram function bins the number of pixels for each possible r_k; for example an 8-bit image histogram has 256 bins to cover the spectrum of gray-levels ranging from 0 to 255. The division by the total number of pixels in the image yields a percentage for each particular pixel value, hence the fact that the image histogram is in fact the discrete PDF of the 2D image function $f(i,j)$. Inspection of the histogram is a reasonable first indicator of the overall appearance of an image, as demonstrated by Figure 3-3, which shows a dark and bright image along with their associated histograms.

This chapter explores three image enhancement algorithms that feature manipulation of the input image's histogram to produce a qualitatively better looking output image: contrast stretching, gray-level slicing, and histogram equalization.

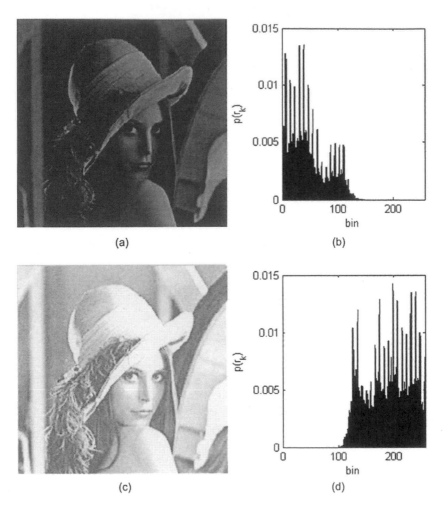

Figure 3-3. Comparison of dark vs. bright images. (a) Dark image. (b) Dark image histogram. (c) Bright image. (d) Bright image histogram.

3.2 CONTRAST STRETCHING

The contrast of an image is a measure of its dynamic range, or the "spread" of its histogram. The dynamic range of an image is defined to be the entire range of intensity values contained within an image, or put a simpler way, the maximum pixel value minus the minimum pixel value. An 8-bit image at full contrast has a dynamic range of $2^8-1=255$, with anything less than that yielding a lower contrast image. Figure 3-4 shows a low-contrast 8-bit x-ray image with a dynamic range of 49, along with its histogram.

The minimum pixel value in Figure 3-4a is 81, and the maximum pixel value is 127. Increasing the dynamic range entails "stretching" the histogram in Figure 3-4b by applying a linear scaling function that maps pixel values of 81 to 0, maps pixel values of 127 to 255, and scales pixel intensities that lie within the range [82-126] accordingly. The contrast-stretch algorithm essentially pulls the boundaries of the original histogram function to the extremes. This process is detailed in Algorithm 3-1.

Algorithm 3-1: Contrast Stretch
    INPUT: q-bit image I

        $MP = 2^q-1$ (*e.g. 255 for 8-bpp image*)
        $a = min(I)$
        $b = max(I)$
        $R = b-a$

        foreach (pixel p in I)
          $p = [(p-a)/R]MP$ (*apply linear scaling function*)
          $p = round(p)$
        end

Contrast stretching the image in 3-4a along the lines of Algorithm 3-1 produces what is shown in Figure 3-5. The image now takes on the full 8-bit range, and correspondingly the new histogram is spread out over the range 0-255, resulting in an image that subjectively looks far better to the human eye. However, the drawback to modifying the histogram of an image in such a manner comes at the expense of greater "graininess." If the original image is of rather low-contrast and does not contain much information (i.e. most of the pixels fall into a rather small subset of the full dynamic range), stretching the contrast can only accomplish so much.

Contrast stretching is a common technique, and can be quite effective if utilized properly. In the field of medical imaging, an x-ray camera that consists of an array of x-ray detectors creates what are known as digital radiographs, or digital x-ray images. The detectors accumulate charge (which manifests itself as a larger pixel intensity) proportional to the amount of x-ray illumination they receive, which depends on the quality of the x-ray beam and the object being imaged. A high-density object means less x-rays pass through the object to eventually reach the detectors (hence the beam is said to be *attenuated*), which results in such higher density areas appearing

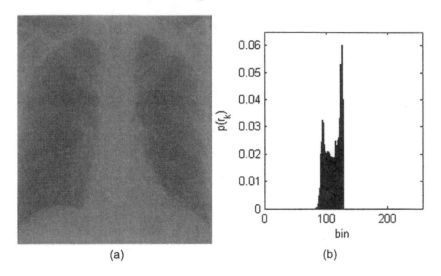

Figure 3-4. X-ray image courtesy of SRS-X, http://www.radiology.co.uk/srs-x. (a) Low contrast chest x-ray image. (b) Low contrast histogram.

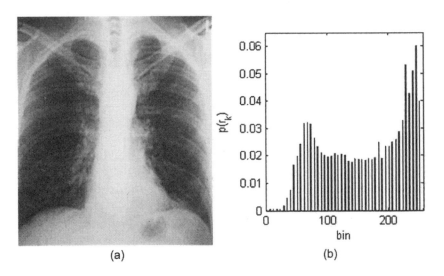

Figure 3-5. (a) Contrast-stretched chest x-ray image. (b) Modified histogram.

darker. Likewise, the beam will not be as attenuated when passing through low-density areas and here the image will appear bright, as the detectors accumulate comparatively more charge. When displaying the acquired image, it is common to reverse the polarity of the image by creating the negative, which is why the dense areas (bone) in Figure 3-4a and 3-5a appear white and the less dense areas (soft tissue and background) appear dark.   The overall quality of the radiograph is directly related to the characteristics of the detectors and the amount of x-ray exposure the detectors absorb. Too little illumination and the image appears dark and of low-contrast. More illumination means greater contrast and a better quality image, with a larger delta between pixel intensities corresponding to the bony areas and the background (compare the quality of 3-4a with that of 3-5a). Too much illumination however, will saturate the image, a condition where most of the pixel values will be clustered in the upper portion of the dynamic range and a loss of information occurs, due to clamping of gray-scale intensities. Such a situation is analogous to using too high of a flash level when taking photographic images, thereby producing a "washed out" image.

In certain medical imaging applications such as fluoroscopy, repeated radiographs are taken of a patient. Because the patient is subjected to larger amounts of x-ray radiation, it is advantageous to acquire the radiographs using the least amount of x-ray exposure as possible, while still obtaining images of the required contrast. Here the contrast stretching technique can be quite useful, as employing this histogram modification technique allows acquisition systems to obtain useful images while reducing the amount of x-ray exposure to the patient. However, one major drawback with the contrast-stretching algorithm is its susceptibility to noise - a single outlier pixel can wreak havoc with the output image. Figure 3-6 shows the same low-contrast chest radiograph and the results of enhancing the radiograph via the contrast stretching process, but this time the image suffers from outlier noise. As exhibited here, the presence of just a few outlier pixels dramatically reduces the overall effectiveness of the linear scaling function - in fact in this case, it is nullified completely. One problem with digital radiograph detectors is so-called "shot" or "salt and pepper" noise, whereby the image is corrupted with certain pixels taking on the maximum or close to maximum pixel value. For example, if even a single detector fails to accumulate charge during the acquisition process, perhaps because of a transient condition or due to insufficient x-ray illumination, the estimated dynamic range is artificially high, and the linear scaling function fails to remap the pixels properly, as demonstrated in Figure 3-6b.

Applying a small amount of intelligence during estimation of the dynamic range of the input image can help ameliorate this problem. Instead of blindly computing the dynamic range in Algorithm 3-1 using just the minimum and maximum pixel values, a more robust and adaptive technique is to use the 5th and 95th percentiles of the input values when deriving the

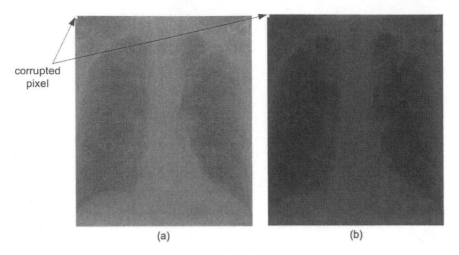

corrupted
pixel

(a)   (b)

Figure 3-6. Noise reducing the effectiveness of contrast-stretching. (a) Corrupted low-contrast x-ray image. (b) The presence of the single noisy pixel prevents the scaling function from stretching the histogram. In fact, in this case it actually shrinks the histogram, thereby reducing the contrast!

dynamic range of the input image. Yet another strategy is to locate the histogram peak and then march in both directions until only a small number of pixel intensities are rejected. These boundary points are then used as the dynamic range of the input. The percentile scheme is outlined in Algorithm 3-2.

Algorithm 3-2: Adaptive Contrast Stretch
    INPUT: q-bit image I, percentile range plo and phi)

    P = 2^q-1
    hist[] = int(MP+1)   (*array to store histogram bins*)
    cdf[] = int(MP+1)   (*array to store cumulative dist function*)

    for (b = 0 … MP)
      hist[b] = 0
    end

    foreach (pixel p in I) (*compute histogram, or PDF*)
      hist[p] = hist[p] + 1
    end

Algorithm 3-2 (continued)
   (*from PDF compute CDF*)
   cdf[0] = hist[0]
   for (b = 1...MP)
    cdf[b] = cdf[b-1] + hist[b]
   end

   a = find(cdf $\geq$ plo)   (*bin corresponding to lower percentile*)
   b = find(cdf $\geq$ phi)   (*bin corresponding to upper percentile*)
   R = b-a

   foreach (pixel p in I)
    p = [(p-a)/R]MP   (*apply linear scaling function*)
    p = round(p)
   end

3.2.1 MATLAB Implementation

A MATLAB implementation of Algorithm 3-2, with parameterized percentile inputs, is given in Listing 3-1. This M-file does not use any functions from the Image Processing Toolbox.

Listing 3-1: contrast_stretch.m

```
function Istretched = contrast_stretch(I, percentiles)

    if (~strcmp('uint8', class(I))) % validate input arg
        error(' contrast_stretch only works with uint8 images');
    end

    if 1==nargin % default percentiles
        percentiles = [ .05 .95 ];
    end


    N = length(I(:)); % total # of pixels
    I = double(I); % many MATLAB functions don't work with uint8
    H = histc(I(:), 0:255); % image histogram

    % find the 2 bins that correspond to the specified percentiles
    cutoffs = round(percentiles .* N);
    CH = cumsum(H);
    % lower bound
```

```
a = find(CH >= cutoffs(1));
a = a(1) - 1; % remember, 0-based pixel values!
% upper bound
b = find(CH >= cutoffs(2));
b = b(1) - 1;

% apply scaling function to each pixel
Istretched = I;
R = b - a; % dynamic range
scale = 255/R;
Istretched = round((I-a) .* scale);
Istretched = uint8(Istretched);
```

Figure 3-7 demonstrates the effectiveness of this more robust algorithm, using the same corrupted chest x-ray image from Figure 3-6a. This image's first pixel is completely saturated, simulating the situation of a detector in an x-ray camera failing to collect charge. Applying Algorithm 3-1 fails to yield an improved image, as illustrated in Figure 3-6b. However, the improved algorithm, using the default percentiles (5% and 95%) in Listing 3-1, does manage to return an enhanced image, albeit at the cost of some bony structure near the patient's ribs.

Displaying contrast enhanced gray-scale images using the Image Processing Toolbox is extremely easy. Assuming that the variable I refers to an image matrix, the following MATLAB command can be used to display a stretched image:

```
imshow(I, [])
```

Note that MATLAB assumes all pixel values are normalized to lie within the range [0, 1] if I is of type double. For integer "classes" (MATLAB parlance for data types) such as uint8, uint16, and so on, the range is defined to be $[0, 2^{bpp}-1]$. For uint8, bpp is 8 and for uint16, bpp is 16. The imshow command only displays the contrast-stretched image; it does not modify the pixels. To actually apply the scaling function to the pixel values, you can use the Image Processing Toolbox functions stretchlim and imadjust, as so:

```
Ienhanced = imadjust(I, stretchlim(I), [])
```

The stretchlim function can be used to determine the limits for contrast stretching an image (variables *a* and *b* in Algorithms 3-1 and 3-2). The imadjust function adjusts the pixel intensities by applying a scaling

corrupted pixel

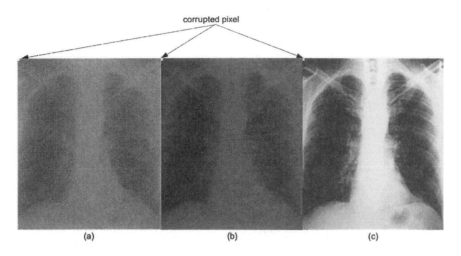

(a)   (b)   (c)

Figure 3-7. Adaptive contrast-stretching algorithm. (a) Corrupted low-contrast x-ray image.
(b) Result from applying algorithm 3.1. (c) Result from MATLAB adaptive contrast
stretching function, an implementation of Algorithm 3-2.

function, mapping pixel values specified by the second input argument to the pixel values specified by the third input argument. stretchlim(I) returns an array with the minimum and maximum pixels in I, and if the third argument is empty, imadjust defaults to [0 1]. So the above MATLAB statement maps the minimum pixel value to 0, and the maximum pixel value to the largest possible intensity the image can take on (255 for uint8, 65536 for uint16, and 1.0 for double) – in other words Algorithm 3-1.

3.2.2 TI C67xx Implementation and MATLAB Support Files

One of the many challenges when developing embedded systems is getting test data into the system. This problem is particularly acute with image processing applications, as the sheer size of the input data oftentimes precludes certain methods like compiling data directly into the executable (e.g. directly initializing an array containing the image pixels). In our first C6x implementation, we use some of Code Composer Studio's built-in file I/O facilities to bring image data into our program and also to save the processed data. While CCStudio does provide traditional C file I/O functionality through its standard C library, like fopen, fwrite, and their ilk, these functions are so terribly slow that it is not viable to use them for transferring even moderate amounts of data. This section describes how to use one such means of circumventing the standard C library – subsequent chapters describe the usage of other technologies, including Real-Time Data Exchange (RTDX) and High-Performance Interface (HPI).

On the CD-ROM in the Chap3\contrast_stretch directory is a project named, appropriately enough, contrast_stretch.pjt. This project can be used as a template for other simple EVM projects, as it is about as bare bones as a CCStudio project can be, consisting of just a "linker command" (.cmd) file and a single C source file. The project does not utilize any DSP/BIOS facilities, and links to four other libraries: the runtime support library (rts6701.lib), image processing library (img62x.lib), peripheral support library (dev6x.lib), and driver library (drv6x.lib). DSP/BIOS projects, which we introduce in Chapter 5, need only explicitly link to the appropriate IMGLIB library (e.g. img64x.lib for the C6416 DSK) if imaging functions are used - the CCStudio DSP/BIOS configuration tools take care of generating code and/or implicitly linking to the other aforementioned support libraries.

The linker command file, contrast_stretch.cmd, is shown in Listing 3-2. This file contains directives for the TI linker, and serves a few purposes. If the project uses DSP/BIOS, then the CCStudio configuration tool creates this file automatically, and the developer should not manually modify it. However, since this application does not utilize DSP/BIOS, a hand-crafted linker file is necessary.

Listing 3-2: C6701 EVM linker command file.

```
/* Memory map definition */
MEMORY
{
  IPRAM :   origin = 0x00000000 length = 0x00010000
  IDRAM :   origin = 0x80000000 length = 0x00010000
  SBSRAM : origin = 0x00400000 length = 0x00040000
  SDRAM0 : origin = 0x02000000 length = 0x00400000
  SDRAM1 : origin = 0x03000000 length = 0x00400000
}


/* Bind various sections into the different areas of memory */
SECTIONS
{
  .vec:      load = 0x00000000    /* Interrupt vector table */
  .text:     load = IPRAM         /* Code */
  .const:    load = IDRAM         /* Variables defined with const */
  .bss:      load = IDRAM         /* Global variables */
  .data:   load = IDRAM
```

```
.cinit      load = IDRAM
.pinit      load = IDRAM
.stack      load = IDRAM      /* Stack (local variables) */
.far        load = SDRAM0     /* Variables defined with far */
.sysmem     load = SDRAM0     /* Heap: malloc, free, etc */
.cio        load = IDRAM
.ipmtext    load = IPRAM
}
```

The MEMORY section of the linker command file delineates the memory map, and the nominal values for the location and size of the memory banks come directly from the documentation[1,2]. The first two memory sections, the internal program RAM (IPRAM) and internal data RAM (IDRAM), are located on the chip, while the last three sections define off-chip memory segments accessed via the external memory interface (EMIF) controller. For this example, we will be using the synchronous burst static RAM (SBSRAM) to store buffers that do not fit in internal RAM. The SECTION part of the linker file binds different sections of the program to the memory sections previously mentioned.

Listing 3-3 is the full contents of the main source module for this project, contrast_stretch.c. This C program is a direct port of the MATLAB function given in Listing 3-1.

Listing 3-3: contrast_stretch.c

```c
#include "board.h" /* EVM library */
#include "limits.h"
#include "stdio.h"
#define _TI_ENHANCED_MATH_H 1
#include "math.h"
#include "img_histogram.h" /* IMGLIB */

/* image dimensions */
#define X_SIZE 128
#define Y_SIZE 128
#define N_PIXELS X_SIZE*Y_SIZE

unsigned char in_img[N_PIXELS];

/* Input & output buffer both will not fit in
   internal chip RAM, this pragma places the
   output buffer in off-chip RAM */
#pragma DATA_SECTION (out_img, "SBSRAM");
```

```
unsigned char out_img[N_PIXELS];

/* image histogram goes here, the pragma aligns the buffer on
   a 4-byte boundary which is required by IMGLIB */
#pragma DATA_ALIGN (hist, 4)
unsigned short hist[256];

/* scratch buffer needed by IMGLIB */
unsigned short t_hist[1024];

void compute_range(unsigned char *pin, int N,
                   float percentile1, float percentile2,
                   unsigned char *pbin1, unsigned char *pbin2)
{
    unsigned short cumsum = 0, /* running tally of cumulative sum */
        T1 = round((float)N * percentile1), /* threshold for 1st bin */
        T2 = round((float)N * percentile2); /* threshold for 2nd bin */
    int ii, jj;

    /* calc image histogram */

    /* buffers must be initialized to zero */
    memset(t_hist, 0, sizeof(unsigned short)*1024);
    memset(hist, 0, sizeof(unsigned short)*256);
    IMG_histogram(pin, N, 1, t_hist, hist);

    /* find location for 1st (lower bound) percentile in histogram */
    for (ii=0; ii<256; ++ii) {
        cumsum += hist[ii];
        if (cumsum >= T1) {
            *pbin1 = ii;
            break;
        }
    }

    /* find location for 2nd (upper bound) percentile in histogram */
    for (jj=ii+1; jj<256; ++jj) {
        cumsum += hist[jj];
        if (cumsum >= T2) {
            *pbin2 = jj;
            break;
        }
```

```
        }
    }

void contrast_stretch(unsigned char *pin,
                      unsigned char *pout,
                      int N)
{
    unsigned char a, b; /* lower & upper bounds */
    float scale, pixel;
    int ii;

    /* estimate dynamic range of the input */
    compute_range(pin, N, .05, 0.95, &a, &b);

    /* apply linear scaling function to input pixels,
       taking care to handle overflow & underflow */
    scale = 255.0/(float)(b-a);
    for (ii=0; ii<N_PIXELS; ++ii) {
        pixel = round( (float)(pin[ii] - a) * scale );
        /* clamp to 8 bit range */
        pout[ii] = (pixel<0.f) ? 0 : ((pixel>255.0) ? 255 : pixel);
    }
}

int main(void)
{
    evm_init(); /* initialize the board */
    contrast_stretch(in_img, out_img, N_PIXELS);
    printf("contrast stretch completed");
}
```

The program uses flattened arrays to store images, an idiom employed in most of the other C and C++ examples presented in this book. As explained in 1.5, in contrast to the MATLAB convention of storing 2D data in column-major format, in C this data is stored in row-major format. Therefore, indexing into a flattened image buffer requires an understanding of some basic pointer arithmetic. To access a pixel at row i and column j of an image with N columns, the following statement is used:

```
pixel_value = image[i*N + j];
```

The image dimensions for this project are hard-coded to 128x128. Due to the size of the resultant image buffers, not all of the image data fits in internal RAM. The DATA_SECTION pragma, or compiler directive, tells CCStudio to allocate space for the symbol in the specified section, which in the case of out_img is the SBSRAM, or external RAM, portion of the memory map. The linker must know about the specified memory bank, so either the hand-crafted linker command file contains the requisite definitions or in the case of DSP/BIOS, the memory segment must be configured appropriately through the IDE.

There are two other global buffers, both of which are used for computing the histogram of the image. The IMGLIB function IMG_histogram requires two buffers: one to contain the histogram bin counts upon return (hist), and another temporary scratch buffer of length 1024 (t_hist). Moreover, the documentation[3,4] states that the histogram buffer must be aligned on a 4-byte boundary, which is specified in the source code via the DATA_ALIGN pragma.

All of the grunt work is performed within two functions: compute_range and contrast_stretch. The compute_range function starts by calculating the discrete PDF of the image data via IMG_histogram. IMG_histogram is used in lieu of a C-coded histogram function for performance reasons - all of the IMGLIB functions are heavily optimized and will offer better performance metrics than C code. The next step is to use the discrete PDF to compute bounds of the linear scaling function that stretches the contrast of the image. To that end, we use the cumulative distribution function (CDF). The CDF is closely related to the probability density function; simply put, the CDF of a random variable (in this case the set of pixel values in the input image) is a nondecreasing function yielding the probability that a random variable (pixel intensity) takes on a value less than or equal to some number. This code stores the current value of the CDF in cumsum, and in conjunction with the specified percentile ranges calculates the lower and upper bounds needed to stretch the histogram. Finally, contrast_stretch, which initially invoked compute_range, uses the returned bounds a and b to apply the scaling function to each image pixel.

To test the program, we need to first load pixel data onto the DSP and then extract the processed image data. To do this, we will use the data I/O facilities of CCStudio to send a block of image data stored in a text file located on the host directly into a memory location on the target (EVM). The MATLAB function save_ccs_data_file, shown in Listing 3-4, generates an ASCII text file from an image matrix that can be used by CCS to import data into the DSP[5].

Listing 3-4: MATLAB function to save image matrix to CCS data file.

```
function save_ccs_data_file(I, filename)

fid = fopen(filename, 'w+');
if -1 == fid
    error(['Failed to open ' filename]);
end

if strcmp('uint8', class(I))
    bpp = 8;
elseif strcmp('uint16', class(I))
    bpp = 16;
else
    error('Only 8-bit and 16-bit images are supported');
end

fprintf(fid, '1651 1 0 0 0\r\n');
[M N] = size(I);
I = double(I);
for ii=1:M
    if 8==bpp
        for jj-1:N/4
            fprintf(fid, '0x%.2x%.2x%.2x%.2x\r\n', ...
                    I(ii, 4*jj:-1:4*(jj-1)+1));
        end
    elseif 16==bpp
        for jj=1:N/2
            fprintf(fid, '0x00%.2x00%.2x\r\n', ...
                    I(ii, 2*jj-1), I(ii, 2*jj));
        end
    end
end % for each row

fclose(fid);
```

The CCStudio file format dictates that each line contains a single word (32 bits) of data. The `save_ccs_data_file` function supports both 8-bit (`uint8`) and 16-bit (`uint16`) pixel formats. The `in_img` buffer in the EVM program is defined as an array of unsigned chars, so to use `save_ccs_data_file` to generate an input file for use with the EVM program, the string 'uint8' should be passed in as the second argument. Since each word consists of four bytes (32 bits), each line in the output text

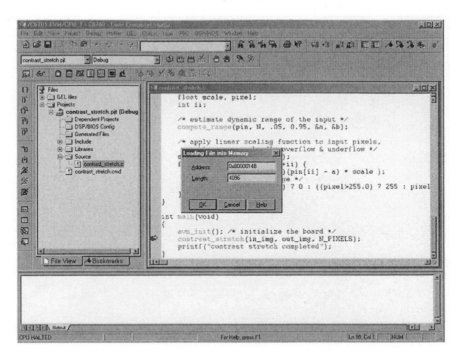

Figure 3-8. Code Composer Studio **File|Load** facility. After typing in_img for the address, CCStudio automatically replaces the string with the actual address. Notice the address maps to the SBSRAM portion of the memory map defined by the linker command file.

file contains four pixels worth of data. The function assumes little-endian ordering - that is, bytes are numbered from right to left within a word, with the more significant bytes in a word having a larger address. The C6x architecture is unique in that it is bi-endian, meaning that depending on the build options it can be either little or big endian. In order for save_ccs_data_file to work correctly, the build options for the project must be set to little endian (which is the default setting). A sample data file with the corrupted x-ray image pixels from Figure 3-6a can be found in the project directory under the name xray.dat, and this input data is used in the following screen captures to illustrate the usage of the program.

After building the project and loading the executable program onto the DSP, set a breakpoint in main where contrast_stretch is called. Upon running the program, the debugger will stop at this line. By using the **File|Data|Load** menu item, we can send a block of data to the DSP, given a CCStudio data file containing the raw data, the memory address of where to

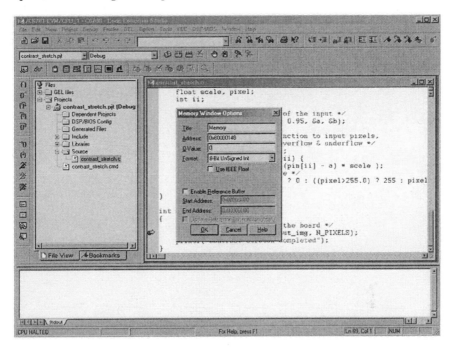

Figure 3-9. Code Composer Studio view memory configuration. As in the case with the File I/O dialog, CCStudio automatically replaces `in_img` with the actual address.

place the block of data, and the length of that data block. When the "Load Data" dialog comes up, choose the "Hex" file format, as this is the format that `save_ccs_data_file` uses when it formats the input data. The next dialog allows the user to specify where in memory the data should be placed, by asking for the destination memory address and data length. Figure 3-8 is a CCStudio screen capture illustrating how to send image data into the `in_img` array, assuming an image with dimensions 128x128. Since each word contains four bytes, the data file actually contains 128x128/4 = 4096 elements.

You can verify that the pixel data transferred correctly into the `in_img` array by using one the most helpful debug facilities offered by the CCStudio IDE – the ability to peer into memory and display array data in a large variety of formats. Selecting **View|Memory** brings up the dialog box shown in Figure 3-9. Figure 3-10 shows what happens after setting the memory window options to display the contents of `in_img` in an 8-bit unsigned integer format. Note that the first pixel value is 255, corresponding to the top-left corrupted pixel from Figure 3-6a. The CCStudio memory viewer is

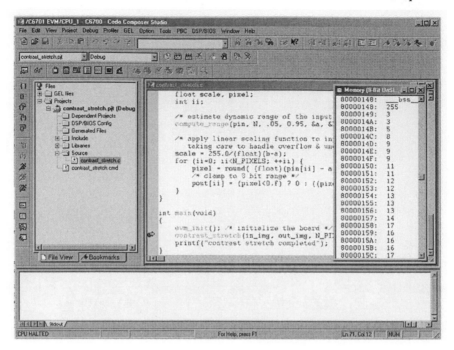

Figure 3-10. Contents of `in_img` input buffer. The corrupted x-ray image from Figure 3-6a was used as input data, and the first value in the array reflects the obvious outlier pixel intensity (255).

an important debugging tool in your arsenal – it is far easier to inspect the contents of data arrays in this type of block format, as opposed to the "quick watch" option, which is more geared towards examination of scalar variables. In fact, one can visualize image data directly within CCStudio, as we shall see in 4.3.1.

By stepping over `contrast_stretch` in the debugger we execute the function and proceed to process the image. Upon return, we can save the contrast stretched image data to a file on the host machine using the **File|Data|Save** menu selection. Specify the memory address as `out_img` and the length as 4096. The MATLAB function `read_ccs_data_file`, shown in Listing 3-5, can then be used to parse this CCStudio data file (hex format) and resize the data into an image matrix. This matrix can then be viewed using standard MATLAB commands like `imagesc`, `image`, or `imshow`.

Listing 3-5: `read_ccs_data_file.m`

```
function I = read_ccs_data_file(filename, type)

if strcmp(type, 'uint8') | strcmp(type, 'int8')
    nbytes = 4; % each word has 4 bytes
elseif strcmp(type, 'uint16') | strcmp(type, 'int16')
    nbytes = 2; % 2 bytes per word
else
    error('Please specify a desired data-type, [u]int8 or [u]int16');
end

fid = fopen(filename, 'r');
if -1 == fid
    error(['Could not open ' filename]);
end

tline = fgetl(fid);
% last token in header gives us the length
header_data = sscanf(tline, '%x');
len = header_data(length(header_data));
if (0 == len)
    % data length not defined, so compute it
    % from the # of text lines in the data file.
    fclose(fid); % will reopen afterwards
    len = how_many_lines(filename);
    % get back to where we were prior to this
    % diversion.
    fid = fopen(filename, 'r');
    tline = fgetl(fid);
end
N = sqrt(len*nbytes); % assumes square image matrix!
I = zeros(N, N);
for ii = 1:N
    if 2==nbytes % 16 bpp
        for jj = 1:N/2
            tline = fgetl(fid);
            I(ii, 2*jj-1:2*jj) = [sscanf(tline(3:6), '%x') ...
                                  sscanf(tline(7:10), '%x')];
        end
    else % 8 bpp
        for jj = 1:N/4
            tline = fgetl(fid);
```

```
        % little endian format
        p1 = tline(9:10);  p2 = tline(7:8);
        p3 = tline(5:6);  p4 = tline(3:4);
        I(ii, 4*(jj-1)+1:4*jj) = [sscanf(p1,'%x') ...
                                  sscanf(p2,'%x') ...
                                  sscanf(p3,'%x') ...
                                  sscanf(p4,'%x')];
      end
    end
  end % for each image row

  fclose(fid);

  if 'u' ~= type(1) % unsigned-to-signed integer conversion
    halfway = 2^(32/nbytes-1)-1; % e.g., 127 for 8 bpp
    % find and replace those values that are actually negative
    knegative = find(I>halfway);
    offset = zeros(size(I));
    offset(knegative) = bitshift(1,32/nbytes); % (1<<8) or 256
    I = I-offset;
  end

  eval( sprintf('I=%s(I);', type) );

  function N = how_many_lines(filename)

  fid = fopen(filename, 'r');
  tline = fgetl(fid);
  N = 0;
  while 1
    tline = fgetl(fid);
    if ~ischar(tline), break, end
    N = N + 1;
  end
  fclose(fid);
```

The `read_ccs_data_file` function supports both signed and unsigned 8 or 16 bit integer formats, and is specified in the second argument (type). This M-file actually consists of two functions: the main entry function `read_ccs_data_file`, and a MATLAB *subfunction* `how_many_lines`. Subfunctions may only be called from functions residing in the same M-file, and there can be multiple subfunctions within a single M-file. In this fashion, MATLAB subfunctions are reminiscent of static functions in the C language, where functions marked with the `static` keyword are only visible to those functions residing in the same C file, or more strictly speaking, the same translation unit.

More interesting to the implementation of `read_ccs_data_file` is the use of MATLAB's `eval` facility at the tail end of the function to perform double-to-integer and unsigned-to-signed data conversion. The built-in `eval` function offers incredible convenience and takes advantage of the MATLAB interpreter to execute a string containing a MATLAB expression. It is used in a variety of contexts, for instance as a quasi function pointer or to condense C-style `switch` statements or long if-then-else blocks into a single line of code. For example, in `read_ccs_data_file`, the statement

```
eval( sprintf('I=%s(I);', type) );
```

evaluates to one of:

- `I = uint8(I); % double to unsigned 8-bit integer conversion`
- `I = int8(I); % double to signed 8-bit integer conversion`
- `I = uint16(I); % double to unsigned 16-bit integer conversion`
- `I = int16(I); % double to signed 16-bit integer conversion`

depending on the value of `type`.

3.3 WINDOW/LEVEL

The previous section described a contrast enhancement technique whereby a linear scaling function was applied to all pixels, resulting in an image whose histogram was stretched so as to maximize the contrast. This section describes another point-processing enhancement method that entails applying a gray-scale lookup-table to each pixel in an image. This technique goes by many names, including "window/level", "density/level", "gray-level slicing", or in general "LUT" (a mnemonic for lookup table) processing. Here we will use the term window/level. The general window/level algorithm can be extended to perform a variety of other image transformations that are not covered in this book, such as gamma correction. At the end of this section, two window/level implementations are discussed: one than runs on the Windows platform and another CCStudio program that uses "GEL" files to provide some measure of interactivity, allowing the user to dynamically change certain parameters driving the algorithm.

The basic concept behind window/level is to apply a linear gray-scale transform function, in the form of a lookup table (LUT) specified by two parameters, *window* and *level*. The end result is that pixel intensities corresponding to a subset of the entire dynamic range are highlighted, at the expense of those pixels falling outside the specified range. The LUT processing technique is depicted graphically in Figure 3-11, where each pixel in the input image *f* is mapped to the corresponding pixel in the output image *g*, with the new pixel intensity taking on a value specified by the LUT.

In Figure 3-12, a few examples of gray-scale transformation lookup tables are plotted. Figure 3-12a and 3-12b are linear LUTs, the class of transform functions we concern ourselves with in this section. Figure 3-12b shows a LUT where input pixels of intensity ≤ 73 are mapped to 0, pixels of

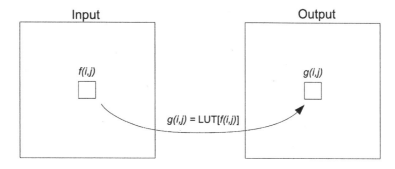

Figure 3-11. Image transformation via gray-scale lookup table.

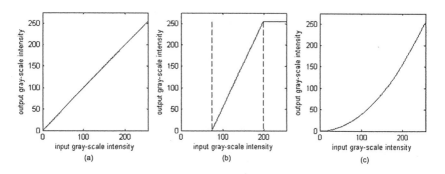

Figure 3-12. Three examples of 8-bit gray-scale LUT transform functions. (a) Nominal identity function, a one-to-one "transform" mapping the intensity range [0-255] to [0-255]. (b) Linear function highlighting pixels lying within the range [74-197]. (c) Non-linear gamma transformation function.

intensity ≥ 198 are mapped to 255, and pixels within the range [74-197] are scaled appropriately depending on the slope of the transform function in the region bracketed by the dotted vertical lines. Thus, those input pixels whose intensities lie within the range of interest are selectively contrast stretched.

The actual LUT values corresponding to Figure 3-12b are enumerated in Table 3-1.

Table 3-1. Actual contents of the LUT from Figure 3-12b

Input	Output
0	0
1	0
2	0
.	.
.	.
.	.
74	2.04 (2)*
75	4.08 (4)*
76	6.12 (6)*
.	.
.	.
.	.
197	252.96 (253)*
198	255
199	255
.	.
.	.
.	.
255	255

* transformed pixel intensities are rounded to fit into 8-bit integers.

One way of specifying a linear gray-scale transform function of the form in Figure 3-12b is to define the width of the LUT where the slope is non-zero (window) and the center of that same segment of the LUT (level) – so holding the window constant while adjusting the level has the effect of moving the non-zero slope portion of the transform to the left or to the right. Likewise, altering the window parameter either broadens or narrows the pixel intensity range to be highlighted. Pseudo code for the construction and application of the window/level LUT algorithm is given in Algorithm 3-3.

Algorithm 3-3: Window/Level LUT
    INPUT: q-bit image I, window, level

    $MP = 2^q - 1$
    LUT[] = int(MP+1) (*array to store gray-scale transform function*)
    a = level – window/2

Algorithm 3-3 (continued)
 b = level + window/2

 foreach (i = 0 ... a-1)   (*1ˢᵗ portion all zeros*)
   LUT[i] = 0
 end

 foreach (i = a...b)   (*2ⁿᵈ portion linear scaling function*)
   LUT[i] = round[(MP/window)(i-a)]
 end

 foreach (i = b+1 ... MP)   (*3ʳᵈ portion max pixel intensity*)
   LUT[i] = MP
 end

One area (among many) where window/level LUT processing is useful is during enhancement of *synthetic aperture radar*, or SAR, images. SAR is a high-resolution satellite imaging technique capable of acquiring useful images at night that is impervious to clouds and inclement weather. These advantages make SAR a more viable technology than traditional photographic imaging modalities in modern satellite applications. By virtue of their capabilities, SAR systems have proven highly useful in geological terrain mapping, environmental applications (such as demarcating oil spill boundaries), and military surveillance and targeting. Figure 3-13 is an unprocessed SAR image of M-47 tanks.

Given an interactive means of creating window/level lookup tables, a user could selectively enhance certain portions of an image using window/level processing, and then send the adjusted gray-scale image downstream to other processing subsystems for further analysis. Such a system would be semi-automated, as an operator would have to be in the loop. A fully automated system, such as one that might be found in a real-time military application, would of course require some means of determining the appropriate window and level parameters. The downstream analysis may come in the form of image segmentation or pattern recognition algorithms, which could potentially be used here to discern the locations of imaged tanks or other battlefield components. For example, Figures 3-14 and 3-15 show the effect of various windows and levels on the SAR image of Figure 3-13. The bottom-most image of Figure 3-14b is particularly interesting, because in comparison to the other processed images, much of the background has been removed, mostly leaving the pixels pertaining to

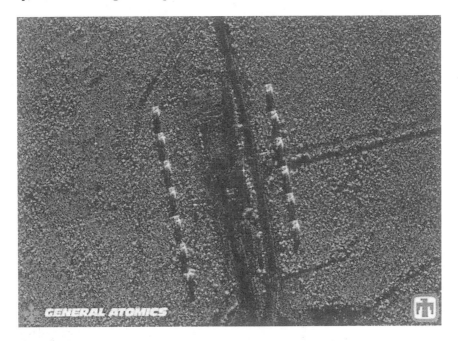

Figure 3-13. Synthetic Aperture Radar image of M-47 tanks at Kirtland Air Force Base, Albuquerque, NM (image courtesy of Sandia National Laboratories).

the 13 tanks. A potential automated targeting application could accept the windowed and leveled gray-scale image as input, and then proceed to bin the connected pixels into disparate objects using morphological operators. The centroids of these objects would serve as an estimate of tank locations.

Window/level processing is also extensively utilized in medical imaging applications, especially with regards to radiographs[9] and three-dimensional imaging modalities such as Computed Tomography (CT) and Magnetic Resonance Imaging (MRI)[10]. A common methodology is to window/level an image, where in the case of a 3D volumetric dataset a 2D image is extracted by pulling out a "slice", or plane, from the volume. For example, it is quite common to use window/level to highlight bony structure at the expense of soft issue, or to highlight lesions, tumors, or other abnormalities at the expense of other structures present within the field-of-view.

3.3.1 MATLAB Implementation

Specifying the pixel ranges used for displaying an image is rather simple using MATLAB. By default the full range is used, so when issuing the command imshow(I) or imagesc(I) MATLAB determines the

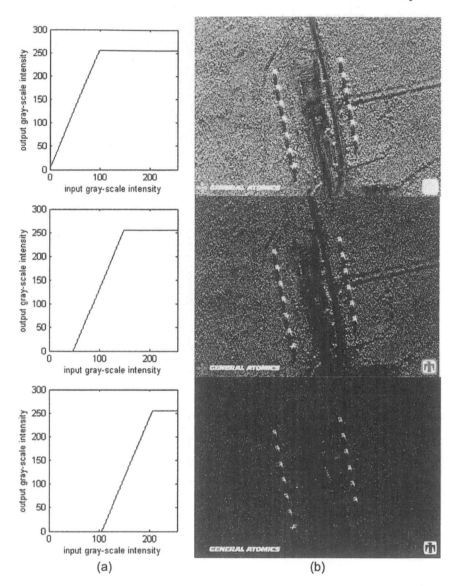

Figure 3-14. Effect of varying the level while holding the window constant (100) on the SAR tank image. (a) LUT transform functions for level=50, 100, and 157 from the top down. (b) Corresponding processed image.

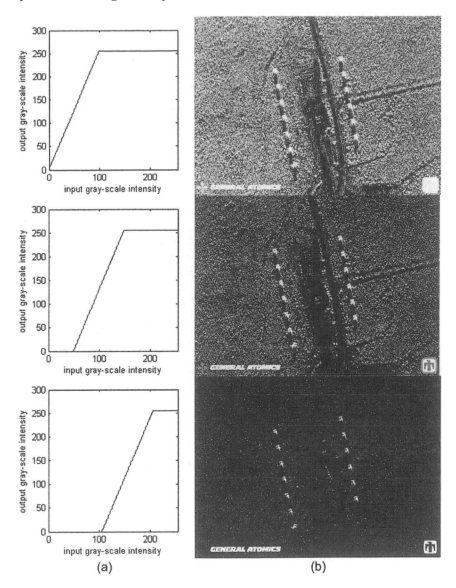

Figure 3-15. Effect of varying the window while holding the level constant (100) on the SAR tank image. The overall effect on this image is not as large as the example shown in Figure 3-14. (a) LUT transform functions for level=79, 100, and 150 from the top down. (b) Corresponding processed image.

minimum and maximum range based on I's type. The imshow and imagesc commands are similar, except that the former is available only if you have the Image Processing Toolbox installed. If you wish to display a gray-scale intensity image using a different range, then this range may be passed in as the second parameter to either of these two image display functions. For example, displaying an image with a window of 125 and level 137 is equivalent to setting the range to [73-198], which can be done using either imshow(I, 73 198]) or imagesc(I, [73 198]).

The aforementioned MATLAB commands will only display images using the specified range. To actually adjust the underlying pixel values, the function window_level (Listing 3-6) may be used.

Listing 3-6: window_level.m

```
function [J, varargout] = window_level(I, win, lev)

if strcmp('uint8', class(I))
    nlevels = 256;
    bpp = 8;
elseif strcmp('uint16', class(I))
    nlevels = 65536;
    bpp = 16;
else
    error('Require 8-bit or 16-bit input image');
end
max_pixel_intensity = nlevels - 1;

% construct gray-scale LUT
LUT = zeros(1, nlevels);

halfwin = floor(win/2);
a = lev - halfwin;
if a<1, a=1; end
b = a + win - 1;
if b>max_pixel_intensity, b=max_pixel_intensity, end

% 1st portion all zeros
% middle portion is a linear scaling function
ratio = max_pixel_intensity / win;
LUT(a:b) = round( ratio .* [1:win] );

% final portion all maximum pixel intensity
LUT(b+1:nlevels) = max_pixel_intensity;
```

```
% apply LUT to input image I
% NOTE: promote to real because certain operations not
%          allowed with integer objects and add 1 because
%          MATLAB is one-based yet pixel intensities are 0-based.
J = LUT(double(I) + 1);

% if client asks for it, return the LUT
if 2==nargout, varargout{1} = LUT; end

eval( sprintf('J=%s(J);', class(I)) );
```

The `window_level` function supports 8-bit or 16-bit images, and returns the processed image as well as the actual LUT, if desired. The innards of this function are a direct port of Algorithm 3-3, but it utilizes some advanced MATLAB features that merit additional discussion. First, the signature of the function, while on the outset appearing normal, actually defines a function that returns an output variable J as well as a list of other variables. The `varargout` output variable is a MATLAB reserved word, and refers to a MATLAB cell array. A MATLAB cell array is a heterogeneous object collection, which may in fact be empty. For example, the following MATLAB command initializes a cell array consisting of a few numbers, a string, and a matrix:

```
my_cell_array = { 1, 2.5, 3, 'hello world', rand(5) }

my_cell_array =

    [1] [2.5000] [3] 'hello world' [5x5 double]
```

With normal MATLAB arrays, parentheses are used to index into the array, whereas for cell arrays the { and } characters are used, and so `my_cell_array{4}` returns the string 'hello world'. The `nargout` reserved word is a variable containing the number of output parameters used in the current invocation of a function. So by using `nargout` in conjunction with `varargout`, as in the MATLAB code below (taken verbatim from `window_level.m`), a MATLAB function can return a variable number of *output* arguments, similar to how C functions may be defined such that they accept a variable number of *input* arguments via the ellipsis (...) operator:

```
if 2==nargout, varargout{1} = LUT; end
```

Thus `window_level.m` has been coded such that it may be invoked in one of two ways:

1. J = window_level(I)
2. [J, LUT] = window_level(I)

So if the caller wishes to obtain the LUT used to transform the pixel intensities, an optional second output parameter may be given to `window_level`. The second unique aspect of the function pertains to the mechanics of how the lookup table is applied to transform the pixel intensities of the input image I. MATLAB is very much a "vectorized" language; in fact, the key to writing efficient MATLAB code is avoiding explicit for loops, although this has been mitigated somewhat with recent advances in the MATLAB just-in-time (JIT) interpreter. So in lieu of a for loop that walks through each pixel in I and replaces that pixel with the value given by the LUT, a single indexing statement suffices. If the variable LUT contains the transformation function, and the variable I is an image matrix of type `double`, then the statement `J = LUT(I)` is a vectorized and more efficient equivalent to the following MATLAB for loop:

```
for  ipixel=1:length(I)
   J(ipixel) = LUT( I(ipixel) );
end
```

In other words, in the statement `J=LUT(I)`, each element (pixel) of I is treated as an index into LUT, thereby transforming I's pixel intensities and assigning them to the processed image matrix J.

3.3.2 A Window/Level Demo Application Using Visual Studio .NET 2003

Our first Visual Studio application is an interactive window/level demo, and it utilizes all of the key libraries that discussed in 2.4. The source code for this program can be found on the CD-ROM in the `Chap3\window_level\WindowLevelVS` directory. The front-end GUI uses the Microsoft Foundation Classes (MFC) framework for the basic user interface elements, GDI+ for image display and file I/O, and the Intel Integrated Performance Primitives (IPP) Library for a highly efficient implementation of the LUT-driven gray-scale transformation function. The application shell was built using the Visual Studio .NET MFC application wizard.

Figure 3-16 shows this application, along with the major MFC objects comprising the main dialog. The user first imports an image into the application using **File|Load**. This program only supports 24-bit color and 8-bit monochrome Windows BMP images, which in the case of 24-bit color are converted to a gray-scale format. The slider bars adjust the window and level, and the processed image is updated in real-time.

A complete listing of the source code is not shown here, but the following explanation pertains to the code in `WindowLevelDlg.cpp`. In order for the slider bars to attain a range appropriate for 8-bit images, the `SetRange` method is called within the main dialog's `OnInitDialog` method, where most of the GUI initialization takes place. In the MFC framework[6], `OnInitDialog` is somewhat analogous to a C++ class constructor. `OnInitDialog` is needed because some MFC initialization

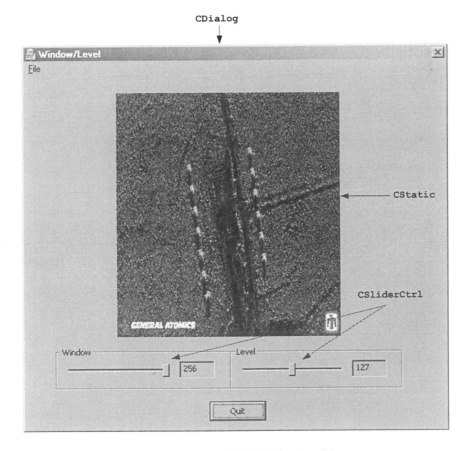

Figure 3-16. Visual Studio .NET 2003 Window/Level demo program.

code, such as that found in SetRange, can not be called from the dialog class constructor. This is because at object construction time, the Windows child controls (CStatic, CSliderCtrl, and so on) have not actually been created yet. For this type of GUI initialization, you must wait until the framework invokes OnInitDialog, which occurs shortly after the dialog is shown. The remaining MFC-centric portion of the code is fairly straightforward. A callback method CWindowLevelDlg::OnFileLoad instantiates two Image8bpp objects (described shortly), passing in the name of the file the user has chosen. The slider callback methods are

- OnNMCustomdrawWindowSlider
- OnNMCustomdrawLevelSlider

In Visual Studio, these are hooked into CWindowLevelDialog from the resource view of the dialog, as illustrated in Figure 3-17. These methods do very little except grab the current window or level value from the slider control and call on a class variable of type WindowLevel8bpp to actually process the image. After the image is processed, the window is "invalidated," which forces Windows to repaint the dialog and display the windowed and leveled image.

Another major component to the program is the Image8bpp class, whose interface is given in Listing 3-7. Image8bpp is an important utility class used in all the other Visual Studio C++ applications in this book. Objects of type Image8bpp encapsulate a gray-scale image compatible with both GDI+ (m_bitmap) and the IPP library (m_pPixels).

Listing 3-7: Image8bpp.h

```
class Image8bpp {

public:
    // ctor that reads in pixel data from a file
    Image8bpp(const CString &filename);

    Image8bpp(const Image8bpp &other);

    ~Image8bpp();

    Image8bpp &operator=(const Image8bpp &rhs);

    // draw the image to the control
    void render(HWND hwnd);
```

```
// accessor method for bitmap width
int getWidth() const { return m_pBitmap->GetWidth(); }

// accessor method for bitmap height
int getHeight() const { return m_pBitmap->GetHeight(); }

// accessor method for bitmap stride, or step-size
int getStride() const { m_stride; }

// accessor method for image pixels
Ipp8u *getPixels() const { return m_pPixels; }

private:

    // helper method called by copy ctor & assignment operator
    void copy(const Image8bpp &src);

    // GDI+ object used for file I/O and rendering purposes
    Gdiplus::Bitmap *m_pBitmap;

    // image pixels aligned to 32-byte boundary, as required by IPP
    Ipp8u *m_pPixels;

    // offset in bytes between one scan line and the next
    int m_stride;

};
```

Normally when using GDI+ it is sufficient to package all of the data associated with the image into an object of type `Gdiplus::Bitmap`. This object contains meta-data associated with the image (image dimensions, bit depth, etc.), as well as a pointer to the underlying buffer used to store the actual pixel data. In this case, the `Image8bpp` class, which wraps the GDI+ object, is designed to interoperate with both GDI+ and IPP. To attain optimal performance when using IPP library functions, the API dictates that for an 8-bit gray-scale image, pixel data must be stored as a word-aligned array of `Ipp8u` elements, similar to how the image buffer was aligned in the TI C67 contrast stretching example using the `DATA_ALIGN` pragma. The `Image8bpp` class also performs color to gray-scale conversion, placing the gray-scale pixels into this aligned buffer pointed to by `m_pPixels`. The initial step is to construct a temporary GDI+ `Bitmap` object with the file

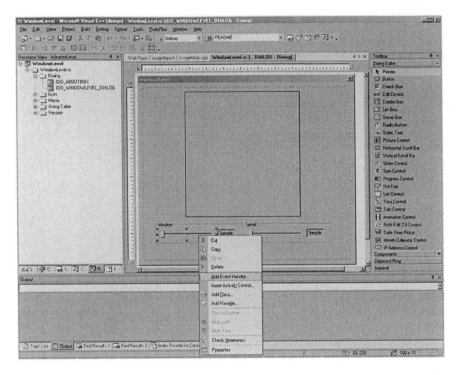

Figure 3-17. Adding an event handler for an MFC slider bar in Visual Studio .NET 2003. After selecting the "Add Event Handler" option, use the Event Handler wizard to add a callback method on your dialog class. In this project, the NM_CUSTOMDRAW Windows event results in CWindowLevelDlg::OnNMCustomdrawWindowSlider being called, which processes and refreshes the image based on the current parameter value.

name of the input image, leaving GDI+ to deal with the tedious machinations of reading in the image data from disk. If the image is a 24-bit color image, a pointer to the GDI+ allocated buffer is obtained, and the color planes are mapped to gray-scale using the following standard formula:

$$\text{gray-scale pixel intensity} = (0.299)(\text{red}) + (0.587)(\text{green}) + (0.114)(\text{blue})$$

The above relation transforms red, green, and blue pixels to "luminance," or monochrome pixels. Unfortunately, it is not possible to use the built-in IPP color conversion functions[7] because Microsoft GDI and GDI+ order color pixels in an unconventional blue-green-red (BGR) order instead of the far more common red-green-blue (RGB) format. The final Image8bpp initialization step involves creating another GDI+ Bitmap object, but this time using a form of the constructor whereby the Bitmap object itself is not responsible for deallocating the pixel data. This is necessary because we are sharing a buffer between GDI+ and IPP, and because that buffer is allocated using IPP routines it must also be deallocated using IPP routines. In addition, an 8-bit gray-scale palette is set so that Windows renders the now monochrome image correctly. All of this bookkeeping is performed within the Image8bpp constructor, shown in Listing 3-8.

Listing 3-8: Image8bpp constructor, from Image8bpp.cpp.

```
Image8bpp::Image8bpp(const CString &filename) :
                        m_pBitmap(NULL), m_pPixels(NULL)
{
  // ASCII -> UNICODE conversion
  int len = filename.GetLength();
  vector<wchar_t> wName(len+1);
  mbstowcs (&wName[0], (LPCSTR)filename, len);
  wName[len] = L'\0';

  // use this temporary object purely to deal with
  // the machinations of performing the file I/O
  Status s = Ok;
  Bitmap bmp(&wName[0]);
  if (Ok != (s = bmp.GetLastStatus()))
    throw runtime_error((LPCSTR)GdiplusUtil::getErrorString(s));

  // now allocate aligned memory for use with Intel IPP functions
  if (NULL == (m_pPixels = ippiMalloc_8u_C1(bmp.GetWidth(),
                                            bmp.GetHeight(),
                                            &m_stride)))
    throw runtime_error("Out of memory.");

  BitmapData bmpData;
  Rect rect(0, 0, bmp.GetWidth(), bmp.GetHeight());
  PixelFormat fmt = bmp.GetPixelFormat();
  if (PixelFormat24bppRGB == fmt) { // convert RGB to grayscale

    s = bmp.LockBits(&rect,
                     ImageLockModeRead,
                     PixelFormat24bppRGB,
                     &bmpData);
    if (Ok != s)
      throw runtime_error((LPCSTR)GdiplusUtil::getErrorString(s));

    // color conversion (note that even though we'd like to use the
```

```
// IPP function ippiRGBToGray we can't because Microsoft
// stores pixel values in BGR format)
unsigned char *pInput =
              (unsigned char*)bmpData.Scan0, *pScanIn = NULL;
Ipp8u *pOutput = m_pPixels, *pScanOut = NULL;
for (UINT iRow=0; iRow<bmp.GetHeight(); ++iRow) {
  pScanIn = pInput + iRow*bmpData.Stride;
  pScanOut = pOutput + iRow*m_stride;
  for (UINT iCol=0; iCol<bmp.GetWidth(); ++iCol)
   pScanOut[iCol] = (Ipp8u)(pScanIn[iCol*3]*0.114 +
                            pScanIn[iCol*3+1]*0.587 +
                            pScanIn[iCol+2]*0.299);

  }

} else if (PixelFormat8bppIndexed != fmt) { // no color conversion

  // get raw bits comprising the temporary GDI+ bitmap object
  s = bmp.LockBits(&rect,
                ImageLockModeRead,
                PixelFormat8bppIndexed,
                &bmpData);
  if (Ok != s)
   throw runtime_error((LPCSTR)GdiplusUtil::getErrorString(s));

  // copy from temporary GDI+ object into aligned memory buffer
  unsigned char *pInput =
                (unsigned char*)bmpData.Scan0, *pScanIn = NULL;
  Ipp8u *pOutput = m_pPixels, *pScanOut = NULL;
  for (UINT iRow=0; iRow<bmp.GetHeight(); ++iRow) {
    pScanIn = pInput + iRow*bmpData.Stride;
    pScanOut = pOutput + iRow*m_stride;
    for (UINT iCol=0; iCol<bmp.GetWidth(); ++iCol)
     pScanOut[iCol] = pScanIn[iCol];
  }

} else
  throw runtime_error("Only 8bpp indexed or 24bpp RGB "
                      "images supported.");

// finally, instantiate the GDI+ object that will by used for display
// purposes (note that Bitmap ctor will not perform a deep copy,
// m_pPixels is shared)
```

```
m_pBitmap = new Bitmap(bmp.GetWidth(),
                       bmp.GetHeight(),
                       m_stride,
                       PixelFormat8bppIndexed,
                       m_pPixels);
if (Ok != (s = m_pBitmap->GetLastStatus()))
  throw runtime_error((LPCSTR)GdiplusUtil::getErrorString(s));

// without correct color palette gray-scale image will not display
// correctly
m_pBitmap->SetPalette(GdiplusUtil::get8bppGrayScalePalette());
}
```

The actual image processing occurs within the `WindowLevel8bpp` class, invoked by `CWindowLevelDlg::winLevChanged` (from the main dialog class) to process the image each time one of the slider bars changes value. The `WindowLevel8bpp::process` method, shown in Listing 3-9, is essentially a direct port of Algorithm 3-3, except that it uses the IPP library function `ippiLUT_8u_C1R` to perform the gray-level LUT pixel transformation in a more computationally efficient manner than a straight-forward C/C++ implementation would yield. Intel IPP functions deal with various data types through a well-defined and consistent naming convention, as opposed to function overloading and/or templates, presumably because the library can then be used within a purely C application. The "8u" in the function name indicates that the function expects a pointer to `Ipp8u` elements (which are `unsigned chars`), and "C1R" denotes a single channel (i.e., a monochrome image). For example, if you happen to be dealing with 16-bit gray-scale pixel data, the correct function to use would be `ippiLUT_16u_C1R`[1].

Listing 3-9: The core C++ window/level algorithm, from `WindowLevel8bpp.cpp`.

```
void WindowLevel8bpp::process(const Image8bpp *pIn,
                              Image8bpp *pOut,
                              int window,
                              int level)
{
// construct LUT
int ii=0, jj=0, N=level - window/2;

// 1st portion is all zeros
if (N > 0)
```

```
for (; ii<N; ++ii)
  m_pValues[ii] = 0;

// 2nd portion is a linear relationship
float ratio = (float)(N_LEVELS-1)/(float)window;
for (; jj<window; ++jj)
  m_pValues[ii+jj] = (Ipp32s)( (ratio*jj)+0.5 ); // +0.5 for rounding

// 3rd portion is just all 255s
ii += jj;
for (; ii<N_LEVELS; ++ii)
  m_pValues[ii] = N_LEVELS-1;

// apply LUT
IppiSize roiSize = { pOut->getWidth(), pOut->getHeight() };
IppStatus s = ippiLUT_8u_C1R(pIn->getPixels(),
                             pIn->getStride(),
                             pOut->getPixels(),
                             pOut->getStride(),
                             roiSize,
                             m_pValues,
                             m_pLevels,
                             N_LEVELS);

if (ippStsNoErr != s)
  throw runtime_error("LUT operation failed");
}
```

The final object of interest in this Visual Studio demo program is
GdiplusUtil which, like Image8bpp, shall be utilized extensively in
later applications. This object provides two static member functions that
have previously been encountered in Listings 3-7 and 3-8. The method
getErrorString converts error identifiers (Gdiplus::Status) to
strings, and get8bppGrayScalePalette returns an 8-bit gray-scale
color palette. More importantly however, a global object of type
GdiplusUtil is declared in GdiplusUtil.cpp. The reason why this
is done is that the GDI+ documentation[8] states that GdiplusStartup
must be called prior to using any GDI+ facility. By creating a global instance
of GdiplusUtil, and making the GdiplusStartup call in
GdiplusUtil's constructor, we ensure that GdiplusStartup is
invoked at program startup, prior to any code that may use GDI+ facilities.
While the C++ standard is somewhat vague about this, in practice one can

safely assume that initialization of global objects takes place prior to any user code being called. Whether or not this means that GdiplusUtil's constructor is called prior to main, or sometime shortly thereafter is besides the point – you can assume that the global object's constructor, and thus by extension GdiplusStartup, will be called prior to the main dialog being created (which is all that matters, since all GDI+ code is called after dialog initialization). In GdiplusUtil's destructor, the GdiplusShutdown function is called, ensuring that GDI+ cleanup is performed at program exit, as Microsoft documentation recommends.

3.3.3 Window/Level on the TI C6x EVM

The previous Visual Studio window/level program was interesting in that it illustrated how slider bars allow the user to interactively control the input parameters for an image processing algorithm, and how algorithm output can be visualized in real-time. Suppose as part of an embedded application, you had to implement window/level, perhaps as a component of a larger imaging subsystem. How would you go about testing your C6x implementation? In this example, we will draw upon some of the tools we previously used to feed data into the TI IDE, but will augment it with the ability to dynamically change the window and level parameters from within CCStudio. The end result, while not approaching the level of user interaction provided by the Visual Studio C++ application, nevertheless serves to show how a fair amount of interactivity may be obtained using CCStudio General Extension Language (GEL) files. The alternative is to hard-code the window and level parameters in the source code, and rebuild the application each time one wants to test various window and level settings - hardly a solution conducive to rapid and effective testing.

In the Chap3\window_level\WindowLevelCCS directory a CCStudio project named window_level.pjt may be found. The linker settings, accessed through the linker command file window_level.cmd, are identical to those of the contrast stretching CCStudio project of 3.2.2, and the build settings are also similar except that this project does not utilize any IMGLIB functions, and thus there is no need to link to imglib.lib nor include the IMGLIB include file directory. Listing 3-10 are the contents of the sole C module for this project, window_level.c.

Listing 3-10: window_level.c.
```
#include <board.h> /* EVM library */
#include <limits.h>
#include <stdio.h>
```

```
/* image dimensions */
#define X_SIZE 128
#define Y_SIZE 128
#define N_PIXELS X_SIZE*Y_SIZE

/* image is processed "in-place" */
unsigned char img[N_PIXELS];

/* LUT */
#define N_LEVELS 256 /* 8 bpp */
int window = N_LEVELS;
unsigned char level = 127,
                  LUT[256] = {0};

void make_lut()
{
   int ii=0, jj=0, N=level-window/2;
   float ratio = (float)(N_LEVELS-1)/(float)window;

   // 1st portion is all zeros
   if (N > 0)
      for (; ii<N; ++ii)
        LUT[ii] = 0;

   // 2nd portion is a linear relationship
   for (; jj<window; ++jj)
      LUT[ii+jj] = (unsigned char)( (ratio*jj)+0.5 ); // +0.5 for rounding

   // 3rd portion is just all 255s
   ii += jj;
   for (; ii<N_LEVELS; ++ii)
      LUT[ii] = N_LEVELS-1;
}

void apply_lut()
{
   int ii=0;
   for (; ii<N_PIXELS; ++ii)
      img[ii] = LUT[ img[ii] ];
}
```

```
int main(void)
{
  evm_init(); /* initialize the board */
  make_lut();
  apply_lut();
  printf("window/level completed");
}
```

This program performs its processing in-place, that is the pixel data is fed into the array img and then this same array is subsequently overwritten as the LUT processing takes place. The program's high-level flow can be garnered from the main function, where we first initialize the EVM board, then construct a LUT according to the values contained in the global variables window and level, and finally process the image in apply_lut. In short, this C program is more or less a direct line-by-line port of Algorithm 3-3.

Fair enough, but then how is the image data read into the program and how can we dynamically alter the window and level parameters, a la the Visual Studio C++ application from 3.3.2? For this functionality, we turn to a very powerful add-on feature to CCStudio, the GEL file. GEL files are written in an interpretive language with a C-like syntax that can plug into CCStudio, and are quite useful for automating repetitive tasks – such as feeding data into the DSP and dynamically changing variables. Listing 3-11 is the contents of window_level.gel, which can be found in Chap3\window_level\WindowLevelCCS.

Listing 3-11: window_level.gel
```
menuitem "Window/Level"

hotmenu LoadImage()
{
  GEL_AddInputFile("window_level.c", // which file has probe point
            49,            // line # for probe point
            "tanks.dat",   // location of input data
            1,             // hex format
            "img",         // where to send the data to
            "4096", 0, 1); // length (128x128 image)
}
```

```
slider W(1, 256, 1, 1, win)
{
   window = win;
}

slider L(0, 255, 1, 1, lev)
{
   level = lev;
}

hotmenu SaveImage()
{
   GEL_AddOutputFile("window_level.c",
                     51,
                     "out.dat",
                     1,
                     "img",
                     "4096");
}
```

This GEL file adds four plug-in features into CCStudio, which should be self-explanatory from the function names in Listing 3-11. LoadImage reads in a CCS data file, specified by the third input argument, and sends it to the global variable specified by the fifth argument (in this case img). The test data contained in the CCStudio-formatted data file can be created with the save_ccs_data_file MATLAB function (see Listing 3-4) or through other means. SaveImage performs the converse, taking data contained in img and sending it to a CCStudio data file, which in turn can be parsed using the read_ccs_data_file MATLAB function (see Listing 3-5). The remaining two functions, W and L, implement sliders allowing the user to change window and level, respectively.

To see the application in action, build and load the program window_level.out onto the DSP. Set a breakpoint in the main function, at the line where make_lut is called. Run the program, and when execution halts at the breakpoint, load window_level.gel from the **File|Load GEL** menu selection. After the GEL file has been loaded into CCStudio, a new menu appears under the main GEL menu, "Window/Level", whose name is specified by the first line of Listing 3-11. Underneath this new menu is a sub-menu with the four previously described selections. You can now show the window and level slider bars by using the **GEL|Window\Level|W** and **GEL|Window\Level|L** menu options.

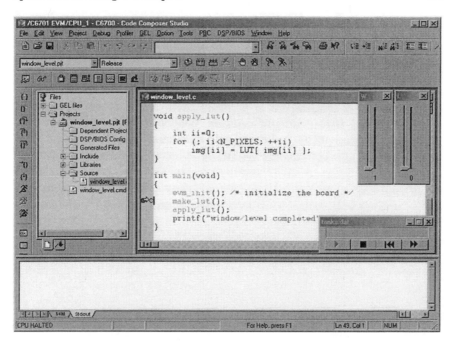

Figure 3-18. Code Composer Studio running `window_level.out`, after having selected the **GEL|Window\Level|W, GEL|Window\Level|L,** and **GEL|Window\Level|Load Image** menu options. Pressing the ">>" button on the tape-player control sends the data in `tanks.dat` into the global `img` array.

Selecting **GEL|Window\Level|Load Image** displays a CCStudio tape-player control, through which you feed the input data to the DSP. Note that you can change the location of the input data file to match whatever happens to be on your local disk by modifying the third argument to `GEL_AddOutputFile` in the `LoadImage` GEL function, and subsequently reloading the GEL file. After selecting these three menu options, the CCStudio IDE will look similar to the screen-shot shown in Figure 3-18.

The forward arrow in the tape-player control fills `img` with the pixel values stored in the CCStudio data file, and the slider bars update the window and level global variables. After window and level have been set, step past `apply_lut` in the debugger, and then select **Gel|Window\Level|Save Image**. This menu selection brings up another tape player control, which is then used to send the contents of `img` to the file specified in the third argument to `GEL_AddOutputFile`.

3.4 HISTOGRAM EQUALIZATION

In both the contrast stretching and window/level algorithms, monotonic and linear (and in the case of window/level, piece-wise linear) gray-level transform functions are applied to an image whose processed histogram yields an image of enhanced quality. In this section we consider a non-linear transform that takes into account the morphology of the input histogram. Consider the image shown in Figure 3-19a. One way to interpret the image's discrete PDF (histogram) is to analyze its shape. Underutilized pixels correspond to valleys or shallow portions of the histogram, and if there were an algorithm that shifted pixel intensities such that the histogram was more evenly distributed, is it reasonable to assume that such a method would produce a better looking image? Figure 3-19c is the result of applying just such an operation, known as histogram equalization.

There are instances when histogram equalization can dramatically improve the appearance of an image, and this algorithm also forms a building block leading to a more general algorithm, histogram specification, that is used in image fusion applications. Histogram equalization alters the input histogram to produce an output whose histogram is *uniform*, where the various pixel intensities are equally distributed over the entire dynamic range. This situation is depicted in Figure 3-20. In the ideal continuous case, histogram equalization normalizes the pixel distribution such that the number of pixels in the image with pixel intensity p is MN/Q, where M is the number of image rows, N is the number of image columns, and Q is the number of gray-levels – in other words, a completely uniform probability density function whose value is constant. However, because digital image processing algorithms operate in the discrete domain, digitization effects come into play and prevent a purely uniform distribution, as evidenced by the shape of Figure 3-19d, where the equalized histogram is close to, but not exactly constant, over the interval 0-255.

So the goal of the histogram equalization algorithm is to find a gray-level transformation, of the now familiar form $s = T(r)$, which makes the histogram $p(r_k)$ for k=0,1,...,Q-1 flat. The question now becomes how to find $T(r)$? For the purpose of this discussion, we normalize the allowable pixel range such that all pixel intensities are between 0 and 1, i.e. $0 \leq r \leq 1$. We then assume the following two conditions:

1. $T(r)$ is a monotonically increasing function that preserves the intensity ordering.
2. $T(r)$ maps the range [0,1] to [0,1].

Figure 3-19. Histogram equalization. (a) Original image of birds at San Francisco Zoo. (b) Unprocessed histogram. (c) Histogram equalized birds image. (d) Equalized histogram, exhibiting an approximation to a uniform distribution across the range [0-255].

These two conditions are present for both the contrast stretching and window/level transforms, and if they hold, it is possible to define the inverse mapping $s = T^{-1}(r)$ because T is monotonic and single-valued. This inverse transform will in turn satisfy both (1) and (2).

All of the pixels in the input image whose gray-levels lie within the range $[r, r+\Delta r]$ will have their pixel intensities reassigned such that they lie within the range $[s, s+\Delta s]$. If $p_f(r)$ and $p_g(s)$ denote the PDFs of the input and output images, respectively, then the total number of input pixels with gray-level values lying between r and Δr is $p_f(r) \Delta r$. Likewise, the number of output pixels with gray-level values lying between s and Δs is $p_g(s) \Delta s$. Knowing what we do about $T^{-1}(s)$, the following condition must be true:

$$p_g(s)\Delta s = \left\lfloor p_f(r)\Delta r \right\rfloor_{r=T^{-1}(s)}$$

This equation basically states that for a given interval, the area underneath the two probability density functions $p_f(r)$ and $p_g(s)$ is the same (see Figure 3-21). If $p_f(r)$ is known, and T and T^{-1} both satisfy condition 1, then it follows from probability theory that we can take the limit as both Δr and Δs go to zero and obtain

$$\text{Limit } \Delta r, \Delta s \to 0 : p_g(s) = \left[p_f(r)\frac{dr}{ds} \right]_{r=T^{-1}(s)}$$

Since we desire a uniform distribution for the processed image pixels, the transform function T should adjust the gray-levels so that the output histogram is constant:

$$p_g(s) = C, \ 0 \le s \le 1$$

Plugging the above equation into the previous one yields:

$$C = p_f(r) \ dr/ds$$

$$Cds = p_f(r)dr$$

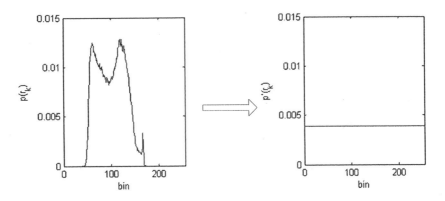

Figure 3-20. Histogram equalization in the idealized (continuous) case.

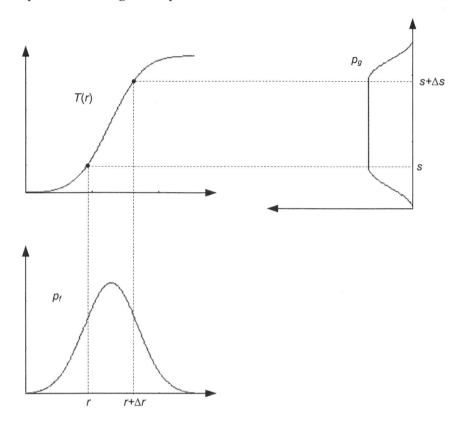

Figure 3-21. Histogram equalization in action. The transform function (top left) is such that the surface area under the curve p_f from r to $r+\Delta r$ is the same as the surface area under the curve p_g from s to $s+\Delta s$. The plot of the output histogram p_g is rotated by 90 degrees.

From the fundamental theorem of calculus, we can take the integral of both sides:

$$s = \int_0^r p_f(r')dr' = T(r)$$

This integral is in fact the CDF of the distribution function $p_f(r)$ – so to construct the gray-scale transform function $T(r)$ producing an output image whose histogram is approximately uniform, we first compute the CDF of the input image and then use it to remap the pixel intensities. As an example, consider Table 3-2 where we prepare to perform histogram equalization on a 10x10 3 bpp image. Such an image consists of 100 pixels, with pixel intensities ranging from 0-7. Starting with the discrete PDF (histogram) of the image (first three columns of Table 3-2), we next compute the CDF (cumulative histogram) of the image (last two columns of Table 3-2).

Table 3-2. Calculation of the CDF for histogram equalization.

Pixel Intensity	Histogram Count	PDF $p_r(r)$	Cumulative Histogram	CDF
0	7	.07	7	.07
1	0	0	7	.07
2	13	.13	20	.2
3	15	.15	35	.35
4	6	.06	41	.41
5	15	.15	56	.56
6	32	.32	88	.88
7	12	.12	100	1.0
Total	100	1.0		

In the next step, shown in Table 3-3, we use the CDF in conjunction with the maximum pixel intensity (in this case 7), to remap the pixel intensities. Note the properties of this mapping, namely that:

- $s = T(0) = 0$
- $s = T(2^{bpp}-1) = 2^{bpp}-1$.

Table 3-3. Remapping pixels using histogram equalization.

Pixel Intensity	$T(r) = 7*CDF$	$s = \text{round}[T(r)]$
0	.49	0
1	.49	0
2	1.4	1
3	2.45	2
4	2.87	3
5	3.92	4
6	6.16	6
7	7	7

The complete histogram equalization procedure is given in pseudo-code form in Algorithm 3-4.

Algorithm 3-4: Histogram Equalization
INPUT: MxN q-bit image I

$MP = 2^q-1$
hist[] = int(MP+1)   (*array to store histogram bins*)
cdf[] = int(MP+1)   (*array to store cumulative distribution f*)

Algorithm 3-4 (continued)
```
foreach (b = 0...MP)
   hist[b] = 0
end

(compute histogram, or PDF)
foreach (pixel p in I)
   hist[p] += 1
end

(from PDF compute CDF)
cdf[0] = hist[0]
for (b = 1...MP)
   cdf[b] = cdf[b-1] + hist[b]
end

(normalize CDF)
for (b = 0...MP)
   cdf[b] = cdf[b] / MN
end

(remap pixels)
foreach (pixel p in I)
   p = MP*cdf[p]    (apply non-linear transform function)
   p = round(p)
end
```

3.4.1 Histogram Specification

Histogram equalization is used as an intermediate step in a closely related and more useful spatial processing technique, histogram specification. As we have just seen, in histogram equalization a single output is generated – a transformed image with a flat, or uniform, probability distribution. In histogram specification, we generalize this algorithm so that we can produce an output image whose histogram matches some desired probability distribution. For example, suppose you have an image and wish to apply a gray-level pixel mapping so that the altered image's histogram takes on a Gaussian distribution – or in general, if you want to remap an image's pixels such that the transformed image has the same histogram as some target image.

Let $p_f(w)$ denote the input image probability distribution, $p_d(w)$ the specified (desired) pixel distribution, and g the processed image. Then we can obtain $T(r)$ which equalizes $p_f(w)$ using the histogram equalization equation:

$$s = T_f(r) = \int_0^r p_f(w)\, dw$$

Likewise, because we specified $p_d(w)$, we can equalize the desired output histogram:

$$v = T_d(r) = \int_0^r p_d(w)\, dw$$

Histogram equalization produces a single output – a uniform PDF – and hence it follows that:

$$s = T_f(r) = \int_0^r p_f(w)\, dw$$

$$s = T_d(r) = \int_0^r p_d(w)\, dw$$

$$\rightarrow T_d(r) = T_f(r)$$

Assuming the inverse transform T_d^{-1} exists and is single-valued and monotonically increasing, inverting T_d and feeding in s (the equalized levels of the input image) yields:

$$g = T_d^{-1}(r) = T_d^{-1}\left[T_f(r)\right]$$

To summarize, we first transform the input histogram to a uniform distribution via the application of T_f, and then transform this uniform distribution to a set of pixels characterized by the desired distribution by feeding the intermediate result into T_d^{-1}. Thus, to perform histogram specification, it is not necessary to explicitly histogram equalize the input. Rather, we merely need to compute the equalization gray-level transform and combine it with the inverse transform T_d^{-1}. In the continuous domain, obtaining analytical expressions for $T_f(r)$ and T_d^{-1} is difficult, if not impossible. In the digital domain, we are able to circumvent this problem by approximating these functions using discrete values. An optimized histogram specification algorithm that does not explicitly equalize the input is given in Algorithm 3-5.

Algorithm 3-5: Histogram Specification
INPUT: M_1xN_1 q-bit image I, M_2xN_2 q-bit target image T

```
MP = 2^q-1
Ihist[] = int(MP+1)  (discrete PDF of I)
Thist[] = int(MP+1)  (discrete PDF of T)
Icdf[] = int(MP+1)  (cumulative histogram of I)
Tcdf[] = int(MP+1)  (cumulative histogram of T)
LUT[] = int(MP+1)  (histogram specification pixel re-mapping function)

for each (b = 0...MP)
   Ihist[b] = Thist[b] = 0
end

for each (pixel p in I)              (compute histograms of both images)
   Ihist[p] += 1
end
foreach (pixel p in T)
   Thist[p] += 1
end

for (bin = 0...MP)                        (normalize both histograms)
   Ihist[b] /= M_1N_1
   Thist[b] /= M_2N_2
end
(compute CDFs, to find fraction of pixels with values ≤p)
Icdf[0] = Ihist[0]
Tcdf[0] = Thist[0]

for (p = 1 ... MP)
   Icdf[p] += Icdf[p-1]
   Tcdf[p] += Tcdf[p-1]
end
```

Algorithm 3-5 (continued)
   (*histogram specification gray-level transform function: scan along input CDF, stopping when cumulative histogram exceeds target CDF*)

```
newval = 0 (the remapped pixel value)
for (bin = 0...MP)
   while Tcdf[bin]<Icdf[bin] && newval<MP
      newval += 1
   end
   LUT[bin] = newval
end

(remap pixels)
foreach (pixel p in I)
   p = LUT[p]       (apply transform function)
end
```

3.4.2 MATLAB Implementation

The Image Processing Toolbox has a function that performs both histogram equalization as well as histogram specification, histeq. The first argument to this function is the input image, and if a second argument, a histogram vector, is provided histeq performs histogram specification. On the CD, in the Chap3 directory, there are three M-files which may be used in lieu of histeq for those who do not have access to the Image Processing Toolbox. Listing 3-12 is the implementation for hist_equalize, which is based on Algorithm 3-4 and equalizes an image.

Listing 3-12: MATLAB function for histogram equalization.

```
function J = hist_equalize(I)

if strcmp('uint8', class(I))
   nlevels = 256;
elseif strcmp('uint16', class(I))
   nlevels = 65536;
else
   error('Require 8-bit or 16-bit input image');
end

% normalized histogram (discrete PDF) and from that get CDF
hgram = histc(I(:), 0:nlevels-1) ./ length(I(:));
CDF = cumsum(hgram);
```

```
max_pixel = nlevels - 1;
LUT = round(max_pixel .* CDF);
```

```
% apply LUT to input image I
% NOTE: must promote to real because certain operations
%       now allowed with integer objects and add 1 because
%       MATLAB is one-based yet pixel intensities are 0-based.
J = LUT(double(I) + 1);
```

```
% retain original type
if 256 == nlevels
    J = uint8(J);
else
    J = uint16(J);
end
```

It is sometimes difficult to gauge the effectiveness of histogram equalization on gray-scale images, but the procedure often shines when employed on color images. Color images are typically represented as multi-channel two-dimensional signals, whereas the gray-scale images we have encountered have just a single (monochrome) channel. Color images in MATLAB can be stored as three-dimensional RGB matrices, with each "slice" or sub-matrix containing the red (R), green (G), and blue (B) channels. In the `images` directory on the CD, a 24-bit (8 bits per color channel) Windows BMP image file `hills.bmp` may be found. This image file may be read into MATLAB using the built-in `imread` function:

```
I = imread('hills.bmp');
```

The variable `I` has dimensions 492x489x3, and the command `I(:,:,c)` for c=1,2,3 returns the red, green, and blue channels respectively. How does one go about equalizing a color image? While it may be tempting to simply treat each channel as its own gray-scale image and proceed to separately equalize them and subsequently combine them back together to form a processed RGB image, this is not advised. What will happen is that separate equalization of the three color channels will result in artificial color shifts, producing a psychedelic effect of sorts. Rather, the recommended course of action is to apply a color transformation to the RGB channels that maps R, G, and B to the hue, saturation, and value (HSV) color space – this process is detailed in Figure 3-22. The value channel in this space is roughly equivalent to intensity, so histogram equalization may be safely performed on this channel. The inverse color transform is then used to

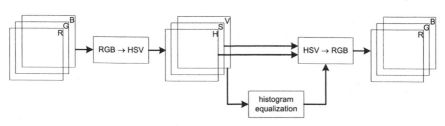

Figure 3-22. Histogram equalizing an RGB image.

map the H, S, and equalized V channels back to RGB space, which can then be displayed. The MATLAB Image Processing Toolbox functions rgb2hsv and hsv2rgb can be used to perform these sorts of color transformations.

Unfortunately, the Image Processing Toolbox histeq function does not handle RGB images – Listing 3-13 is a function histeqRGB which can be used to perform the aforementioned color equalization. Figure 3-23 illustrates the effect histogram equalization can have on a digital photograph. This figure is shown in monochrome, however the histogram equalized color image, processed via histeqRGB, may be found on the CD under the images directory.

Listing 3-13: MATLAB function to histogram equalize an RGB image.
```
function J = histeqRGB(I)

Ihsv = rgb2hsv(I); % convert from RGB space to HSV space

% V is roughly equivalent to intensity, bear in mind that
% V is now of type double, with values in the range [0-1]
V = Ihsv(:,:,3);

% perform histogram equalization on the V channel,
% overwriting the original data
Ihsv(:,:,3) = histeq(V);

J = hsv2rgb(Ihsv); % back to RGB space

% we're definitely now of type double, here we retain I's original type
if strcmp('uint8', class(I))
    J = uint8(255.*J);
elseif strcmp('uint16', class(I))
    J = uint16(65535.*J);
end
```

(a)   (b)

Figure 3-23. Histogram equalizing hills.bmp via histeqRGB (photo courtesy of
http://philip.greenspun.com). (a) Original 24-bit RGB image. (b) Result of equalizing the
"value" channel of the HSV transformed image, and then applying the HSV-to-RGB
mapping.

Algorithm 3-5 delineated an efficient means of performing histogram
specification that does not explicitly equalize the input image. A MATLAB
function, hist_specification, that implements this algorithm is given
in Listing 3-14. If asked for, this function also returns the gray-level
transformation function used to perform the histogram matching, a la
window_level from 3.3.1. When using hist_specification, it is
imperative to pass in a *normalized* histogram (that is, divide each frequency
bin by the total number of pixels, such that the summation of the entire
vector is 1.0). A common mistake is forgetting to normalize the target
histogram (hgram in hist_specification). In this case, the gray-
level transform function is erroneously constructed because the scale of the
two discrete PDFs are no longer the same. The hist_specification
function does perform a sanity check on the desired histogram, by verifying
that the maximum element is not greater than 1.0.

Listing 3-14: MATLAB function for histogram specification.

```
function [J, varargout] = hist_specification(I, hgram)

if strcmp('uint8', class(I))
    nlevels = 256;
elseif strcmp('uint16', class(I))
    nlevels = 65536;
```

```
else
    error('Require 8-bit or 16-bit input image');
end


% this function expects a normalized target histogram
if max(hgram(:) > 1)
    error('Please divide target hist by # pixels in desired image!');
end


% calculate normalized histogram of input image
H = histc(I(:), 0:nlevels-1);
H = H ./ length(I(:));
% cumulative histograms of both input and desired distributions
cdf_orig = cumsum(H);
cdf_desired = cumsum(hgram);


% construct gray-level transformation lookup table here:
% scan along input CDF, stopping when cumulative histogram
% exceeds target CDF.
p = 0;
for ii=0:255
    while cdf_desired(p+1)<cdf_orig(ii+1) && p<255
        p = p + 1;
    end
    LUT(ii+1) = p;
end


% apply LUT to input image I
% NOTE: must promote to real because certain operations
%       not allowed with integer objects, and add 1 because
%       MATLAB is one-based yet pixel intensities are 0-based.
J = LUT(double(I) + 1);


% if client asks for it, return the LUT
if 2==nargout
    varargout{1} = LUT;
end
```

3.4.3 Histogram Specification on the TI C6x EVM

A histogram equalization implementation targeting the C6701 EVM, located in `Chap3\histogram_equalization`, is instructive in a number of ways. The obvious is that it is a C port of Algorithm 3-4, again making use of the IMGLIB function IMG_histogram. In addition, this program is illustrative of the fact that substantial performance increases may be obtained via high-level optimizations, which in this particular algorithm entail eschewing floating-point operations in favor of far more efficient fixed-point operations.

While it is often tempting to focus on low-level optimizations (e.g., rewriting tight loops in assembly language) when attempting to increase the performance of a computationally expensive algorithm, the adage that "90% of the work is done in 10% of the code" definitely holds true most of time, and even more so in the case of DSP and image processing applications. Developers will obtain far greater "bang for their buck" by initially focusing their efforts and time on higher-level optimization opportunities first. Then, if the performance of the algorithm is still unsatisfactory, lower-level optimizations begin to take on added importance. This discussion serves two purposes: one, to present the C implementation of a histogram equalization routine running on the DSP and two, to describe how the CCStudio profiling tools can be used to measure the performance of C code running on a TI DSP[11].

The histogram equalization EVM project is modeled closely after the contrast stretching and window/level projects. The same linker command file is used, and the only substantial infrastructure change is that the in_img buffer is now pre-initialized with the pixel data from Figure 3-19 in the image.h header file. As described in previous sections, the processed image can be sent back to the host PC using the CCStudio file I/O facilities and pointing them to the out_img array. Listing 3-15 is the contents of histeq.c, which contains both a floating-point as well as a fixed-point implementation of Algorithm 3-4.

Listing 3-15: histeq.c

```
#include <board.h> /* EVM library */
#include <limits.h>
#include <stdio.h>
#define _TI_ENHANCED_MATH_H 1
#include <math.h>
#include <img_histogram.h> /* IMGLIB */
#include "image.h" /* birds image, along with image dimensions */

/* Input & output buffer both will not fit in
   internal chip RAM, this pragma places the
   output buffer in off-chip RAM */
#pragma DATA_SECTION (out_img, "SBSRAM");
```

```
unsigned char out_img[N_PIXELS];

/* image histogram goes here, the pragma
   aligns the buffer on a 4-byte boundary
   which is required by IMGLIB */
#pragma DATA_ALIGN (hist, 4)
unsigned short hist[256];

/* scratch buffer needed by IMGLIB */
#pragma DATA_ALIGN (t_hist, 4)
unsigned short t_hist[1024];

/* pixel mapping function */
#pragma DATA_ALIGN (T_r, 4)
unsigned short T_r[256];

#define FIXED_POINT 1 /* 0=floating point, 1=fixed point */
#if FIXED_POINT
unsigned int cumsum; // 32-bit accumulator
#else // floating-point
const float num_pixels = (float)N_PIXELS;
float cumsum;
#endif /* FIXED_POINT */

void histeq()
{
  int ii = 0;
  cumsum = 0;

  /* buffers must be initialized to zero */
  memset(t_hist, 0, sizeof(unsigned short)*1024);
  memset(hist, 0, sizeof(unsigned short)*256);

  /* use IMGLIB to compute image histogram */
  IMG_histogram(in_img, N_PIXELS, 1, t_hist, hist);

  /* form gray-level transform function T(r) = CDF */
#if FIXED_POINT
  for (ii=0; ii<256; ++ii) {
    cumsum += (hist[ii] << 2);
    T_r[ii] = (255*cumsum) >> 16;
  }
```

```
#else
  for (ii=0; ii<256; ++ii) {
    cumsum += hist[ii];
    /* +0.5 followed by integer truncation accomplishes rounding */
    T_r[ii] = (255.f*cumsum/num_pixels) + 0.5;
  }
#endif /* FIXED_POINT */

  /* apply histogram equalization transform function */
  for (ii=0 ; ii<N_PIXELS; ++ii)
    out_img[ii] = T_r[in_img[ii]];
}
int main(void)
{
  evm_init(); /* initialize the board */
  histeq();
  printf("histogram equalization completed\n");
}
```

Toggling between the two variants of the algorithm is accomplishing through the C preprocessor directive `FIXED_POINT` (if defined as 0 then the `histeq` function is floating-point, else the function utilizes a fixed-point implementation). In both scenarios, the code proceeds in three phases, where the first and third phases are identical in both fixed and floating-point:

1. Estimate the discrete PDF of the input image using the IMGLIB `IMG_histogram` function.
2. Construct the `T_r` array, which holds the lookup table that is the basis for the pixel remapping function.
3. Remap pixels from `in_img` to `out_img`, using `T_r`.

The floating-point version of phase 2 is straightforward. The `cumsum` variable serves as an accumulator storing the current value of the CDF, and in the subsequent statement `T_r[ii]` is set to a scaled version of this value, thereby mapping the normalized range [0-1] to the 8-bit range [0-255]. At first glance, the fixed-point version of this loop is not as straightforward. Fixed-point numbers were introduced in 2.1.2, and this program utilizes what could be called a Q2 format. For comparison purposes, Listing 3-16 is a side-by-side comparison of the floating-point and fixed-point histogram equalization loop.

Listing 3-15: floating-point loop (left) versus fixed-point loop (right).

```
for (ii=0; ii<256; ++ii) {              for (ii=0; ii<256; ++ii) {
  cumsum += hist[ii];                     cumsum += (hist[ii]<<2);
  T_r[ii] =                               T_r[ii] =
    (255.f*cumsum/num_pixels)+0.5;          (255*cumsum/N_PIXELS)>>2;
}                                       }
```

Note that in the actual code of Listing 3-15, the second line in the fixed-point loop replaces the division by N_PIXELS (which is known to be $2^{14} =$ 16384) with a corresponding right shift of 14 bits. Thus the right-hand side of that statement, (255*cumsum/N_PIXELS)>>2, reduces to (255*cumsum)>>16. We can use the profiling tools included with Code Composer Studio to quantify exactly how much is gained by using fixed-point representations of numbers on a floating-point processor like those of

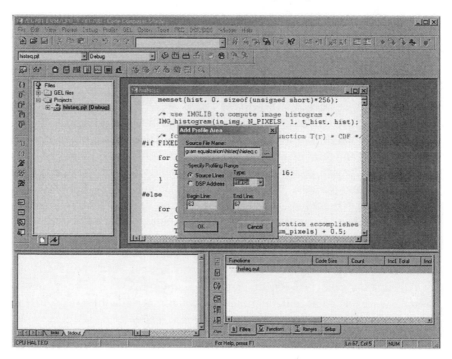

Figure 3-24. Setting a CCStudio profile area to benchmark histeq performance. Lines 62-68 contain the code to construct the floating-point T_r array. To benchmark the fixed-point implementation, set the FIXED_POINT preprocessor symbol to 1, recompile, reload the program, and use the range 55-60.

the C67x series (the difference will be even more pronounced on a fixed-point processor). Of course, in order to make a fair comparison aggressive compiler optimizations should be enabled, which is done by selecting the '−o3' optimization level within the Build tab of the dialog from the **Project|Build Options** menu selection. Note that the '−g' option (debug symbols) must be enabled to use the Code Composer Studio profiler.

After loading the program onto the DSP, and selecting **Debug|Go Main** (which loads the symbols), start a profiler session via **Profiler|Start New Session**. After starting the session, a profile window appears where a section of code may be marked for profiling by clicking on the **Create Profile Area** button in this new window (the fourth button from the top). This button brings up an **Add Profile Area** dialog where the section of code to profile is set. Enter the lines of code to profile, and take care to select the "Range" type as opposed to "Function" type if you wish to profile just a portion of the function in question, as we do here (see Figure 3-24). Also note that to measure accurate cycle times, the processor should be reset prior to each benchmark, so as to clear out the cache and make a truly fair comparison (we wish to avoid the situation where a warm cache skews the measurements). Resetting the DSP is done by selecting **Debug|Reset**.

The floating-point version clocks in at 35,702 cycles on a C6701 EVM, while the fixed-point version takes 528 cycles, as measured by the CCStudio profiler. However, the CCStudio profiler, while quite convenient, is unable to tell the full story, the reason being that it only works if the program is compiled with the −g option. The inclusion of debug symbols increases code size and disables certain optimizations, so in order to get a true idea of how long each implementation takes the −g option should be disabled, thereby precluding the use of the CCS profiler. Another C6x EVM program can be found in the Chap3\histogram_equalization_profile directory. This program is fundamentally the same as its predecessor, except that profiling statements using the clock function have been interspersed throughout the code, as described in [12]. Listing 3-17 shows the relevant portions of the source code.

Listing 3-17: histeq.c

```
clock_t histeq()
{
  int ii = 0;
  cumsum = 0;

  /* estimate discrete PDF via histogram as before */

  /* form gray-level transform function T(r) = CDF */
```

```
#if FIXED_POINT

    start = clock(); /* begin "profile area" */
    for (ii=0; ii<256; ++ii) {
        cumsum += (hist[ii] << 2);
        T_r[ii] = (255*cumsum) >> 16;
    }
    stop = clock(); /* end "profile area" */
    t = stop-start-overhead;
    printf("Fixed point compute T_r cycles: %d\n", t);

#else

    start = clock();
    for (ii=0; ii<256; ++ii) {
        cumsum += hist[ii];
        T_r[ii] = round(255.f*cumsum/num_pixels);
    }
    stop = clock();
    t = stop-start-overhead;
    printf("Floating point compute T_r cycles: %d\n", t);

#endif /* FIXED_POINT */

    /* perform pixel re-mapping as before */

    return t;
}

int main(void)
{
    const int N = 10; /* how many times to profile */
    int ii = 0;
    clock_t profile = 0;

    evm_init(); /* initialize the board */
    start = clock(); /* calculate overhead of calling clock*/
    stop = clock();  /* and subtract this value from the results*/
    overhead = stop - start;

    for (; ii<N; ++ii)
        profile += histeq();
```

```
printf("average time to compute T_r is %.2f cycles\n",
    (float)profile/(float)N);
}
```

To use this program properly, enable aggressive optimizations as before, however do not set the '–g' compile option. Alternatively, changing the build configuration to Release should by default enable fairly aggressive optimizations, although it is always recommended to ensure that the most appropriate build options for your particular application are set. Prior to running the program, the CCStudio clock facility must be enabled via the **Profiler|Enable Clock** menu option. This program calls the histeq function N times, prints the number of cycles taken to execute each separate invocation of histeq, and at the conclusion of the program prints the average cycle count. The program now reports approximately 34,833 cycles (versus 35,702) for the floating-point loop, and approximately 331 cycles (versus 528) for the fixed-point loop. Another area one might tackle to optimize this program's performance is in its memory usage. The out_img array is stored in off-chip memory, and accessing off-chip memory is enormously prohibitive. In the next chapter, we investigate techniques to ameliorate this bottleneck.

REFERENCES

1. Texas Instruments, *TMS3206201/6701 Evaluation Module Technical Reference* (SPRU305.pdf), Chapter 1, section 1.4, *External Memory.*
2. Spectrum Digital, *TMS3206416 DSK Technical Reference* (505945-001 Rev. A), Chapter 1, section 1.4, *Memory Map.*
3. Texas Instruments, *TMS320C62x Image/Video Processing Library Programmer's Reference* (SPRU400.pdf), Chapter 5, section 5.2.4, *Histogram Computation.*
4. Texas Instruments, *TMS320C64x Image/Video Processing Library Programmer's Reference* (SPRU023.pdf), Chapter 5, section 5.2.4, *Histogram Computation.*
5. Code Composer Studio online help, see **Using CCS IDE|Debugger|File Input/Output|Data File Formats.**
6. Microsoft Developers Network (MSDN), see "MFC Overview." Retrieved October 2004 from:
   http://msdn.microsoft.com/library/default.asp?url=/library/en-us/vccore/html/vcorimfcoverview.asp
7. Intel Corp., *Intel® Integrated Performance Primitives for Intel® Architecture (Part 2: Image and Video Processing)*, 2004.
8. Microsoft Developers Network (MSDN), see "GDI+." Retrieved October 2004 from:
   http://msdn.microsoft.com/library/default.asp?url=/library/en-us/gdicpp/GDIPlus/GDIPlus.asp

9. Bushberg, J., Seibert, J., Leidholdt, E., and Boone, J., *The Essential Physics of Medical Imaging* (Williams & Wilkins, 1994).

10. Udupa, J., and Herman, G., *3D Imaging in Medicine, 2nd Edition* (CRC Press, 2000).

11. Texas Instruments, *Code Composer Studio IDE 2.2 Profiler* (SPRA905.pdf).

12. Texas Instruments, *TMS3206000 Programmer's Guide* (SPRU198.pdf), Chapter 2, section 3.1.2, *Using the Clock() Function to Profile.*

Chapter 4

IMAGE FILTERING

Chapter 3 dealt with the class of image processing algorithms that enhance images through pixel intensity remapping via lookup tables that discretize a gray-level transform function $T(r)$. In this chapter we move on to a different but related set of algorithms, broadly classified as image filtering. The topic of filtering digital waveforms in one dimension or images in two dimensions has a long and storied history. The basic idea in digital image filtering is to post-process an image using standard techniques culled from signal processing theory. An analogy may be drawn to the type of filters one might employ in traditional photography. An optical filter placed on a camera's lens is used to accentuate or attenuate certain global characteristics of the image seen on film. For example, photographers may use a red filter to separate plants from a background of mist or haze, and most professional photographers use a polarizing filter for glare removal[1]. Optical filters work their magic in the analog domain (and hence can be thought of as analog filters), whereas in this chapter we implement digital filters to process digital images.

4.1 IMAGE ENHANCEMENT VIA SPATIAL FILTERING

A common means of filtering images is through the use of "sliding neighborhood processing", where a "mask" is slid across the input image and at each point, an output pixel is computed using some formula that combines the pixels within the current neighborhood. This type of processing is shown diagrammatically in Figure 4-1, which is a high-level illustration of a 3x3 mask being applied to an image.

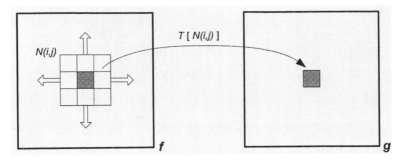

Figure 4-1. Image Enhancement via "sliding window", or neighborhood processing. As the neighborhood currently centered about the shaded pixel *f(i,j)* is overlaid around the image *f*, the pixel intensities combine to produce filtered pixels *g(i,j)*.

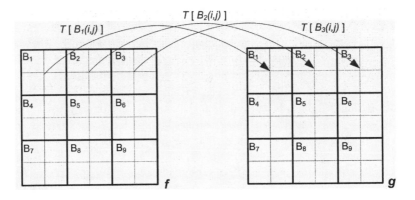

Figure 4-2. "Block processing" of images. Processed image *g* is created by partitioning *f* into distinct sections, an operation T is applied to all the pixels in each block $f(B_k)$, and the corresponding block $g(B_k)$ receives this output. The Discrete Cosine Transform (DCT), the transform that forms the heart of the JPEG image compression system, is performed in this manner using 8x8 blocks.

The sliding window operation - where *N(i,j)* is larger than just a single pixel - is used to perform *spatial filtering* of images. The mask, also referred to as a filter *kernel*, consists of weighting factors that depending on its makeup, either accentuates or attenuates certain qualities of the processed image. Basically what happens is that each input pixel *f(i,j)* is replaced by a function of *f(i,j)*'s neighboring pixels. For example, an image can be smoothed by setting each output pixel in the filtered image *g(i,j)* with the weighted average of the corresponding neighborhood centered about the input pixel *f(i,j)*. If the neighborhood is of size 3x3 pixels, and each pixel in

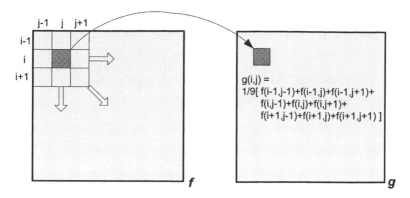

Figure 4-3. Sliding window in action. As the mask is shifted about the image for each position *(i,j)*, the intensity *g(i,j)* is set to the average of *f(i,j)*'s nine surrounding neighbors.

the neighborhood is given equal weight, then Figure 4-3 shows the formula used to collapse each neighborhood to an output pixel intensity. The filter kernel shown in Figure 4-3 is typically represented in matrix form, i.e.

$$mask = \frac{1}{9}\begin{bmatrix} 1 & 1 & 1 \\ 1 & 1 & 1 \\ 1 & 1 & 1 \end{bmatrix}$$

Each element in the matrix kernel is a filter coefficient, sometimes referred to as a filter "tap" in signal processing parlance. The neighborhood pixels need not be weighted evenly to smooth the image. For example, it may make sense to give a larger weight to pixels in the center of the neighborhood, and use smaller weights for those pixels located on the neighborhood's periphery, as in the following mask:

$$\begin{bmatrix} 1/16 & 2/16 & 1/16 \\ 2/16 & 4/16 & 2/16 \\ 1/16 & 2/16 & 1/16 \end{bmatrix}$$

It is important to note that this kernel is both symmetric and more importantly, the sum of its weights is 1. If the sum of an averaging kernel's weights is not exactly 1, then the processed image will not have the same gain (overall brightness) as the input image and this is usually not desired. The application of these masks can be computationally expensive. If the input gray-scale image is of size MxN pixels, and the mask is a square kernel of size PxP (all of the kernels we use in this book are square), then spatially filtering the image requires MNP^2 multiplication/addition operations. Using reasonable values for M, N, and P (640x480 image dimensions, and a 5x5 mask) yields over 7.5 million operations to filter an image! The good news is that DSP cores are specifically architected so that a core element in digital filtering, the multiply and accumulate (MAC) operation, almost always can be performed in a single clock cycle – a statement not necessarily true when speaking of general-purpose CPUs (see 1.6).

4.1.1 Image Noise

Spatial filters are very often used to suppress noise (corrupted pixels) in images – this is known as the "image restoration" or "image denoising" problem, and figures quite prominently in the image processing field. Image noise may be the result of a variety of sources, including:

- Noisy transmission channels, for example in space imaging where the noise density overcomes any forward error detection capabilities.
- Measurement errors, such as those emanating from flawed optics or artifacts due to motion jitter. In the former case, a classic example in astronomical imaging was the initial Hubble telescope, where imperfections in the main mirror caused sub-standard image acquisition. In the latter case, early satellite imagery provides plenty of case studies, as the motion of the spacecraft and/or vibration of the acquisition machinery introduced distortions in the acquired images.
- Quantization (the act of sampling an analog signal and converting it to a digital form) effects manifest themselves as high-frequency noise.
- Camera readout errors cause impulsive noise, sometimes referred to as "shot" or "salt and pepper" noise.
- Electrical noise arising from interference between various components in an imaging system.
- The deterioration of aging films, such as the original 1937 "Snow White and the Seven Dwarfs," many of which have been digitally restored in recent years.

Restoring a degraded image entails modeling the noise, and utilizing an appropriate filter based on this model. Image noise comes in many flavors, and as a consequence the appropriate model should be employed for robust image restoration. Noise may be additive, which can be expressed as

$$g(x, y) = f(x, y) + n(x, y)$$

where f is the original 2D signal (image), n is the noise contribution, and g is the corrupted image. Image noise can also be multiplicative, where

$$g(x, y) = f(x, y)n(x, y)$$

Noise may be highly correlated with the signal f, or completely uncorrelated. If the noise is highly structured, it may be due to defects in the manufacturing process of the detector, such as when faulty pixels produce images with pixel dropouts. Here, one possible model is to replace any offending pixels with the average of its neighbors. However, since the exact location of any faulty pixel is presumably known a priori, it follows that a far more efficient and effective means of reducing this highly structured noise is to generate a list of image locations that should be filtered, and apply an averaging mask over just those locations. Yet another form of structured noise is the "dark current" associated with Charged Coupled Device (CCD) cameras. This type of noise is very dependent on temperature and exposure time, however if these two variables are fixed it is possible to remove this type of noise via image subtraction. The activation energy of a CCD pixel is the amount of energy needed to excite the constituent electrons to produce a given pixel intensity. Pixels with low activation energies need comparatively less energy to produce equivalent pixel counts and are said to run "hot." Removing this form of noise is best accomplished through judicious calibration – capturing a so-called "dark field" image, one acquired without any illumination, at nominal temperature and exposure times and subsequently subtracting this calibration image from each new acquisition.

Spatial filtering is best used when dealing with unstructured or stochastic noise, where the location and magnitude of the noise is not known in advance, but development of a general model of the degrading process is feasible. Averaging filters are sometimes utilized in x-ray imaging, which suffer from a grainy appearance due to a well-known effect that goes by the moniker "quantum mottle." Reducing the amount of radiation a patient is exposed to during an x-ray acquisition is always beneficial for the patient – the problem is that with fewer photons hitting the image acquisition device, the statistical fluctuation of the x-ray begins to take on added effect, resulting in a non-uniform image due to the increased presence of additive film-grain noise. The problem gets progressively worse when image intensifiers are used, as boosting the image unfortunately also boosts the effects of the noise. By "smoothing" the radiograph using a procedure that will be discussed in great detail in 4.1.2, one can often enhance the resultant image, as demonstrated in Figure 4-4.

The image smoothing operation does succeed in reducing noise, but concomitant with the filtering is a loss of spatial resolution which in turn leads to an overall loss of information. Through careful selection of the size of the averaging kernel and the filter coefficients, a balance is struck between reduction of quantum mottle and retention of significant features of the image. In later sections of this chapter we explore other types of filters that are sometimes better suited for the removal of this type of noise.

Other types of stochastic noise are quite common. Digitization devices, such as CCD cameras, always suffer from readout noise as a result of the uncertainty during the analog-to-digital conversion of the CCD array signal. Any image acquisition system using significant amounts of electronics is susceptible to electronic noise, especially wideband thermal noise emanating from amplifiers that manifests itself as Gaussian white noise with a flat power spectral density. The situation is exacerbated when the system gain of the acquisition device is increased, as any noise present with the system is amplified alongside the actual signal.

It should be emphasized that noise present within a digital image most likely does not arise from a single source and could very well contain multiple types of noise. Each element in the imaging chain (optics, digitization, and electronics) contributes to potential degradation. An image

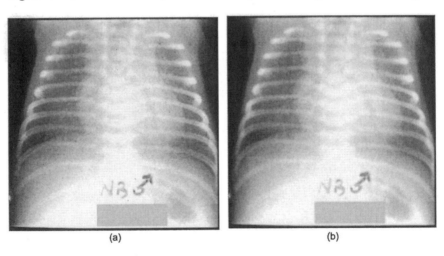

(a)   (b)

Figure 4-4. Reduction of quantum mottle noise by image smoothing (image courtesy of Neil Trent, Dept. of Radiology, Univ. of Missouri Health Care). (a) Original chest radiograph of infant. Note the speckled and grainy nature of the image. (b) Processed image, using an 11x11 Gaussian low-pass filter ($\sigma=1$). The grainy noise has been reduced, at the cost of spatial resolution.

may suffer from both impulsive shot noise and signal-dependent speckle noise, or an image may be contaminated with blurring due to motion during acquisition, as well as some electronic noise. In short, for image filtering to be effective, these factors must be taken into account – no filter exists that can robustly handle all types of noise.

4.1.2 Two-Dimensional Convolution, Low-Pass and High-Pass Filters

The smoothing operation, which is really nothing more than a moving average, is an example of two-dimensional linear filtering (non-linear filters are covered in 4.5). The filter kernel is multiplied by the neighborhood surrounding each pixel in the input image, the individual products summed, and the result placed in the output image, as depicted in Figure 4-3. This process may be described in mathematical form as

$$g(i,j) = \sum_{i'=-\frac{NH}{2}}^{\frac{NH}{2}} \sum_{j'=-\frac{NH}{2}}^{\frac{NH}{2}} f(i+i', j+j') h(i', j')$$

where f and g are MxN images, so that $0 \le i < M$, $0 \le j < N$, and h is a square kernel of size NHxNH pixels. The above expression is the *convolution* equation in two dimensions. Convolution is of fundamental importance in signal processing theory, and because this topic is covered so well in many other sources[2,3], a mathematically rigorous and thorough discussion is not given here. Suffice it to say that for the most part, the theory behind one-dimensional signal processing extends easily to two dimensions in the image processing field. In 1D signal processing, a time domain signal x_n is filtered by *convolving* that signal with a set of M coefficients h:

$$y_n = \sum_{k=0}^{M-1} h_k x_{n-k}$$

These types of filters, and by extension their 2D analogue, are called *finite impulse response* (FIR) filters because if the input x is an impulse ($x_n=1$ for $n=0$ and $x_n=0$ otherwise), the filter's output eventually tapers off to zero. The output of a filter obtained by feeding it an impulse is called the filter's impulse response, and the impulse response completely characterizes the filter. FIR filters are more stable than *infinite impulse response* (IIR)

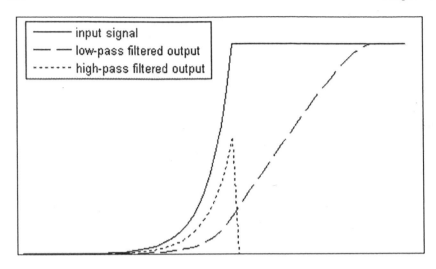

Figure 4-5. The effect of low-pass and high-pass filters in one dimension. In one dimension, the analogue to an image smoothing kernel is a filter that produces a running average of the original signal (the low-pass filtered output). A one-dimensional high-pass filter emphasizes the differences between consecutive samples.

filters, where the impulse response never tapers off to zero. FIR filters are simply weighted averages, and the 2D convolution operator we use in image processing to spatially filter images is in fact a 2D FIR filter.

The makeup of the filter coefficients, or filter taps, determines how the filter behaves. As we have already seen, various types of averaging kernels produce a blurred, or smoothed, image. This type of output is an example of using a *low-pass* filter, which diminishes high spatial frequency components, thereby reducing the visual effect of edges as the surrounding pixel intensities blur together. As one might expect, there are other types of linear filters, for example *high-pass* filters which have the opposite effect of low-pass filters. High-pass filters attenuate low-frequency components, so regions in an image of constant brightness are mapped to zero and the fine details of an image emphasized. Figure 4-5 illustrates the effects of low-pass and high-pass filters in a single dimension.

High-pass filters are considerably more difficult to derive than low-pass filters. While averaging filters always consist of purely positive kernels, that is not the case for high-pass kernels. A common 3x3 high-pass kernel is

$$h = \frac{1}{9} \begin{bmatrix} -1 & -1 & -1 \\ -1 & 8 & -1 \\ -1 & -1 & -1 \end{bmatrix}$$

In general, a high-pass kernel of size NHxNH may be generated by setting the center weight to $(NH^2-1)/NH^2$ and the rest to $-1/NH^2$. Upon passing an image through a high-pass filter of this form, the average pixel intensity (DC level) is zero, implying that many pixels in the output will be negative. As a consequence a scaling equation, perhaps coupled with clipping to handle saturation, should then be used as a final post-processing step. In addition, since most noise contributes to the high frequency content of an image, high-pass filters have the unfortunate side effect of also accentuating noise along with the fine details of an image.

Better looking and more natural images can be obtained by using different forms of high-pass filters. One means of performing high-pass filtering is to first repeatedly low-pass filter an image, and then subtract the smoothed image from the original, thereby leaving only the high spatial frequency components. In simpler terms,

high-pass filtered image = original image – smoothed imaged

The high-pass filtered image in the above relation contains only high-frequency components. However, it is usually desirable to retain some of the content of the original image, and this can be achieved through application of the following equation:

$$g = \alpha f - \beta [f * h]_n$$

where h is a low-pass kernel, $\alpha - \beta = 1$, the $*$ symbol denotes convolution, and n signifies the fact that we blur the image by convolving it with the low-pass kernel n times. This procedure is known as *unsharp masking*, and finds common use in the printing industry as well as in digital photography, astronomical imaging, and microscopy. By boosting the high-frequency components at the expense of the low-frequency components, instead of merely completely removing the low-frequency portions of the image, the filter has the more subtle effect of sharpening the image. Thus the unsharp mask can be used to enhance images that may appear "soft" or out-of-focus. The unsharp mask is demonstrated in Figure 4-6.

4.1.3 Fast Convolution in the Frequency Domain

Two-dimensional convolution is an expensive operation, and some image processing techniques call for repeatedly passing an image through a filter, such as the unsharp masking example of Figure 4-6. Even with efficient code running on architectures explicitly designed for fast digital filtering,

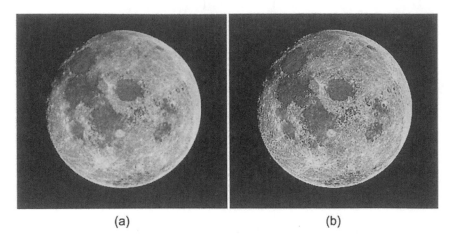

(a)   (b)

Figure 4-6. Unsharp mask on moon image (courtesy of the National Space Science Data Center). (a) Original image. (b) Processed image, using a 20x20 Gaussian low-pass filter (σ=1), with α=1.5 and β=0.5.

such as DSPs, as the image and kernel sizes increase the price of operating purely in the time-domain eventually becomes prohibitive. At this point, it is advantageous to migrate the process over to the "frequency domain", which necessitates incorporating the Fourier Transform and its digital equivalent, the Discrete Fourier Transform (DFT). The DFT is calculated highly efficiently not through direct evaluation of its mathematical formulation, but rather through a well-known algorithm known as the Fast Fourier Transform (FFT). The details behind the Fourier Transform, the DFT, and how the FFT performs its magic are beyond the scope of this book. For our purposes, it is sufficient to say that 2D convolution can be performed in the frequency domain by exploiting one of the central tenets of signal processing theory, the *convolution theorem*, which states that convolution in the time domain is equivalent to multiplication in the frequency domain.

Figure 4-7 show how to perform 2D convolution in the frequency domain. Both the image and kernel are transformed using the 2D FFT, which in general produces two complex numbered matrices containing the discrete Fourier coefficients for both the image and filter kernel. Because the kernel and image matrices are probably not of the same size, prior to applying the 2D FFT the smaller of the two is expanded through an operation known as zero-padding, where zeros are appended to the matrix (see Figure 4-7b). The coefficients are multiplied together and then the inverse transform is performed, thereby sending the result back to the time domain. The result of the inverse transform is the convolved image, or equivalently, the result of passing the original image through the filter defined by the kernel.

(a)

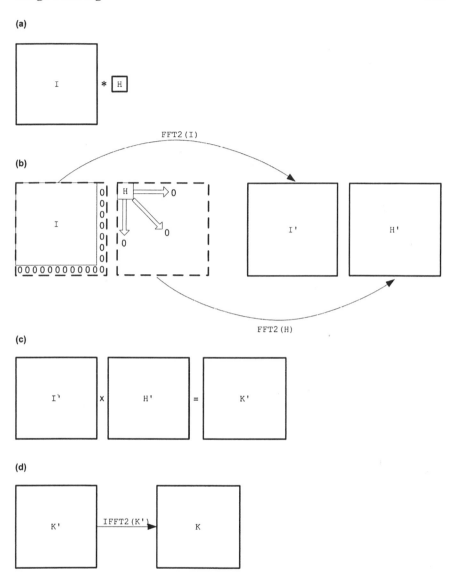

Figure 4-7. 2D linear convolution in the frequency domain. (a) I is an image matrix of size MxM, H is the NxN filter kernel matrix. (b) I and H are zero-padded such that they are of size (M+N-1)x(M+N-1), and the 2D FFT is performed on both augmented matrices. (c) FFT matrices are multiplied together, in an element-wise fashion (not matrix multiplication). (d) The result of the multiplication is transformed back to the time domain via the 2D inverse FFT.

It should be emphasized that 2D convolution in the frequency domain is not necessarily faster than direct application of the 2D convolution equation in the time domain. Typically, performing the operation in the manner shown in Figure 4-7 is faster when large kernel sizes are used with large images, the exact sizes being dependent on the speed of the FFT routine and the particular architecture on which the code is running on. Eventually, the cost of transforming two matrices to the frequency domain via the 2D FFT and one matrix back to the time domain via the 2D inverse FFT is balanced out by the fact that in between these two transforms, element-wise multiplication occurs, as opposed to sliding a mask across an image and computing the sum of products at each step. However, this means of performing filtering in the frequency domain only works for linear filters. Non-linear filters are precluded from taking advantage of this scheme, as the convolution theorem does not apply to them. A general rule of thumb is that linear convolution in the frequency domain should be considered if:

$$N_1 N_2 > 6 \log_2 (M_1 M_2) + 4$$

where the filter kernel is of size N_1xN_2 and the image is of size M_1xM_2.

4.1.4 Implementation Issues

Algorithm 4-1 is the pseudo-code for linear filtering of images in the time domain, that is, via direct application of the 2D convolution equation.

Algorithm 4-1: 2D Convolution
    INPUT: MxN image I, NHxNH kernel H, where NH is odd
    OUTPUT: filtered image J

    b = floor(NH/2)   (*half-width of kernel*)

    (*initialize output image to zero*)
    for (r = 0...M-1)
      for (c = 0...N-1)
        J[r][c] = 0
      end
    end

    (*filter interior of image, by sliding mask across the image*)
    for (r = b...M-2b+1) (*1st b and last b rows not filtered*)
      for (c = b...N-2b+1) (*1st b and last b cols not filtered*)

Algorithm 4-1 (continued)
    for (ih = -b...+b)
      for (jh = -b...+b)

          filter_coeff = h[ih+b][jh+b]
          input_pixel = I[r+ih][c+jh]
          J[r+ih][c+jh] += (filter_coeff)(input_pixel)

      end
    end

    end (*for each column*)
  end (*for each row*)

    Algorithm 4-1 brings up a few practical considerations that any image filtering implementation must take into account, namely what to do about edge pixels. Figure 4-8 illustrates the situation for a 5x5 kernel situated at the top-left corner of an image. The neighborhood for this pixel, and all others centered about a pixel in the first and last two rows or columns have to be handled in a different manner than the interior region of the image.

    There are a few different ways in which to handle the edge pixels. One is to pad the image on both sides with zeros, and another is to simply extend the image by replicating the first and last row or column however many times is needed (as dictated by the half-width of the kernel). Both of these

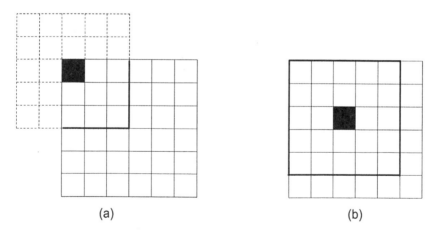

(a)   (b)

Figure 4-8. Edge effects issues with 2D convolution. (a) With the neighborhood centered around the upper-left pixel, the mask extends out past the boundaries of the image by the half-width of the kernel (in this case 2). (b) By the time the neighborhood has shifted down to the 3rd row and 3rd column, the entire mask tiles completely over a valid portion of the image.

solutions have their drawbacks, especially the first as the overall intensity levels of the outer regions of the processed image will be less than their neighbors. Most of the time the edge pixels are simply ignored and set to zero, and this is the solution specified in Algorithm 4-1 as well as the C code that follows in this chapter. When performing 2D convolution in the frequency domain, edge artifacts will be present as the FFT assumes a periodic signal and so in essence, the image is extended such that it wraps around at all four of its edges (i.e. the right-most column is contiguous with the left-most column, the bottom-most row is contiguous with the top-most row, and so on).

4.2 LINEAR FILTERING OF IMAGES IN MATLAB

The Image Processing Toolbox includes a function `imfilter` that can be used to filter images. The first argument to this function is the image matrix, and the second argument is a matrix containing the filter coefficients. The command

J = imfilter(I, ones(3,3)./9);

passes the image I through a 3x3 averaging filter. The `ones` function returns a constant matrix of ones. There is another similar MATLAB function `zeros` that returns the zero matrix - both of these functions are useful for preallocating space for matrices and initialization of matrix values. The ./ operator specifies element-wise division, so the second argument above passes into `imfilter` a 3x3 matrix where each element is 1/9. The `imfilter` command also allows one to specify how to handle edge pixels with an optional third argument, and a variety of other options are available (refer to the online help for more information).

The Image Processing Toolbox function `fspecial` can be used to generate 2D kernels for use with `imfilter`. This function is especially helpful, as it knows about a slew of common filters, all of which can be tuned according to kernel size and additional filter-specific parameters. The first argument to `fspecial` is a string indicating which class of filter to generate (refer to the online help for a full list of supported filter types), the second argument is the size of the kernel, and the rest are optional arguments for tuning the filter[4], the exact specifications of which depend on the type of filter. For example, the command

fspecial('gaussian', [20 20], 1)

returns the low-pass filter used in Figure 4-6 to implement the unsharp mask image sharpening procedure. The image filtering C programs that follow in sections 4.3.1-4.3.5 and 4.4.1-4.4.2 for the most part utilize two-dimensional arrays to store filter coefficients. To experiment with different filters, all that is required is to generate the coefficients via fspecial and copy the coefficients into the source code. For fixed-point programs, care must be taken to use coefficients in Q15 format. Conversion from floating-point to Q15 format is quite easy in MATLAB - the following snippet of code shows how to generate the fixed-point filter coefficients for a 5x5 Gausssian low-pass filter:

```
coeffs = fspecial('gaussian', [5 5], 1)

coeffs =

    0.0030    0.0133    0.0219    0.0133    0.0030
    0.0133    0.0596    0.0983    0.0596    0.0133
    0.0219    0.0983    0.1621    0.0983    0.0219
    0.0133    0.0596    0.0983    0.0596    0.0133
    0.0030    0.0133    0.0219    0.0133    0.0030

coeffsQ15 = uint16(coeffs .* 2^15 + 0.5)

coeffsQ15 =

      97     436     719     436      97
     436    1954    3222    1954     436
     719    3222    5312    3222     719
     436    1954    3222    1954     436
      97     436     719     436      97
```

If the Image Processing Toolbox is not available, conv2 can be used as a substitute for imfilter to perform 2D linear filtering of images. This function has a smaller set of arguments compared to imfilter and differs in some subtle ways. The first and second arguments are the image matrix and filter kernel. The third optional argument is a string indicating how much of the output to return[5]. The conv2 function always assumes zero padding of edge pixels, and by passing in the string 'valid', conv2 returns just the portion of the convolution that did not require zero-padding. The other difference to be aware of is the class of the returned type. If imfilter receives a matrix of type uint8, it will return back a uint8 matrix, in contrast to conv2 which always returns a double matrix. Thus, displaying the result of conv2 using image or imshow typically results in a completely white display if the image is not first converted back to the uint8 type, as shown below:

J = uint8(conv2(uint8(I), H, 'same') + 0.5);

In the above statement, the addition of 0.5 following the integer truncation via uint8 aids in the correct rounding. The matrix J can now be displayed in the appropriate scale using either of image or imshow. Alternatively, since both of the aforementioned image display functions expect image matrices of the double type to be scaled between [0,1.0], J could be divided by the maximum allowable pixel value for this data type (255 for uint8) prior to display:

J = conv2(uint8(I), H, 'same') ./ 255;

4.3 LINEAR FILTERING OF IMAGES ON THE TI C62XX/C67XX

C6701 EVM image filtering programs may be found on the CD-ROM under the directory Chap4\LinearFilter\C62xxC67xx. The image dimensions defined in these programs are more realistic and larger than those encountered in Chapter 3 (256x256 vs. 128x128). As a consequence, neither the input image buffer nor the output image buffer fits in internal on-chip RAM, a situation that if not dealt with appropriately has deleterious effects on the efficiency of the code. Thus the implementations presented in this chapter serve two purposes; one, to provide tested code that implements 2D convolution and two, to illustrate strategies for handling the situation where data does not fit in the DSP internal RAM. All of these programs are fixed-point implementations, using a Q15 data format.

The image dimensions, 16-bit short integer input buffer in_img, and kernel size NH are defined in the header file image.h, which is shared amongst all of the programs in the C62xxC67xx directory. The in_img buffer is initialized with pixel data from the ubiquitous "Lenna" image, which is used as a test image in many image-processing texts. Of course, alternate data can be fed into the in_img array if so desired, using techniques covered in Chapter 3. The size of the kernel may also be changed by adjusting the definition of NH, however NH must be odd. The description of the code often makes reference to the half-width of the kernel, and this parameter is referenced in the code by the BOUNDARY preprocessor symbol.

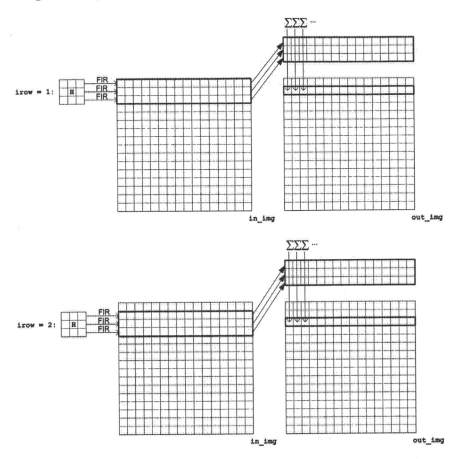

Figure 4-9. `IMG_corr_gen` in action. This diagram depicts two iterations of the 2D image filtering algorithm in `filter_imglib.c`. Labels in **bold, `courier`** font refer to actual variables in the implementation. A series of 1D FIR filtering operations is performed on each scan-line of `in_img`, and then these 1D results are combined via a summation in the other direction and placed in `out_img`. This process continues as we march down `in_img`.

The linker command file `image_filter.cmd` is shared amongst the various C62xx/C67xx programs. It is identical to the linker command files used in the CCStudio projects for Chapter 3, except that the ".cinit" section (the portion of the memory map reserved for constants and other data for the initialization of global C variables) is now mapped to external RAM (SBSRAM) instead of internal RAM. This is necessary if the 64k (256x256) element array `in_img` is to be pre-initialized with pixel data, as the initialization data will not fit in DSP internal RAM.

Finally, there is another shared source file, vecs.asm, used for some of the later implementations presented in this section. This file contains the vector interrupt table and will be explained in more detail when Direct Memory Access (DMA) is introduced.

4.3.1 2D Filtering Using the IMGLIB Library (filter_imglib)

A program set up to use the IMGLIB 2D filtering function IMG_corr_gen can be located in the filter_imglib subdirectory. IMG_corr_gen can be used to implement a generalized 2D filtering process by repeatedly invoking it for every sliding window position down the rows of an image, each time passing into the function the current row of the mask and the current row of the image[6]. This in effect ends up passing rows of in_img through the 1D FIR filter given by the ith row of mask. NH filtered rows are then combined via summation across the orthogonal (column-wise) direction, thereby producing pixels in the output image that have been filtered in a 2D fashion. This process is illustrated in Figure 4-9 for the case of a 3x3 kernel (NH=3).

Listing 4-1 is the contents of filter_imglib.c. Note that this and all other programs set the first and last kernel half-width columns and rows to zero. Thus if the C preprocessor symbol NH is 5, the first and last two rows and columns of out_img are set to zero.

Listing 4-1: filter_imglib.c

```
#include <board.h>          /* EVM library */
#include <stdio.h>          /* printf() */
#include <string.h>         /* memset() */
#include <img_corr_gen.h>
#include <time.h>           /* clock() */

#include "..\image.h"
#pragma DATA_ALIGN (in_img, 4);
#pragma DATA_SECTION (in_img, "SBSRAM");

#pragma DATA_ALIGN (out_img, 4);
#pragma DATA_SECTION (out_img, "SBSRAM");
unsigned char out_img[N_PIXELS]; /* the filtered image */

/* filter coefficients in Q.15 format */
#define NH 5
#define BOUNDARY (NH/2) /* kernel half-width */
```

```
#pragma DATA_ALIGN (H, 4)
short H[NH][NH] = {
/* Averaging filter, 1/25 */
  {1310, 1310, 1310, 1310, 1310},
  {1310, 1310, 1310, 1310, 1310},
  {1310, 1310, 1310, 1310, 1310},
  {1310, 1310, 1310, 1310, 1310},
  {1310, 1310, 1310, 1310, 1310}
};

/* temporary buffers for IMG_corr_gen */
#pragma DATA_ALIGN (out_corr, 4)
short out_corr[NH][Y_SIZE-2*BOUNDARY];
#define N_OUT_CORR NH*(Y_SIZE-2*BOUNDARY)

#define N_PIXELS_2_FILTER Y_SIZE-2*BOUNDARY

/*
 * This function was adapted from the behavorial C code
 * given in img_corr_gen.h and SPRU400 (5.3.2)
 */
void corr_gen (short *in_data, short *h,
               short *out_data, int m, int cols)
{
  /*
   * For all columns compute an M-tap filter. Add
   * correlation sum to value, to allow for a generalized 2-D
   * correlation to be built using several 1-D correlations
   */

  int i, j, sum; // sum is a 32-bit accumulator
  for (j = 0; j < cols; j++) {
    sum = out_data[j];
   for (i = 0; i < m; i++)
    sum += in_data[i + j] * h[i]; /* Q.15 multiplication */
    out_data[j] = (short) (sum>>15); /* cast output to 16 bits */
  }
}

void filter_image() {
  unsigned char *p; /* ptr into output image buffer */
  int ii, irow, icol, ih, sum;
```

```
/* set 1st BOUNDARY rows to zero */
memset(out_img, 0, BOUNDARY*Y_SIZE*sizeof(char));

/* filter the interior region of the image matrix */
for (irow=BOUNDARY; irow<X_SIZE-BOUNDARY; ++irow) {
  /* zero out the correlation output array */
  memset(out_corr, 0, N_OUT_CORR);

  /* IMG_corr_gen doesn't work, corr_gen does  */
  for (ih=0; ih<NH; ++ih)
   /*IMG_*/corr_gen(&in_img[(irow-BOUNDARY+ih)*Y_SIZE],
                    H[ih],
                    out_corr[ih],
                    NH,
                    N_PIXELS_2_FILTER);

  /* 1st BOUNDARY cols are zero */
  p = out_img+irow*Y_SIZE;
  for (ii=0; ii<BOUNDARY; ++ii) *p++ = 0;

  /* sum up correlation results */
  for (icol=0; icol<N_PIXELS_2_FILTER; ++icol) {
   sum = 0;
   for (ih=0; ih<NH; ++ih)
    sum += (out_corr[ih][icol]);
   *p++ = sum;
  }

  for (ii=0; ii<BOUNDARY; ++ii)
   *p++ = 0; /* BOUNDARY cols = 0 */
 }

 /* last BOUNDARY rows are zero */
 memset(out_img+(X_SIZE-BOUNDARY)*Y_SIZE,
        0,
        BOUNDARY*Y_SIZE*sizeof(char));
}

int main(void) {
  clock_t start, stop, overhead, t = 0; /* timing */
  const int N = 10; /* how many times to profile */
```

```
int ii = 0;

evm_init(); /* initialize the board */
start = clock(); /* calculate overhead of calling clock*/
stop = clock();  /* and subtract this value from The results*/
overhead = stop - start;

for (; ii<N; ++ii) {
  start = clock(); /* begin "profile area" */
    filter_image();
  stop = clock(); /* end "profile area" */
  t += stop-start-overhead;
  printf("# cycles to filter image: %d\n", stop-start-overhead);
}

printf("avg time is %.2f cycles.\n", (float)t/(float)N);
}
```

Unfortunately, as of version 1.02 of the C62x/C67x IMGLIB, IMG_corr_gen does not work as advertised! Most of the TI library reference manuals contain "behavioral C code" translations of their documented functions. Since the functions are more often than not written in hand-optimized assembly code, this behavioral C code (which can usually also be found in the relevant algorithm header file) can be invaluable when attempting to understand exactly what an algorithm is intended to accomplish. In this program, a local function corr_gen has been defined which is basically the behavioral C code for IMG_corr_gen copied and pasted from img_corr_gen.h into filter_imglib.c. This function does work, but serves the purpose of providing a completely unoptimized (aside from using fixed-point arithmetic) implementation of image filtering. The corr_gen function in filter_imglib.c does not handle saturation correctly, and there are numerous inefficiencies in the loop, many of which could be tackled by using special features of the TI C compiler (so-called intrinsics, introduced in 4.4.1 and covered extensively in Appendix B) and/or rewriting the loop in linear assembly. Using the profiling technique described in 3.4.3, this implementation takes an average of 46,972,076 cycles to filter a 256x256 image with a 5x5 kernel, with no debug symbols and the −o3 compiler option selected. The fact that memory usage, specifically the cost incurred by accessing pixels stored in external RAM, is not optimized at all in this program contributes greatly to the cycle count – a solution to this issue will be presented in subsequent image filtering programs. Incidentally, using IMG_corr_gen instead of the locally defined `corr_gen` yields an average cycle count of 10,775,445 cycles, for a speedup of over 4x. However it is difficult to draw any clear conclusions from this measurement as the output is incorrect. In the next section we shall see how an alternative function from the DSPLIB library that works can be used as a replacement for `IMG_corr_gen`.

Code Composer Studio provides numerous data visualization tools. In fact, it is such a powerful development environment that it even offers the ability to render buffers as images, a feature that is obviously of great utility when developing image processing applications. To use this feature, build the `filter_imglib` program with debug symbols enabled and set a breakpoint in `main` sometime after the call to `filter_image`. Then run the program, and when the debugger stops at the breakpoint, select the **View|Graph|Image** menu item. This brings up a dialog that configures how to render the data in an image buffer. There is no monochrome color space option, so to display gray-scale images, select the RGB color space option and point the R, G, and B channels to the same buffer (in this case `out_img`). Figure 4-10 is a CCStudio screen-shot showing the output of the low-pass filtered Lenna image, after passing it through an averaging 5x5 kernel. The image visualization feature of CCStudio is quite full-featured, in fact if you click within the image window the pixel coordinates and the intensity is displayed in the status bar at the bottom of the display. MATLAB provides similar functionality through its `pixval` command.

4.3.2   Low-Pass Filtering Using DSPLIB (`blur_dsplib`)

Through judicious use of the DSPLIB function `DSP_fir_gen`, which performs one-dimensional filtering of signals[7], one can implement a two-dimensional filtering algorithm. This particular implementation is not as general as `filter_imglib`, as the variable H in this case is a 1D array of length NH, and as a consequence the only meaningful filters that can be represented in this fashion are low-pass averaging filters (where every filter coefficient is the same). This requirement leads to a simplification and optimization of the algorithm, and is useful to consider for that reason. A more general variant of the 2D filtering algorithm, where H is a 2D array and thus can represent any convolution, a la `filter_imglib`, is developed in the subsequent sections.

Figure 4-11 illustrates how this program goes about low-pass filtering an image, for a 3x3 kernel. Essentially what is going on is that since we store the input image as a flattened 2D array, if every row of the input image is run through the same FIR filter, we can simply FIR filter every row in the image with a *single* call to `DSP_fir_gen`. There remain a few caveats:

Figure 4-10. Visualizing the results of the `filter_imglib` program by rendering `out_img` as a 256x256 RGB image.

1. We must discard some of the output from the FIR filter, because during the transition from row r to row r+1 (the last pixel in row r and the first pixel in row r+1), DSP_fir_gen assumes contiguous samples, and consequently this portion of the output has meaningless values.

2. DSP_fir_gen requires the input array to be slightly larger than the image buffer, to account for the last few samples in the input array. Specifically, we must define in_img to be of length (# rows)(# columns) + (NH) - 1.

3. DSP_fir_gen expects the filter coefficient vector to be in reverse order, but since the averaging filter is symmetric (in fact it is constant) that requirement does not matter here.

Listing 4-2 is the contents of blur_dsplib.c, sans main, which remains identical to the version in filter_imglib.c. The memory inefficiencies are still present, but this program now takes on average 2,199,517 cycles to low-pass filter a 256x256 image with a 5x5 averaging kernel, or 4.7% percent of the clock cycles required for filter_imglib. Keeping in mind that this is not a truly fair comparison, as this program is

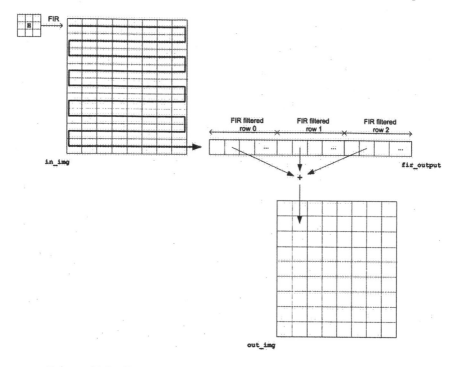

Figure 4-11. 2D filtering using the 1D DSPLIB function DSP_fir_gen. In this diagrammatic representation of how blur_dsplib works for a 3x3 kernel, the steps to produce the pixel at out_img[1][1] are shown. All of the pixels in in_img are passed through the FIR filter in one fell swoop, producing a series of contiguous filtered rows in fir_output. The second element in each of the first three filtered rows are then summed to produce the final 2D filtered pixel.

only capable of a small (however important) subset of spatial filters, it nevertheless represents a substantial performance boost, and serves to illustrate just how important it is to use optimized code, especially when such functions are readily available.

Listing 4-2: portions of blur_dsplib.c

```
#pragma DATA_ALIGN (in_img, 4);
#pragma DATA_SECTION (in_img, "SBSRAM");

#pragma DATA_ALIGN (fir_output, 4);
#pragma DATA_SECTION (fir_output, "SBSRAM");
short fir_output[N_PIXELS]; /* output of DSP_fir_gen */

#pragma DATA_ALIGN (out_img, 4);
```

```
#pragma DATA_SECTION (out_img, "SBSRAM");
unsigned char out_img[N_PIXELS]; /* filtered image */

/* filter coefficients in Q15 format */
#define BOUNDARY (NH/2) /* "kernel offset" or half-width */
short h[NH] = {1310, 1310, 1310, 1310, 1310}; /* (1/25 * 2^15) */

void filter_image() {
  int irow, icol, ii;
  short *pFirBand[NH], /* ptr to the FIR filtered row of pixels */
        sum; /* accumulator */
  unsigned char *p; /* ptr into the output image array */

  /* filter all of the rows */
  DSP_fir_gen(in_img, h, fir_output, NH, N_PIXELS);

  /* 1st BOUNDARY rows are zero */
  memset(in_img, 0, sizeof(char)*BOUNDARY*Y_SIZE);

  /*
   * walk down the column dimension, thereby
   * performing the 2D filtering operation.
   */
  for (irow=BOUNDARY; irow<X_SIZE-BOUNDARY; ++irow) {
    for (ii=0; ii<NH; ++ii)
    pFirBand[ii] = fir_output + (irow-BOUNDARY+ii)*Y_SIZE;

  /* 1st BOUNDARY cols are zero */
  p = out_img + irow*Y_SIZE;
  for (ii=0; ii<BOUNDARY; ++ii) *p++ = 0;

  for (icol=0; icol<X_SIZE-2*BOUNDARY; ++icol) {
    sum = 0;
    for (ii=0; ii<NH; ++ii)
      sum += *(pFirBand[ii])++;
    *p++ = sum; /* >>15 performed within DSP_fir_gen */
  }

  /* last BOUNDARY cols are zero */
  for (ii=0; ii<BOUNDARY; ++ii)  *p++ = 0;
  }
}
```

4.3.3 Low-Pass Filtering with DSPLIB and Paging (`blur_dsplib_paging`)

It has been repeatedly alluded to that failing to consider the memory footprint of an image processing algorithm implementation is a major detriment to performance. Indeed, the external memory interface (EMIF) on the C6701 is slow – it takes between 15-17 cycles to access a pixel stored in external RAM, versus a single cycle for a pixel in on-chip RAM[8]. Such latencies result in the DSP stalling while it waits for data to arrive via the EMIF. This issue of data residing in external off-chip RAM is even more pressing in comparison to the point processing operations of Chapter 3, because here the interior pixels of the input image need to be accessed multiple times. Consider an interior image pixel and a 5x5 kernel. This pixel will be accessed 25 times, for a worst-case access penalty of $(25)(17) = 425$ cycles, versus just $17+25 = 42$ cycles if the pixel is first copied from external RAM and then accessed repeatedly from on-chip RAM. Because image filtering in general is a well-structured algorithm, this spatial locality (once the algorithm has accessed the pixel NH^2 times, it is never needed it again) can be exploited using a memory optimization technique known as *paging*. The `blur_dsplib_paging` program provides the same functionality as `blur_dsplib`, but augments that functionality by paging in blocks of the image to an on-chip input scratch buffer (`input_buf`) prior to passing image pixels through the FIR filter implemented via `DSP_fir_gen`. Additionally, another scratch on-chip buffer, `output_buf`, contains the filtered pixels for the current block, and when this entire block has been filtered (and will no longer be referenced again), the contents of `output_buf` are paged out to `out_img`, which also resides in external RAM. In Listing 4-3 the relevant portions of `blur_dsplib_paging.c` pertaining to this memory optimization is given.

Listing 4-3: portions of `blur_dsplib_paging.c`

```
#define BOUNDARY (NH/2) /* kernel half-width */
#define NUM_SCAN_LINES   16 /* must divide evenly into X_SIZE */
#define NUM_BLOCKS      (X_SIZE/NUM_SCAN_LINES)
/* BLOCK_X_SIZE = # rows in a block */
#define BLOCK_X_SIZE    (NUM_SCAN_LINES+BOUNDARY)
/* how many rows to pass through FIR filter */
#define FIR_BLOCK_X_SIZE (NUM_SCAN_LINES+2*BOUNDARY)
/* filter coefficients in Q15 format */
short h[NH] = {1310, 1310, 1310, 1310, 1310}; /* (1/25 * 2^15) */
short *pFirBand[NH]; /* ptrs into the NH filtered rows */
```

```
/* These are scratch buffers, strategically placed in on-chip RAM:
 *
 * input_buf = input pixels about to be passed through FIR filter
 * fir_buf = output of FIR filter placed in this buffer
 * output_buf = filtered pixels go here
 *
 */
#pragma DATA_ALIGN (input_buf, 4);
short input_buf[BLOCK_X_SIZE*Y_SIZE + NH-1];
#pragma DATA_ALIGN (fir_buf, 4);
short fir_buf[FIR_BLOCK_X_SIZE*Y_SIZE + NH-1];
#pragma DATA_ALIGN (output_buf, 4);
unsigned char output_buf[NUM_SCAN_LINES*Y_SIZE];

void filter_image()
{
  int ii, jj, sum,
     irow = BOUNDARY,
     ifir = 0, /* ptr to filtered row of pixels */
     /* # filtered pels to copy during transition to next block */
     nfir2copy = 2*BOUNDARY*Y_SIZE;
  unsigned char *p = output_buf + BOUNDARY*Y_SIZE;

  /* 'prologue': move 1st block into on-chip RAM and filter */
  DSP_blk_move(in_img, input_buf, BLOCK_X_SIZE*Y_SIZE);
  DSP_fir_gen(input_buf, h, fir_buf, NH, BLOCK_X_SIZE*Y_SIZE);

  /* 1st BOUNDARY rows = 0 */
  memset(output_buf, 0, BOUNDARY*Y_SIZE);

  /* algorithm main loop: filter the individual blocks */
  for (; irow<Y_SIZE-BOUNDARY; ++irow)
  {
    if (0 == irow%NUM_SCAN_LINES)
    {
      /*
       * We just hit the start of the next block, so here we:
       *
       * o page in next block into working buffer
       * o pass these scan-lines through the FIR filter
       * o page out last block to output array
       * o do some prep work for the rest of the loop
```

```
      */
      int d = irow/NUM_SCAN_LINES, /* which block */
          k = (1==d)?3:2; /* very 1st row is a special case */

      /*
       * Move some rows from the bottom to the top of the scratch
       * buffer fir_buf, because they can be reused in the next
       * pass. Variable srow is an index to the start row of the
       * portion of the block that will be "moved"
       */
      int srow2copy = FIR_BLOCK_X_SIZE-k*BOUNDARY;

      /* n is the # of rows to filter */
      int n = (d != NUM_BLOCKS-1) ?
              NUM_SCAN_LINES : NUM_SCAN_LINES-BOUNDARY;

      /* move from the bottom to the top */
      DSP_blk_move(fir_buf+srow2copy*Y_SIZE, fir_buf, nfir2copy);

      /* copy from external to internal RAM */
      DSP_blk_move(in_img+(irow+BOUNDARY)*Y_SIZE,
              input_buf,
              n*Y_SIZE);

      /* run through FIR filter */
      DSP_fir_gen(input_buf,
              h,
              fir_buf+2*BOUNDARY*Y_SIZE,
              NH,
              n*Y_SIZE);
      /* copy processed pixels into the final output array */
      memcpy(out_img+(d-1)*NUM_SCAN_LINES*Y_SIZE,
          output_buf,
          NUM_SCAN_LINES*Y_SIZE);

      ifir = 0;
      p = output_buf;
    } /* end if (we just finished a block) */

    /* setup the pointers to the filtered rows */
    for (ii=0; ii<NH; ++ii)
      pFirBand[ii] = fir_buf + (ifir+ii)*Y_SIZE;
```

```
/* 1st BOUNDARY columns are set to zero */
for (ii=0; ii<BOUNDARY; ++ii) *p++ = 0;

/* here 2D mask is applied from the 1D filtered rows */
for (ii=0; ii<X_SIZE-2*BOUNDARY; ++ii) {
  sum = 0;
  for (jj=0; jj<NH; ++jj)
    sum += *(pFirBand[jj])++;
  *p++ = sum;
}

/* last two BOUNDARY columns are set to zero */
for (ii=0; ii<BOUNDARY; ++ii) *p++ = 0;

ifir++;
} /* end (for each row) */

/* algorithm 'epilogue': handle the final block */

/* zero out final few rows */
memset(p, 0, BOUNDARY*Y_SIZE);

/* copy final block into the output buffer */
memcpy(out_img+(NUM_BLOCKS-1)*NUM_SCAN_LINES*Y_SIZE,
       output_buf,
       NUM_SCAN_LINES*Y_SIZE);

} /* end (filter_image) */
```

This program uses two functions for shuttling blocks of pixels around: `memcpy` and `DSP_blk_move`. `DSP_blk_move` is optimized for word-aligned 16-bit `short` integers[7], and thus is used for moving data between those buffers containing Q15 data : `in_img`, `input_buf`, and `fir_buf`. The standard C library function `memcpy` is used for paging the 8-bit processed pixels out from internal RAM to `out_img` in external RAM. Note that the transition between blocks provides an avenue for further optimization, as the last BOUNDARY rows of the input image scratch buffer (`input_buf`) need not be paged in if the last BOUNDARY rows of the scratch FIR output buffer (`fir_buf`) is moved from the bottom of the buffer to the top in a circular fashion.

With the same 256x256 image, a 5x5 smoothing kernel, and 16 row block, `blur_dsplib_paging` takes on average 2,195,546 cycles to perform the low-pass filtering operation. While this constitutes a savings of 3,971 cycles versus an implementation that does not utilize paging, it represents only a very small savings of .18%. A far more dramatic time savings will be accomplished by using the C62xx/C67xx DMA controller.

4.3.4 Low-Pass Filtering with DSPLIB and Paging via DMA (`blur_dsplib_paging_dma`)

The DMA controller on the C62xx/C67xx allows for the transferring of data between internal memory and external memory and peripherals without intervention by the processor[8,31]. The DMA controller can be used to perform burst transfers of data, where only the initial access incurs the 15-17 cycle penalty, and the remainder entail only 1-2 cycles per word. DMA must be employed for the quickest access to external memory. The project files for this program are to be found in the `blur_dsplib_paging_dma` subdirectory. The overall algorithm is identical to that of `blur_dsplib_paging`, except that DMA support infrastructure is incorporated and all `memcpy` calls and every `DSP_blk_move` call except one are replaced by a call to a local function, `dma_copy_block`. The relevant contents of `blur_dsplib_paging_dma.c` are shown in Listing 4-4.

Listing 4-4: portions of `blur_dsplib_paging_dma.c`

```
/*
 * a couple of macros that convert a group of pixels into
 * the # of elements the DMA transfer function expects.
 */
#define ELEM_COUNT_UCH(N)   N*sizeof(unsigned char)/sizeof(int)
#define ELEM_COUNT_SHORT(N) N*sizeof(short)/sizeof(int)

/* global variables used in DMA interrupt ISR to indicate completion */
volatile int transfer_done = FALSE;

/* reference to the vector table to call the IRQ ISRs hookup */
extern far void vectors();

/* vecs.asm hooks this up to IRQ 09 */x
interrupt void c_int09(void) /* DMA ch1 */
{
  transfer_done = TRUE;
```

```
  return;
}

/* set the interrupts */
void set_interrupts_dma(void)
{
  IRQ_nmiEnable();
  IRQ_globalEnable();
  IRQ_disable(IRQ_EVT_DMAINT1); /* INT9 */
  IRQ_clear(IRQ_EVT_DMAINT1);
  IRQ_enable(IRQ_EVT_DMAINT1);
}

void dma_copy_block(void *psrc, void *pdst, int element_count)
{
  static DMA_Handle hDma1;
  hDma1 = DMA_open(DMA_CHA1, DMA_OPEN_RESET);
   DMA_configArgs(hDma1,
     DMA_PRICTL_RMK(
     DMA_PRICTL_DSTRLD_DEFAULT,
     DMA_PRICTL_SRCRLD_DEFAULT,
     DMA_PRICTL_EMOD_DEFAULT,
     DMA_PRICTL_FS_DEFAULT,
     DMA_PRICTL_TCINT_ENABLE, /* TCINT =1 */
     DMA_PRICTL_PRI_DMA, /* DMA priority over CPU */
     DMA_PRICTL_WSYNC_DEFAULT,
     DMA_PRICTL_RSYNC_DEFAULT,
     DMA_PRICTL_INDEX_DEFAULT,
     DMA_PRICTL_CNTRLD_DEFAULT,
     DMA_PRICTL_SPLIT_DISABLE,
     DMA_PRICTL_ESIZE_32BIT, /* 32-bit element size */
     DMA_PRICTL_DSTDIR_INC, /* incr dest by element size */
     DMA_PRICTL_SRCDIR_INC, /* incr src by element size */
     DMA_PRICTL_START_DEFAULT
     ),
    DMA_SECCTL_RMK(
     DMA_SECCTL_DMACEN_DEFAULT,
     DMA_SECCTL_WSYNCCLR_DEFAULT,
     DMA_SECCTL_WSYNCSTAT_DEFAULT,
     DMA_SECCTL_RSYNCCLR_DEFAULT,
     DMA_SECCTL_RSYNCSTAT_DEFAULT,
     DMA_SECCTL_WDROPIE_DEFAULT,
```

```
                    DMA_SECCTL_WDROPCOND_DEFAULT,
                    DMA_SECCTL_RDROPIE_DEFAULT,
                    DMA_SECCTL_RDROPCOND_DEFAULT,
                    DMA_SECCTL_BLOCKIE_ENABLE,
                DMA_SECCTL_BLOCKCOND_DEFAULT,
                    DMA_SECCTL_LASTIE_DEFAULT,
                    DMA_SECCTL_LASTCOND_DEFAULT,
                    DMA_SECCTL_FRAMEIE_DEFAULT,
                    DMA_SECCTL_FRAMECOND_DEFAULT,
                    DMA_SECCTL_SXIE_DEFAULT,
                    DMA_SECCTL_SXCOND_DEFAULT
                    ),
                DMA_SRC_RMK((Uint32)psrc), /* source buffer */
                DMA_DST_RMK((Uint32)pdst), /* destination buffer */
                DMA_XFRCNT_RMK(
                    DMA_XFRCNT_FRMCNT_DEFAULT,
                    DMA_XFRCNT_ELECNT_OF(element_count) /*)
            );

            /* initialize the interrupts: */
            /* Enable the interrupts after the DMA channels are opened */
            /* as the DMA_OPEN_RESET clears and disables the channel */
            /* interrupt when specified and clears the corresponding */
            /* interrupt bits in the IER. */
            set_interrupts_dma();

            transfer_done = FALSE;
            DMA_start(hDma1); /* start DMA channel 1 */

            /* ISR will flip the value of transfer done */
            while (!transfer_done);

            DMA_close(hDma1);
        }

        void filter_image()
        {
            int ii, jj, sum,
                irow = BOUNDARY,
                ifir = 0, /* ptr to filtered row of pixels */
                /* # filtered pels to copy during transition to next block */
                nfir2copy = 2*BOUNDARY*Y_SIZE;
```

```
unsigned char *p = output_buf + BOUNDARY*Y_SIZE;

/* 'prologue': move 1st block into on-chip RAM and filter */
dma_copy_block(in_img,
               input_buf,
               ELEM_COUNT_SHORT(BLOCK_X_SIZE*Y_SIZE));
DSP_fir_gen(input_buf, h, fir_buf, NH, BLOCK_X_SIZE*Y_SIZE);

memset(output_buf, 0, BOUNDARY*Y_SIZE);

/* algorithm main loop: filter the individual blocks */
for (; irow<Y_SIZE-BOUNDARY; ++irow)
{
  if (0 == irow%NUM_SCAN_LINES)
  {
    /*
     * We just hit the start of the next block, so here we:
     *
     * o page in next block into working buffer
     * o pass these scan-lines through the FIR filter
     * o page out last block to output array
     * o do some prep work for the rest of the loop
     */
    int d = irow/NUM_SCAN_LINES, /* which block */
    k = (1==d)?3:2; /* very 1st row is a special case */

    /*
     * Move some rows from the bottom to the top of the scratch
     * buffer fir_buf, because they can be reused in the next
     * pass. Variable srow is an index to the start row of the
     * portion of the block that will be "moved"
     */
    int srow2copy = FIR_BLOCK_X_SIZE-k*BOUNDARY;

    /* move from the bottom to the top */
    DSP_blk_move(fir_buf+srow2copy*Y_SIZE, fir_buf, nfir2copy);

    /* copy from external to internal RAM */
    dma_copy_block(in_img+(irow+BOUNDARY)*Y_SIZE,
                   input_buf,
                   ELEM_COUNT_SHORT(n*Y_SIZE));
```

```
  /* run through FIR filter */
  DSP_fir_gen(input_buf, h,
                  fir_buf+2*BOUNDARY*Y_SIZE, NH, n*Y_SIZE);

  /* copy processed pixels into the final output array */
  dma_copy_block(output_buf,
                  out_img+(d-1)*NUM_SCAN_LINES*Y_SIZE,
                  ELEM_COUNT_UCH(NUM_SCAN_LINES*Y_SIZE));

    ifir = 0;
    p = output_buf;
  } /* end if (we just finished a block) */

  /* setup the pointers to the filtered rows */
  for (ii=0; ii<NH; ++ii)
    pFirBand[ii] = fir_buf + (ifir+ii)*Y_SIZE;

  /* 1st BOUNDARY columns are set to zero */
  for (ii=0; ii<BOUNDARY; ++ii) *p++ = 0;

  /* here 2D mask is applied from the 1D filtered rows */
  for (ii=0; ii<X_SIZE-2*BOUNDARY; ++ii) {
    sum = 0;
    for (jj=0; jj<NH; ++jj)
      sum += *(pFirBand[jj])++;
    *p++ = sum;
  }

  /* last two BOUNDARY columns are set to zero */
  for (ii=0; ii<BOUNDARY; ++ii) *p++ = 0;

  ifir++;

} /* end (for each row) */

/* algorithm 'epilogue': handle the final block */

/* zero out final few rows */
memset(p, 0, BOUNDARY*Y_SIZE);

/* copy final block of pixels to the output buffer */
dma_copy_block(output_buf,
```

```
        out_img+(NUM_BLOCKS-1)*NUM_SCAN_LINES*Y_SIZE,
        ELEM_COUNT_UCH(NUM_SCAN_LINES*Y_SIZE));
}
```

This program is our first encounter with the C6x Chip Support Library (CSL)[9], used here instead of the EVM library. To use the CSL, the project file should link to the appropriate library (in the case of the C6701 EVM, `csl6701.lib`) and define a preprocessor symbol indicating the DSP architecture. This symbol is of the form `CHIP_xxxx`, where `xxxx` is replaced by the model of the DSP. So for the C6701, the macro `CHIP_6701` is defined prior to inclusion of any of the CSL header files.

The `dma_copy_block` function in Listing 4-4 provides the speedup that this program achieves over its predecessors. Control of the DMA process is achieved by setting bit fields in the DMA registers, which is what occurs in the call to the `DMA_configArgs` function (for the exact details, refer to the CCStudio on-line help or [9]). The DMA transfer is set up so that the callee of `dma_copy_block` must provide the number of *words* to be transferred. Since this algorithm transfers both 16-bit short integers and 8-bit unsigned characters, and a word in the C6x architecture is 32 bits, a simple conversion must take place. The are two macros, `ELEM_COUNT_UCH` and `ELEM_COUNT_SHORT`, that are used towards this purpose inside of `filter_image`.

The DMA transfer is said to be *asynchronous*, meaning because the transfer is transparent to the CPU, the processor is free to perform other duties while the transfer is taking place. Because this program is coded in a serial fashion, we need to wait while the transfer is taking place. At the end of `dma_copy_block` the DSP is put in a busy spin loop until the global variable `transfer_done` is set to 1. The value of `transfer_done` is flipped to 1 in the interrupt service routine (ISR) `c_int09`, which in turn is hooked to `interrupt_9` by the assembly function `vectors` defined in `vecs.asm`. `IRQ_EVT_DMAINT1` is enabled in `set_interrupts_dma`, which is called just prior to initiating the DMA transfer in `dma_copy_block`. This interrupt is mapped to interrupt 9 on the chip and indicates when the current DMA transfer has completed. Thus `dma_copy_block` is turned into a *synchronous*, or blocking, function since it sits in a tight loop until the transfer completes. Finally, note that `transfer_done`'s declaration is decorated with the `volatile` keyword. This keyword is common in embedded applications, and it is important to understand its usage.

Nowhere in the source is `c_int09` ever explicitly called, and `transfer_done`'s value is never modified anywhere else in the code. Of course, the programmer knows that the hardware will cause an interrupt to eventually be generated, resulting in `c_int09` eventually being called, but the compiler has no way of knowing this. Whenever a variable may change its value in ways that the compiler cannot detect, the `volatile` keyword should be used or else an aggressively optimizing compiler may optimize the busy spin loop out of existence, which would indeed be an unfortunate occurrence.

This program takes an average of 1,552,183 cycles to low-pass filter a 256x256 image with a 5x5 kernel, using a block size of 16 rows. This performance time is 70% of `blur_dsplib_paging`, and offers a speedup of 1.42x over the image smoothing program that did not implement any memory optimizations (`blur_dsplib`). Even more performance could be eked out of this program by more sophisticated methods. The serialization in `dma_copy_block`, where the program sits in a busy spin loop waiting for the DMA transfer to complete, simply cries out for further optimization, and there is no reason why the DSP could not be performing useful work during the DMA transfer. This optimization leads to a "ping-pong" implementation, an example of which is given in [10], whereby two buffers, referred to as the ping and the pong buffer, are used to interleave data transfer and processing. After the DSP initiates a DMA transfer into the ping buffer, it moves on to processing the data contained in the pong buffer. When that processing has completed, the DSP initiates another DMA transfer into the pong buffer and moves on to processing the data in ping buffer, which contains the next set of data.

4.3.5   Full 2D Filtering with DSPLIB and DMA (`filter_dsplib_paging_dma`)

This implementation is a variation on the previous three programs, and is an example of a generic image filtering algorithm, capable of handling kernels other than low-pass filters with constant filter coefficients. This program relies on `DSP_fir_gen` like the others and also incorporates the DMA paging optimization we just introduced. Thus the performance this program offers provides a fair comparison with `filter_imglib`, as they both offer the same functionality. The source code is not listed here, as it is quite similar to the code in Listing 4-4; in fact the DMA code, ISR, and `main` are identical. The full project files may be found on the CD in the `filter_dsplib_paging_dma` subdirectory.

Essentially, this program operates in a fashion reminiscent of Figure 4-9, where each row in the image is passed through a potentially different 1D FIR filter multiple times, as the filter mask marches down the image. However, this process is in turn segmented in the block-wise pattern depicted in Figure 4-11, to enable the DMA paging optimization. The cycle count for filtering a 256x256 image using a 5x5 kernel and 16-row blocks is 3,669,928 cycles. The memory optimized blurring program from the preceding section (`blur_dsplib_paging_dma`) filters the same-sized image in 42% of the time it takes this program, so if an application calls for an averaging kernel that program should be used instead. However, most 2D filters do not consist of kernels where each row is the same, so this program can be used for the more general case. And finally, even in this program we have implemented only first-order optimizations, and there are many additional low-level optimizations that could be performed on the code as well. For example, fixing the kernel size and thereby getting rid of many of the loops mitigates some looping overhead, although at the cost of some flexibility – this function has been written so that it could be used with both a 3x3 and a 5x5 kernel, for example. As a rule of thumb, the more general an algorithm implementation, the less opportunity there is for fine-tuning towards further optimization. Table 4-1 summarizes the performance results for the various C62xx/C67xx 2D filtering programs.

Table 4-1. Performance results for the various C62xx/C67xx image filtering programs, profiled using the `clock` function as described in [14]. Cycle counts are the average of ten 10 runs, with the –o3 compiler optimization level and no debug symbols.

Program	Comments	Number Cycles
`filter_imglib*`	Uses C function instead of `IMG_corr_gen`.	46,972,076
`blur_dsplib`	Each row in kernel identical.	2,199,517
`blur_dsplib_paging`	Each row in kernel identical.	2,195,546
`blur_dsplib_paging_dma`	Each row in kernel identical.	1,552,183
`filter_dsplib_paging_dma`		3,669,928

* With `IMG_corr_gen`, 2D filtering takes 10,775,445 cycles, but output is incorrect.

4.4 LINEAR FILTERING OF IMAGES ON THE TI C64X

As described in 2.1, the C6416 fixed-point DSP is a newer member of the C6000 DSP family that offers higher performance than the C62xx series (it is not truly fair to compare the C64x to the C67x, as the C67x is a floating-point architecture). With regards to the CCStudio projects and C source code accompanying this book, all of the C62xx/C67xx projects were built and tested using the EVM development environment. The C64xx projects in this book, on the other hand, were built and tested using the C6416 DSK.

This section describes how to optimize a low-pass filtering program targeting the C6416. As one would expect for a DSP marketed to the imaging community, there is a version of IMGLIB optimized specifically for the C64xx line of DSPs[11]. Starting with a core IMGLIB convolution routine (that works!), an initial fixed-point implementation of a 3x3 low-pass filtering operation will be optimized in much the same fashion as in sections 4.3.3-4.3.5, by paging in blocks of the image as they are needed.

The source code for these programs are found in the Chap4\LinearFilter\C64xx directory. As in the C62xx/C67xx case, there are some common project files: the image.h header file containing default pixel data for the Lenna image and the linker command file image_filter.cmd. The image.h header file remains much the same as its C62xx/C67xx counterpart, except that in_img buffer is defined to be exactly N_PIXELS long, as we do not need to pad it with extra samples because we are no longer using DSP_fir_gen to filter the image. The linker command file for the projects is similar to that of the C62xx/C67xx projects, but the memory map for the C6416 DSK is different from that of the C62xx/C67xx EVM, meaning the MEMORY section is tailored for the DSK. Listing 4-5 shows the contents of the C64x DSK image_filter.cmd. Note that normally the ".cinit" section would map to the IRAM section, but due to the large amount of initialization data it is mapped to an external RAM segment.

Listing 4-5: Example C6416 DSK linker file, image_filter.cmd.

```
MEMORY
{
      VECS:    origin = 0h            length = 200h
      IRAM:    origin = 00000200h     length = 0000ffe0h
      SDRAM:   origin = 80000000h     length = 01000000h
}


SECTIONS
{
      .vectors > VECS
      .text    > IRAM, align(64)
      .bss     > IRAM
      .cinit   > SDRAM, align(64)
      .const   > IRAM
      .stack   > IRAM, align(64)
      .cio     > IRAM
      .sysmem  > IRAM
      .data    > IRAM
      .far     > IRAM
      .tables  > IRAM
}
```

The supporting infrastructure also changes slightly to accommodate changes necessitated by switching over to a different processor and the DSK environment. In particular, these programs link to C6416 versions of the runtime support library (rts6400.lib), chip support library (csl6416.lib), and a library we have yet to encounter, the board support library (dsk6416.lib). And of course, the programs link to a different IMGLIB static library (img64x.lib). Finally, the CHIP_xxxx preprocessor symbol is set to CHIP_6416.

4.4.1 Low-Pass Filtering with a 3x3 Kernel Using IMGLIB (blur3x3_imglib)

Consider the following 3x3 smoothing kernel:

$$\begin{bmatrix} 1/16 & 2/16 & 1/16 \\ 2/16 & 4/16 & 2/16 \\ 1/16 & 2/16 & 1/16 \end{bmatrix}$$

Similar to a Gaussian kernel, this kernel gives the center pixel more of a contribution than its surrounding neighbors and the sum of the weights is 1, a requirement for an averaging kernel that maintains the gain of the input image. A naive implementation would be to simply apply the convolution equation directly, but a far more efficient fixed-point implementation is to factor out the division by 16, so that for a neighborhood centered about pixel $f(i,j)$ the output pixel $g(i,j)$ is:

$$g(i,j) = 1/16[f(x-1,y-1) + 2f(x-1,y) + f(x-1,y+1) + \\ 2f(x,y-1) + 4f(x,y) + 2f(x,y+1) + \\ f(x+1,y-1) + 2f(x+1,y) + f(x+1,y+1)]$$

A further optimization is to replace the division by 16 with the equivalent operation of bit shifting to the right by 4 bits. This fixed-point convolution algorithm can be simulated in MATLAB with the following code:

```
H = [1 2 1; 2 4 2; 1 2 1];
J = imfilter(uint16(I), H); % I is a uint8 image matrix
J = uint8(bitshift(J), -4); % divide by 16
```

In the second line, I is promoted from 8 to 16 bits because the imfilter command returns a matrix of the same type as the input matrix, and thus without this type promotion J will consist of mostly saturated (255) values if I is originally of type uint8. If conv2 is used instead of imfilter then the uint16 qualifier is not required, as conv2 promotes everything to the double type. The third statement is a vectorized bit shift operation – the entire matrix is divided by 2^4.

IMGLIB includes a function IMG_conv_3x3 that can be used to pass an image through a 3x3 kernel consisting of signed 8-bit coefficients[12]. Internally, the function uses three 16-bit accumulators that sum intermediate values during the convolution operation, and so the caller must provide a shift value, which for this particular kernel is 4. Listing 4-6 is the contents of blur_3x3_imglib.c, which is the C source file for the C6416 program located in Chap4\LinearFilter\C64xx\blur3x3_imglib.

Listing 4-6: blur3x3_imglib.c

```
#define CHIP_6416
#include <dsk6416.h>
#include <stdio.h>   /* printf() */
#include <string.h>  /* memset() */
#include <img_corr_3x3.h>
#include <csl_timer.h>

#include "..\image.h"
#pragma DATA_ALIGN (in_img, 8);
#pragma DATA_SECTION (in_img, "SDRAM");

#pragma DATA_ALIGN (out_img, 8);
#pragma DATA_SECTION (out_img, "SDRAM");
unsigned char out_img[N_PIXELS]; /* filtered image */

/* filter dimensions and coefficients */
#define NH 3 /* NHxNH kernel (needs to be 3x3 for this program) */
#define BOUNDARY (NH/2) /* kernel half-width */
#pragma DATA_ALIGN (H, 8)
char H[NH*NH] = {
   1, 2, 1, /* 1/16 2/16 1/16 */
   2, 4, 2, /* 2/16 4/16 2/16 */
   1, 2, 1, /* 1/16 2/16 1/16 */
};
#define SHIFT 4 /* right-shift by 4 (div by 16) */

#define N_COLS_FILTERED Y_SIZE-2*BOUNDARY
```

```
/*
 * Faster than memset(), count must be a multiple of
 * 8 and greater than or equal to 32
 */
void memclear( void * ptr, int count )
{
  long *lptr = ptr;
  _nassert((int)ptr%8==0);
  #pragma MUST_ITERATE (32);
  for (count>>=3; count>0; count--)
    *lptr++ = 0;
}


void filter_image()
{
  unsigned char *p = out_img+BOUNDARY*Y_SIZE;
  int ii, irow;

  /* set 1st BOUNDARY rows to zero */
  memclear(out_img, BOUNDARY*Y_SIZE);

  /* filter the interior region of the image matrix */
  for (irow=BOUNDARY; irow<X_SIZE-BOUNDARY; ++irow)
  {
    /* 1st BOUNDARY cols are zero */
    for (ii=0; ii<BOUNDARY; ++ii) *p++ = 0;

    /*
     * IMG_conv_3x3 requires 3rd arg to be a multiple of 8,
     * that's why we pass in Y_SIZE instead of N_COLS_FILTERED
     * (last few filtered pixels are ignored)
     */
    IMG_conv_3x3(&in_img[(irow-BOUNDARY)*Y_SIZE],
                 p,
                 Y_SIZE,
                 H,
                 SHIFT);

    /* last BOUNDARY cols are zero */
    p += N_COLS_FILTERED;
    for (ii=0; ii<BOUNDARY; ++ii) *p++ = 0;
  }
```

```
/* last BOUNDARY rows are zero */
memclear(out_img+(X_SIZE-BOUNDARY)*Y_SIZE,
        BOUNDARY*Y_SIZE);
}

int main(void)
{
  TIMER_Handle hTimer;
  unsigned int start, stop, overhead, total = 0, t; /* timing */
  const int N = 10; /* how many times to profile */
  int ii = 0;
  DSK6416_init(); /* initialize the DSK board support library */
  /* configure timer */
  hTimer = TIMER_open(TIMER_DEVANY,0);
  TIMER_configArgs(hTimer, 0x000002C0, 0xFFFFFFFF, 0x00000000);
  /* compute overhead of calling the timer. */
  start=TIMER_getCount(hTimer); /* called 2x to avoid L1D miss. */
  start=TIMER_getCount(hTimer);
  stop=TIMER_getCount(hTimer);
  overhead=stop-start;

  for (; ii<N; ++ii) {
    start = TIMER_getCount(hTimer); /* begin "profile area" */
      filter_image();
    stop = TIMER_getCount(hTimer); /* end "profile area" */
    t = (stop-start-overhead) * 8;
    total += t;
    printf("# cycles to filter image: %d\n", t);
  }
  printf("avg time is %.2f cycles.\n", (float)total/(float)N);
}
```

One difference between this program and the previous C62xx/C67xx EVM programs is the timing mechanism. Unless otherwise stated, the C6416 DSK programs in this book use the CSL timer API as described in [12], whereas the EVM programs use the `clock` function for profiling code. Another difference is the addition of a new local function `memclear`, which is used as a faster substitute for `memset`.

The implementation of `memclear` lends insight into how low-level code optimizations can provide major performance enhancements. Staying with the idea that more generality sometimes means a decrease in speed, the `memclear` function leverages certain known characteristics of the structure of the surrounding code to gain a performance edge over the standard C library function `memset`, which obviously has to be general-purpose in order to maintain its contract with the programmer. This function offers advantages over the general-purpose `memset` function, which in C might be defined as so:

```
void memset( void *ptr, int x, int count ) {
    char *uch = ptr;
    for (; count>0; count--) *uch++ = x;
}
```

By stipulating that the number of iterations through the `memclear` loop (commonly referred to as the *trip count*) is a multiple of 8, the input pointer casted to a long (64-bit) type, and guaranteeing alignment of `lptr` to a 64-bit boundary via `_nassert`, the compiler is given numerous pieces of information so that it can generate a loop that will run faster than a `memset`-like function. In the definition of `memclear`, what appears to be a function call to `_nassert` is actually an example of a TI-specific compiler *intrinsic*[13]. Intrinsics are extensions to ANSI C that either map to inline C6x assembly instructions that cannot be expressed in a pure ANSI-compliant C translation unit, or as in the case of `_nassert`, provide extra information to the compiler. Here, the statement `_nassert((int)lptr%8==0)` asserts that the address of `lptr` is double-word aligned. Consequently, the compiler is free to use the LDDW/STDW (load/store aligned double-word) instructions to initialize the 64-bit number pointed to by `lptr`. LDDW/STDW are special instructions that operate on a data stream lying on an aligned memory address[14], and these aligned instructions are more efficient than their unaligned counterparts (the C64x DSP has non-aligned double word instructions, LDNDW/STDNW, or load/store non-aligned double word). Wherever possible, unaligned stores and loads should be avoided as the DSP can only perform a single unaligned load per clock cycle, whereas multiple aligned loads can occur in a single cycle. The more conservative code compiles to assembly language using LDB/STB (load/store byte) instructions to initialize the 8-bit number pointed to by `lptr`, and a series of these instructions is not as efficient as a series of LDDW/STDW instructions due to the lessened throughput of the data flowing through the DSP. This type of code optimization is explained in greater detail in Appendix B.

The MUST_ITERATE pragma directive in `memclear` is a means of providing the compiler information about a loop, and is analogous to the `.trip` directive in linear assembly code[14]. Through this directive the programmer can specify the exact number of times a loop will execute, if the trip count is a multiple of some number, the minimum number of iterations through the loop, and so on. This pragma should be used wherever possible - especially when the minimum trip count is known as this information allows the compiler to be more aggressive when applying loop transformations. The form of the MUST_ITERATE pragma used in memclear specifies that the loop is guaranteed to execute at least 32 times, and armed with this information the compiler can proceed to unroll the loop. Loop unrolling is a technique where the loop kernel (not to be confused with a filter kernel) is expanded by a factor X - and the loop stopping condition adjusted to N/X – with the intent of reducing the number of branches. By reducing the branch overhead, the efficiency of the loop is increased, and it also creates an opportunity for better scheduling of instructions contained within the loop kernel. However, it is not always the case that it is advantageous to unroll a loop. If a loop is unrolled too much, the code size may increase such that it overflows the instruction cache, which essentially defeats the purpose of the loop transformation. This particular loop however, given its small size (a single assignment statement) is not at risk of this problem. A further holdback is if the loop does not execute enough times to even warrant unrolling. By stipulating that the minimum number of times through this loop is 32, the compiler knows that it should proceed with unrolling the loop.

Getting back to the 2D filtering algorithm, the implementation in Listing 4-6 is more concise and frankly far simpler than earlier incarnations, for a couple of reasons. Because IMG_conv_3x3 is explicitly designed to perform 2D filtering with a fixed kernel size, there are no intermediate 1D results that need to be combined to form 2D filtered pixels. In addition, this particular program does not consider the memory bottlenecks, and simply accesses pixel data stored in external RAM. Figure 4-12 is a CCS screenshot of this program, halted just after the call to filter_image. Both the original image (in_img) and the processed image (out_img) are "graphed" side-by-side. With the –o3 compiler optimization level and no debug symbols, this program takes on average 19,141,824 cycles.

4.4.2 A Memory-Optimized 2D Low-Pass Filter (blur3x3_imglib_paging_dma)

In previous sections, we proceeded through a step-by-step dissection of an optimization of 2D image filtering by reducing memory latencies – now that we have seen this once, we can cut to the chase. Listing 4-7 are the relevant portions of the blur3x3_imglib_paging_dma.c file, which can be found in the CCStudio project directory located in Chap4\LinearFilter\C64xx\blur3x3_imglib_paging_dma.

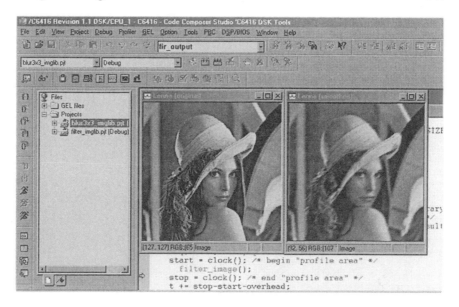

Figure 4-12. Low-pass filtering of the Lenna image on the C6416 DSK. The image on the left is the original image, stored in the in_img array. On the right is the low-pass filtered image, stored in the out_img array.

Listing 4-7: portions of blur3x3_imglib_paging_dma.c.

```
/*
 * For block processing, segment image into individual chunks
 * that are subsequently paged in and out of internal memory.
 */
#define NUM_SCAN_LINES   16
#define NUM_BLOCKS       (X_SIZE/NUM_SCAN_LINES)
/* how many rows each block is */
#define BLOCK_X_SIZE     (NUM_SCAN_LINES+2*BOUNDARY)

/*
 * These are scratch buffers, strategically placed in on-chip RAM:
 *
 * input_buf = input pixels, passed to IMG_conv_3x3
 * output_buf = filtered pixels
 */
#pragma DATA_ALIGN (input_buf, 8);
unsigned char input_buf[BLOCK_X_SIZE*Y_SIZE];
#pragma DATA_ALIGN (output_buf, 8);
/*
```

```
 * NOTE: pad output_buf with 2*BOUNDARY pixels because
 * IMG_conv_3x3 required # cols arg to be multiple of 8.
 * If this wasn't done you'd write past the end of the array.
 */
unsigned char
          output_buf[NUM_SCAN_LINES*Y_SIZE + 2*BOUNDARY];


/* Filter block of an image, returns row index for next block */
int filter_block(int irow, int nrows,
                 const unsigned char *restrict pin,
                 unsigned char *restrict pout)
{
  int jj = irow, kk;
  for (; jj<irow+nrows; ++jj) {

    /* 1st few cols are 0 */
    for (kk=0; kk<BOUNDARY; ++kk) *pout++ = 0;
    /*
     * Even though we only care about N_COLS_FILTERED pixels,
     * in pass Y_SIZE because this function requires # pixels
     * to be filtered be a multiple of 8.
     */
    IMG_conv_3x3(pin, pout, Y_SIZE, H, SHIFT);

    /* last few cols are 0 */
    pout += N_COLS_FILTERED;
    for (kk=0; kk<BOUNDARY; ++kk) *pout++ = 0;

    pin += Y_SIZE; /* incr scan-line in prep for next iteration */
  }
  return jj-1;
}
/* March down the image block-by-block, filtering along the way */
void filter_image()
{
  Uint32 id_EDMAin  = DAT_XFRID_WAITNONE,
         id_EDMAout = DAT_XFRID_WAITNONE;
  int irow = BOUNDARY;
  unsigned char *pout_img = out_img,
                *pin_img =
                &in_img[(NUM_SCAN_LINES+BOUNDARY)*Y_SIZE],
```

```
/*
 * We reuse the bottom-most portion of input_buf by shifting it to the
 * top, prior to the beginning of the subsequent block filtering.  The 1st
 * time through the "interior of the image" loop, the the pointer to
 * bottom-most portion (pinput_buf_row2move_1) is different
 * from the subsequent iterations (pinput_buf_row2move_n).
 */
*pinput_buf_row2move_1 =
           input_buf + (BLOCK_X_SIZE-3*BOUNDARY)*Y_SIZE,
*pinput_buf_row2move_n =
           input_buf + NUM_SCAN_LINES*Y_SIZE,
*pinput_buf_row2move = pinput_buf_row2move_1,
*pinput_buf_row2copy_into = input_buf+2*BOUNDARY*Y_SIZE;

/************************************************************
 * Algorithm 'prologue': filter the 1st block
 ************************************************************/
id_EDMAin = DAT_copy(in_img, input_buf,
                     (BLOCK_X_SIZE-BOUNDARY)*Y_SIZE);
irow = filter_block(irow, NUM_SCAN_LINES,
                    input_buf, output_buf + BOUNDARY*Y_SIZE);

/************************************************************
 * Filter the interior of the image
 ************************************************************/
for (; irow<X_SIZE-NUM_SCAN_LINES; ++irow) {

    /* page out the most recently processed block */
    id_EDMAout = DAT_copy(output_buf, pout_img,
                          NUM_SCAN_LINES*Y_SIZE);
    pout_img += NUM_SCAN_LINES*Y_SIZE;

    /* page in the next block of pixel data */
    /*
     * 1st shift the scan-lines we can reuse from the bottom
     * to the top of the scratch buffer.
     */
    memcpy(input_buf,
           pinput_buf_row2move,
           2*BOUNDARY*Y_SIZE);
    pinput_buf_row2move = pinput_buf_row2move_n;
```

```
/* DMA in next set of scan-lines */
id_EDMAin = DAT_copy(pin_img, pinput_buf_row2copy_into,
                     NUM_SCAN_LINES*Y_SIZE);
pin_img += NUM_SCAN_LINES*Y_SIZE;

/* wait now for both xfers to complete before proceeding */
DAT_wait(id_EDMAout);
DAT_wait(id_EDMAin);
irow = filter_block(irow, NUM_SCAN_LINES,
                    input_buf, output_buf);
}


/**************************************************************
 * Algorithm 'epilogue': filter the last block
 **************************************************************/
/* page out the most recently processed block of image data */
id_EDMAout = DAT_copy(output_buf, pout_img,
                      NUM_SCAN_LINES*Y_SIZE);
pout_img += (NUM_SCAN_LINES)*Y_SIZE;

/* page in the last block of data (shift scan-lines 1st) */
memcpy(input_buf, pinput_buf_row2move, 2*BOUNDARY*Y_SIZE);
id_EDMAin =
  DAT_copy(pin_img, pinput_buf_row2copy_into,
           (NUM_SCAN_LINES-BOUNDARY)*Y_SIZE);

/* must wait now for both xfers to complete before proceeding */
DAT_wait(id_EDMAout);
DAT_wait(id_EDMAin);
filter_block(irow, NUM_SCAN_LINES-BOUNDARY,
             input_buf, output_buf);

/* last few rows are zero */
memclear(output_buf + (NUM_SCAN_LINES-BOUNDARY)*Y_SIZE,
         BOUNDARY*Y_SIZE);

/* we're done, page out this final block of pixel data */
id_EDMAout = DAT_copy(output_buf, pout_img,
                      NUM_SCAN_LINES*Y_SIZE);
DAT_wait(id_EDMAout);
}
```

This program utilizes the paging memory management "design pattern", to borrow a term from software engineering, to circumvent the problems associated with accessing data located in slow off-chip RAM. As image data is needed, blocks are paged in and out via DMA channels. There are a few subtle changes in the C64xx implementation that do warrant additional discussion. For starters, the DMA specific portion of the implementation is more concise than its C62xx/C67xx counterpart (see Listing 4-4), because here we do not use an ISR to wait for the DMA transfer to complete. Instead, this code relies on the CSL API function DAT_wait to wait for a memory transfer to complete (internally DAT_wait more than likely utilizes an ISR but this implementation detail is well hidden from the programmer). The DMA mechanism in the C6x1x line of DSPs (e.g. C6211, C6711, C6713, and C6416) differs from that of the C6x0x series (C6201, C6205, C6701) in that an enhanced DMA (EDMA) peripheral supplants the older DMA peripheral. The EDMA controller offers certain advantages over the legacy DMA controller, such as a larger number of channels (64), high concurrency, and automatic re-arming of trigger events so that a sequence of transfers can be triggered from a single event[9]. As we shall soon see, the abstraction accorded to us by the CSL API allows the program to take advantage of a small amount of parallelism, without having to implement a complicated double-buffering or ping-pong scheme.

The general structure of the filtering algorithm closely follows that of Listing 4-4. The blocking function DAT_wait is used to block until DMA transfers are complete. There is some amount of parallelism in this implementation - note the call to DAT_wait is made after the call to memclear in the algorithm prologue section. Similar scheduling is also used during the filtering of the image interior, with the basic premise being to continue on with other independent tasks while DMA transfers are taking place, since the processor is free while the EDMA controller is handling the data transfer. This type of parallelism that DMA offers is a major benefit, in addition to the fact that DMA provides burst transfers for fast access to off-chip RAM and other peripherals.

There is a version of this program (blur3x3_imglib_paging.c) in the same directory that uses memcpy to page in blocks of memory, and with the standard profiling compiler options, that program takes on average 10,408,890 cycles to filter our 256x256 image with the 3x3 smoothing kernel, or roughly half the time it takes the completely memory-unoptimized version to execute. The EDMA-enabled version of this algorithm takes only 1,646,381 cycles on average to execute, for a speedup of over 11.5x versus the memory-unoptimized version! Table 4-2 summarizes the performance benchmarks for the three C6416 DSK image filtering programs.

Table 4-2. Performance results for three C4616 DSK image filtering programs, profiled using the CSL timer API as described in [12]. Cycle counts are the average of ten runs, with no debug symbols and the –o3 compiler optimization level.

Program	Comments	Number Cycles
blur_3x3_imglib	No memory optimization.	19,141,824
blur_3x3_imglib_paging	Paging via memcpy function.	10,408,890
blur_3x3_imglib_paging_dma	Paging via EDMA.	1,646,381

4.5 NON-LINEAR FILTERING OF IMAGES

Linear filters have the property that their output is a linear combination of their input. Filters that do not satisfy this property are non-linear. Non-linear filters are not employed in image processing as frequently as linear filters, because in many cases they are based on heuristics closely tied to the particular application. The general idea in non-linear image filtering is that instead of using the spatial mask in a convolution process, the mask is used to obtain the neighboring pixel values, and then ordering mechanisms produce the output pixel. That is, as the mask is shifted about the image, the order of the pixels in the windowed section of the image is rearranged, and the output pixel is generated from these rearranged input pixels. Figure 4-13 illustrates this procedure, for three types of non-linear filters.

The primary application with which we shall explore the use of non-linear filters will be in the enhancement of noisy images. While non-linear filters are sometimes used to perform other tasks, removal of noise is by and large the most common usage pattern. The topic of noise was introduced in 4.1.1, and to properly demonstrate the effectiveness of these filters in removing noise, it is necessary to first take a small detour to explain how to quantitatively assess their performance. Once we have introduced the theory behind assessing image fidelity, we can go back and present actual implementations of non-linear filters, and evaluate their performance.

4.5.1 Image Fidelity Criteria and Various Metrics

Here we deal with answering the question of how effective a particular filter is in enhancing the quality of an image, relative to some other image. Without answering the image fidelity question it is impossible to ascertain a filter's performance. The answer to this question is also of fundamental importance in other image processing fields, such as when considering the efficacy of lossy compression algorithms, where tradeoffs must be made between the amount of compression and the loss of image quality in the compressed image. In data compression, the "other image" is simply the

•

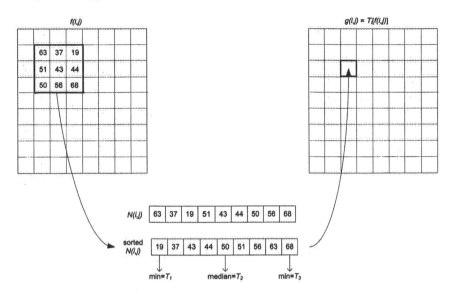

Figure 4-13. Examples of non-linear filters. $g(i,j) = 50$ for T_2 and is referred to as a median filter, and is one of the more common non-linear filters used in image processing. T_1 and T_3 correspond to the minimum and maximum intensity in the 3x3 neighborhood surrounding $f(i,j)$. Minimum filters form the basis of image erosion, and maximum filtering the basis of image dilation.

original, uncompressed image. When dealing with image enhancement algorithms, the situation is not so straightforward. After running a noisy image through a spatial filter, what do we compare the (hopefully) enhanced image against, so as to gauge the filter's performance?

Since the spatial filters in this chapter are going to be presented in the context of removing some type of noise from an image, it follows that a reasonable course of action is to simulate a noisy system by injecting certain forms of noise into an image, running the corrupted image through a filter, and then comparing the processed image to the original pristine image. So how to perform this comparison? Theoretically speaking, the best means is in fact modifying the old adage that "beauty is in the eye of the beholder" to "image enhancement is in the eye of the beholder." In other words, since images are visualized by humans, the human eye provides the best feedback system. Subjective assessment of image quality is prevalent in medical imaging (less so in image compression systems), where it is imperative that highly trained clinicians determine whether or not there is a perceived gain or loss in diagnostic information. Practically speaking, relying on human observers to score the performance of image enhancement algorithms is not a viable strategy; first because it is not economical, and secondly the subjective nature precludes incorporation into mathematical models.

Thus an objective measurement, some means of quantifying image fidelity, is required. The first metric that comes to mind is to consider the difference between the filtered pixels and the original pixels. If the original MxN image is *f(i,j)* and the filtered MxN image is *g(i,j)*, the delta image *d(i,j)* is

$$d(i, j) = g(i, j) - f(i, j)$$

The total delta between the original *f* and processed *g* is then simply the sum:

$$delta = \sum_{i=0}^{M-1} \sum_{j=0}^{N-1} [g(i, j) - f(i, j)]$$

This sum could conceivably be used as a fidelity criterion – however the most commonly used criterion is the *mean squared error (MSE)*, which is the average of the squared delta measurement:

$$mse = \frac{1}{MN} \sum_{i=0}^{M-1} \sum_{j=0}^{N-1} [g(i, j) - f(i, j)]^2$$

The *root mean square error (RMS)* is the square root of the MSE:

$$rms = \left[\frac{1}{MN} \sum_{i=0}^{M-1} \sum_{j=0}^{N-1} [g(i, j) - f(i, j)]^2 \right]^{1/2}$$

Related to these is the *signal-to-noise ratio (SNR)*, roughly defined as

SNR = signal/noise

Essentially, all image enhancement processes aim to improve the SNR. SNR is a term used in a multitude of engineering disciplines, and is oftentimes expressed in decibels (dB). A large SNR indicates a relatively clean image; a low SNR indicates that there is enough noise present within the image such that it impairs one's ability to discriminate between the noise floor and the actual signal. The SNR metric is a very seductive one to use, as it makes intuitive sense. For example, when it comes to digital photography, the bottom line is that increasing the light gathering capability of the lens increases signal and therefore increases SNR, thus producing better looking photos. But care must be taken when quoting actual SNR results, as there are so many different definitions for SNR that proper comparison of results can be difficult. If SNR measurements are quoted, the best strategy is to clearly define exactly how SNR is being calculated, or to rely on RMS or MSE metrics, which everyone agrees on. The mean-squared SNR is defined as

$$SNR_{ms} = \frac{\sum_{i=0}^{M-1}\sum_{j=0}^{N-1} g(i,j)^2}{\sum_{i=0}^{M-1}\sum_{j=0}^{N-1} [g(i,j) - f(i,j)]^2}$$

SNR_{rms}, the RMS value of the SNR, is simply the square root of the above equation. The final metric we consider is the peak signal-to-noise ratio (PSNR), which relates the magnitude of the noise to the peak value in the image. The PSNR, in decibels, is defined as

$$PSNR = 10\log_{10}\left(\frac{p}{RMS}\right) dB$$

where p is the peak value of the signal (255 in an 8-bit image) and *RMS* is the root-mean-square delta between the original image and the output image. It should be emphasized that basing image fidelity and algorithm performance solely on quantitative means can be misleading. It is not uncommon for two images to have the same or approximately the same PSNR, but for one of the images to exhibit characteristics that subjectively jump out at the human observer. The classic example is when one image differs from another by 1 pixel count across the board (i.e. *imageB* = *imageA* + 1), hence *imageB* is ever so slightly brighter than *imageA*. Here the human eye will not perceive any distortion, but a numerical fidelity criterion will have a large magnitude. Now imagine the situation where one image differs from another in that a small block of the image is missing. This blemish, a type of "shot noise," will immediately be noticed yet standard error measures will not bear this out. This situation is illustrated in Figure 4-14.

4.5.2 The Median Filter

The median filter is conceptually similar to an averaging kernel. An averaging kernel produces an output pixel $g(i,j)$ that is the mean intensity of a neighborhood around $f(i,j)$. The median filter is the same operation, except that instead of the mean intensity the median intensity is chosen. In other words, the reordering mechanism that defines this non-linear filter is a sorting operation, and the output is simply the middle value of the sorted list. Figure 4-13 shows an example of how the median filter is implemented.

The median filter is ideal for removing impulsive noise, sometimes referred to as "salt and pepper" or "shot" noise (see Figure 4-14c). This type of noise manifests itself as outlier pixel intensities, completely independent of signal content. Because of the need to sort the values within the sliding window, the median filter can be quite costly to implement – however they are preferred to a smoothing filter in these cases because they are far better at preserving the edges in an image. Figure 4-15 demonstrates this behavior for a small 5x5 test image. Notice how the blurring filter spreads the error around, and reduces the effect of the sharp vertical edge in the input image, while the median filter does a superb job of reconstructing the input image.

Figure 4-16 is a more realistic example of the median filter in action. Figure 4-16b shows the Lenna image corrupted with 40% impulse noise, and Figures 4-16c and 4-16d show the effect of passing the corrupted image through a 3x3 and 5x5 median filter, respectively. This is an example where the quantitative fidelity metrics mesh with the qualitative metrics – not only does Figure 4-16d look better than 4-16c, but the PSNR fidelity measure reflects this observation.

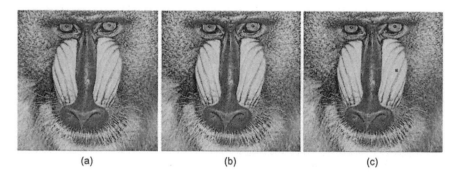

(a)   (b)   (c)

Figure 4-14. The problems with relying solely on numerical image fidelity metrics. (a) Original 256x256 baboon image. (b) Brightened image, with each pixel changed by 1 intensity level, $e_{RMS} = sqrt(256^2/256^2) = 1$. (c) Image with a 3x3 blemish, where each corrupted pixel differs from the original by 50 intensity levels. The e_{RMS} of this image is sqrt $((9*50^2)/256^2) = 0.5859$, which is less than the e_{RMS} of the second image, even though this image looks worse by any subjective criteria.

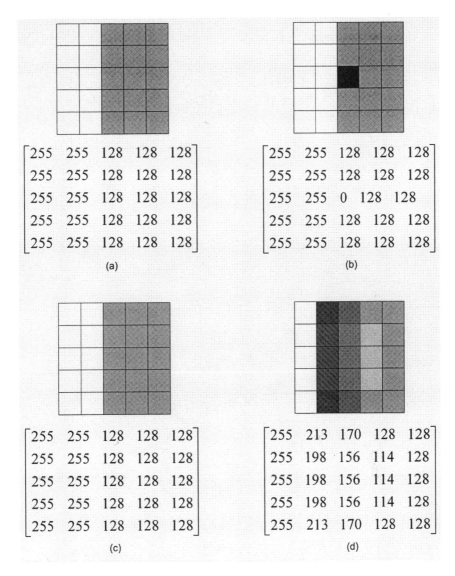

$$\begin{bmatrix} 255 & 255 & 128 & 128 & 128 \\ 255 & 255 & 128 & 128 & 128 \\ 255 & 255 & 128 & 128 & 128 \\ 255 & 255 & 128 & 128 & 128 \\ 255 & 255 & 128 & 128 & 128 \end{bmatrix}$$

(a)

$$\begin{bmatrix} 255 & 255 & 128 & 128 & 128 \\ 255 & 255 & 128 & 128 & 128 \\ 255 & 255 & 0 & 128 & 128 \\ 255 & 255 & 128 & 128 & 128 \\ 255 & 255 & 128 & 128 & 128 \end{bmatrix}$$

(b)

$$\begin{bmatrix} 255 & 255 & 128 & 128 & 128 \\ 255 & 255 & 128 & 128 & 128 \\ 255 & 255 & 128 & 128 & 128 \\ 255 & 255 & 128 & 128 & 128 \\ 255 & 255 & 128 & 128 & 128 \end{bmatrix}$$

(c)

$$\begin{bmatrix} 255 & 213 & 170 & 128 & 128 \\ 255 & 198 & 156 & 114 & 128 \\ 255 & 198 & 156 & 114 & 128 \\ 255 & 198 & 156 & 114 & 128 \\ 255 & 213 & 170 & 128 & 128 \end{bmatrix}$$

(d)

Figure 4-15. Comparison of a median filter vs. an averaging filter. (a) Original 5x5 test image. (b) Test image with impulsive noise added (middle pixel). (c) Output of 3x3 median filter, with boundary values extended symmetrically. (d) Result of 3x3 averaging filter, also with boundary values extended. Notice how this filter, while reducing the effect of the shot noise, spreads the error and greatly attenuates the effect of the sharp edge in the original image.

(a)   (b)

(c) PSNR = 9.2898 dB   (d) PSNR = 12.4450 dB

Figure 4-16. Removing impulse noise from the Lenna image using the median filter. Peak signal-to-noise ratio calculated using provided MATLAB function `psnr.m`, which is based on formula given in the text. (a) Original image. (b) Lenna with 40% additive impulse noise. (c) Processed image using 3x3 median filter. (d) Processed image using 5x5 median filter.

   The median filter is actually a specific form of a *rank filter*, where the ith pixel intensity in the sorted list of neighborhood pixels is chosen as the output. Thus for the case of the median filter, the index i is simply the middle point. If the neighborhood is sorted in ascending order, and the index i is chosen to be the first element, then the filter is a *minimum filter*. Likewise, if the last element is chosen, the filter is a *maximum filter*. Passing an image through a minimum filter is known as *erosion* – dark areas in the image become more pronounced, and oftentimes images are repeatedly passed through erosion filters to aggrandize the effect. One application of the

(a)   (b)

Figure 4-17. Image erosion and dilation. (a) Lenna image after two passes of a 3x3 minimum filter. (b) Image after two passes of a 3x3 maximum filter.

maximum filter is image *dilation*, where bright areas in the image are emphasized. Again, repeated application of dilation is a common technique.

Image erosion and dilation find their way into numerous image-processing systems. The global nature of the image remains the same, while dark (erosion) or bright (dilation) regions dominate, the degree of which is dependent on how many times the filter is applied. Figure 4-17 illustrates this effect on the Lenna image. In astronomical imaging, erosion is useful to bring out an object buried within a dense area – for example a nebula obscured within a star field. The two operations are sometimes coupled, as in image segmentation, where the goal is to deconstruct an image into sections that ideally map to objects of interest. Dilation followed by erosion can be used in image segmentation to enhance the contrast of portions of an image so that a simple brightness threshold can be used as a tool to separate various objects in an image.

4.5.3   Non-Linear Filtering of Images in MATLAB

As its name suggests, the Image Processing Toolbox function `medfilt2` performs two-dimensional median filtering (the `medfilt` function is also available for median filtering of one-dimensional signals). The arguments for `medfilt2` are straightforward: the input image, the size of the mask, and an optional third argument specifying how to handle boundary pixels[15]. The command used to create Figure 4-16c is

```
J = medfilt2(I, [3 3], 'symmetric');
```

The Image Processing Toolbox also contains a very useful function, imnoise, that corrupts an image with various types of noise. This function proves very convenient when performing experiments to determine how effective a filter is in reducing the effects of noise. The imnoise function knows about the following types of noise:

- Gaussian white noise (also white noise whose magnitude tracks the local variance of the image)
- Poisson noise
- Shot noise
- Speckle noise

For each type, imnoise accepts parameters to adjust the characteristics of the noise[16]. For example in the command used to create Figure 4-16b,

J = imnoise(I, 'salt & pepper', 0.4);

the final argument specifies that the number of corrupted pixels should be 40% of the number of pixels in I. Listing 4-8 is a replacement for the above function that does not require the Image Processing Toolbox, for the single case of salt and pepper (a.k.a. shot) noise.

Listing 4-8: MATLAB function to add salt and pepper noise to an image.

```
function J = add_shot_noise(I, how_much)

if how_much > 1.0
    error('Cannot corrupt more than 100% of the pixels');
end
[M N] = size(I);

% rand returns a MxN matrix with uniformly distributed
% #'s in the interval (0.0, 1.0), so by finding how
% many of these are less than or equal to how_much, we
% determine which pixels are to be corrupted.
R = rand(M, N);
icorrupted = find(R <= how_much);
ncorrupted = length(icorrupted);

% generate a normally distributed vector with zero mean,
% so half of them should be negative, and half of them
% positive.
salt_and_pepper = randn(1, ncorrupted);
```

```
% the signum function maps negative #'s to -1 and
% positive #'s to +1.
salt_and_pepper = sign(salt_and_pepper);

% noisy is a temporary matrix to store the pixel
% indices of the corrupted data
noisy = zeros(M, N);
noisy(icorrupted) = salt_and_pepper;
% those locations in noisy that are -1 will be set
% to the 'off' (min) intensity, and those that are
% +1 will be set to the 'on' (max) intensity.
min_noise_val = 0;
if isa(I, 'uint8')
   max_noise_val = 255;
elseif isa(I, 'uint16')
   max_noise_val = 65535;
else
   max_noise_val = 1;
end

J = I;
icorrupted = find(-1 == noisy);
J(icorrupted) = min_noise_val;
icorrupted = find(1 == noisy);
J(icorrupted) = max_noise_val;
```

This function makes extensive use of the built-in MATLAB `find` function, which accepts a predicate and returns the indices where this predicate evaluates to true. Thus the statement

```
icorrupted = find(R <= how_much);
```

returns the indices in the matrix R where the value of R is less than or equal to `how_much`. This begs the question – since R is a 2D matrix, does `icorrupted` somehow return the row and column numbers for indexing into R? The answer is no, as MATLAB indexes into N-dimensional matrices by assuming a flattened array, akin to how the C programs in this book index into flattened image buffers. One significant difference between MATLAB and C is that MATLAB is column-major, whereas C is row-major – so while index 2 is the second column in the first row in C, it is the *first* column in the *second* row in MATLAB. In MATLAB parlance, the row and column indices are matrix *subscripts*, and the function `ind2sub` can then be used to translate between single indices (i.e., the ones returned from `find`) and subscripts, given the dimensions of a matrix. Similarly, there is a function `sub2ind` that translates the other way, from multiple subscripts to linear indices.

Listing 4-9 is a MATLAB function that may be used as a replacement for `medfilt2`, in the event that the Image Processing Toolbox is not available. What is interesting to note about this function's implementation is how MATLAB's "colon" notation allows the programmer great flexibility for indexing into images.

Listing 4-9: MATLAB function to perform median filtering of an image.

```
function J = median_filter_image(I, sz)

if ~mod(sz(1), 2) | ~mod(sz(2), 2)
    error('neighborhood dimensions must be odd!');
end
imedian = ceil(prod(sz)/2);

[MI, NI] = size(I);
J = zeros(MI, NI);

ihw = floor(sz(1)/2); % ihw = i "half-width"
jhw = floor(sz(2)/2); % jhw = j "half-width"

% this double-loop slides the neighborhood around the image
for ii = ihw+1:MI-ihw
    for jj = jhw+1:NI-jhw
        % splice out the portion of the image
        neighborhood = I(ii-ihw:ii+ihw, jj-jhw:jj+jhw);
        % sort the neighborhood (in ascending order)
        n_sorted = sort(neighborhood(:));
        % generate output pixel
        J(ii, jj) = n_sorted(imedian);
    end
end

if isa(I, 'uint8')
    J = uint8(J);
elseif isa(I, 'uint16')
    J = uint16(J);
end
```

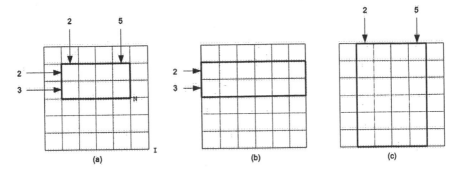

Figure 4-18. MATLAB's colon notation with 2D matrices. (a) N = I(2:3, 2:5) (b) N = I(2:3, :) (c) N = I(:, 2:5)

Given a 2D matrix I, the statement

N = I(sr:er, sc:ec);

returns in N the submatrix of I consisting of rows sr (start row) to er (end row) and columns sc (start column) to ec (end column). This notation is used in median_filter_image to extract the sliding window neighborhood at each iteration within the double loop. If either of the two ranges are replaced with just the colon operator, as in I(sr:er, :) or I(:, sc:ec) then the statement evaluates to every row or column. Figure 4-18 illustrates this situation for the three aforementioned usages of the colon operator.

MATLAB's colon notation is indicative of the language's vectorized nature; in well-written MATLAB code, the amount of looping is always kept to a bare minimum. Another usage of the colon operator is to flatten a multidimensional array, as in the statement

n_sorted = sort(neighborhood(:));

Here, neighborhood(:) serves to flatten the 2D neighborhood into a 1D array, so that we sort a single vector consisting of all the pixels in the current neighborhood. Without this operation, sort would return n sorted vectors, where n is the number of columns in the median filter mask.

It was stated earlier that with various definitions of the peak signal-to-noise ratio floating around, without an exact definition of what is being quoted this particular metric should be taken with a grain of salt. Perhaps for this very reason, there is not an Image Processing Toolbox function to compare two images numerically, but Listing 4-10 is a simple function psnr that accepts two image matrices of the same type (uint8, uint16, or double) and returns the PSNR in decibels.

Listing 4-10: MATLAB function that returns the peak signal-to-noise ratio between two images.

```
function dB = psnr(I, J)

if size(I) ~= size(J)
    error('size of both images must be the same.');
end
N = length(I(:));

% promote to double
if isa(I, 'uint8') & isa(J, 'uint8')
    p = 255;
    I = double(I);
    J = double(J);
elseif isa(I, 'uint16') & isa(J, 'uint16')
    p = 65535;
    I = double(I);
    J = double(J);
elseif isa(I, 'double') & isa(J, 'double')
    p = 1.0;
else
    error('Both images must be of the same type');
end

% RMS between I and J
dIJ = I-J;
dIJsquared = dIJ.^2;
rms = sqrt( mean(dIJsquared(:)) );

% return peak signal-to-noise ratio in decibels
dB = 10 * log10(p / rms);
```

The Image Processing Toolbox function nlfilter is an interesting function that can be used to perform generic non-linear filtering of images. This function can be invoked in a few different manners, one of which is

```
J = nlfilter(I, [3 3], 'min( x(:) )');
```

The above statement performs erosion, by passing the image matrix I through a 3x3 minimum filter. The third argument to nlfilter is a string that is treated as a "function pointer", in a similar fashion to bona fide function pointers in C/C++. As the sliding window moves across the image, nlfilter invokes the function referenced by the string using the built-in MATLAB function eval. The function referenced in the call to nlfilter must accept a single input variable x (the current neighborhood) and return a scalar value. In the above case, each neighborhood matrix is first flattened by the colon operator, and the built-in MATLAB min function returns the minimum pixel value for that neighborhood. It is important to understand that this nlfilter call will fail if the colon operator is omitted, i.e. nlfilter(I, [3 3], 'min(x)'), because min returns a vector of minima from each column if its input is a matrix.

The function pointer string argument to nlfilter is actually shorthand for other forms of invocation. For example the following code snippet implements image dilation:

```
f = inline( 'max(x(:))' );
J = nlfilter(I, [3 3], f);
```

and is equivalent to the following:

```
function J = dilate_image(I)
J = nlfilter(I, [3 3], @max_replacement);

function v = max_replacement(x) % subfunction
v = max(x(:));
```

In the above code, max_replacement is a subfunction passed to nlfilter as a *function handle*, indicated by the @ symbol. Of course, the function handle can also point to a global MATLAB function, whose implementation is provided in a separate M-file. Function handles are a powerful facility, especially when dealing with classes of algorithms whose basic infrastructure remains the same, but where the individual pixel operations differ. Non-linear filtering of images meets this condition perfectly – a skeleton of the process can be coded in a generic fashion and then the generic function specialized through the use of function handles. In some respects, this design pattern closely follows that of the C++ Standard Template Library (STL), where generic data structures and algorithms are specialized through the use of C++ templates.

4.5.4   A Median Filtering Application Built Using Visual Studio .NET 2003

It has been said before but it bears repeating – a working, reference, C/C++ implementation of an algorithm running on the desktop is a major advantage in getting an embedded version of that algorithm running on a DSP. The reasons why are two-fold. Debugging the algorithm in general is substantially easier on the desktop because there are far fewer ancillary issues to deal with (limited memory being one in particular). Moreover, in any real-world system there is more to just getting the image-processing algorithm to work and execute efficiently. The question of whether the particular algorithm works within the application's context must also be considered. This stage of development can usually be done with simulations using test data, and MATLAB or other equivalent IDEs are a perfect tool for this type of software. Microsoft's tools are so good, and so amenable to rapid development, that Visual Studio can sometimes be used in lieu of MATLAB, and then you get the added benefit of having a reference implementation to compare against when porting the code over to the DSP. In this section we discuss a median filtering C++ application (located in `Chap4\median_filter\MedianFilterVS`), that reuses several of the classes first introduced in 3.3.2. A screen-shot of this application is shown in Figure 4.19.

The user loads in a BMP image from the **File|Load** menu item, adds a parameterized amount of shot noise to the image, and cleans up the image by passing the noisy image through a median filter of size 3x3, 5x5, 7x7, or 11x11. The efficacy of the processing may be garnered of course through visual inspection of the right-most image display, and MSE and PSNR metrics are computed and displayed. Border pixels are handled in a similar fashion to the linear filtering programs of 4.3 and 4.4 – they are set to zero. Thus, the MSE and PSNR measurements are computed only for the interior portion of the image where pixel values actually change. So for example, with a 3x3 median filter, a subset of the original and processed images are utilized, where the first and last rows and columns are not considered for purposes of comparing the two images.

Just as coding techniques play an important role in DSP optimization, so do they in the desktop realm. To that end, this particular application implements the median filter using five different methods, each of which shall be explained in further detail. The implementation method is selected by using the bottom-most dropdown list box control, and the time taken to filter the image is displayed on the application window, in microseconds.

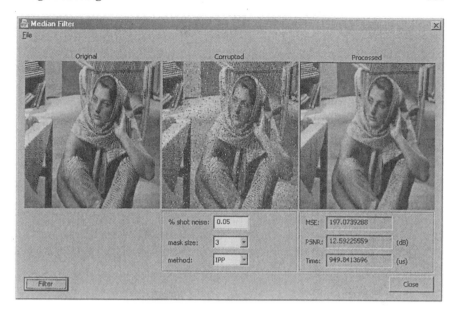

Figure 4-19. Visual Studio .NET C++ median filtering application.

4.5.4.1 Generating Noise with the Standard C Library

This project includes a class named `AddNoise`, which corrupts an image with salt and pepper noise. As we saw in the MATLAB code, generating image noise depends on pseudo-random number generation routines. While the standard C library does not have all of the helpful utility functions that MATLAB has for generating random sequences, with a bit of work the standard C library function `rand` can be used to generate the noise distributions we need in this application. These distributions are used within the static method `SaltAndPepper`, which in turn is called by the GUI code (see `MedianFilterDlg.cpp`) to create a corrupted image. Listing 4-11 is the C++ implementation file for the `AddNoise` class.

Listing 4-11: `AddNoise.cpp`

```
#include "stdafx.h"
#include <numeric>
#include "AddNoise.h"

using namespace std;

Image8bpp *AddNoise::SaltAndPepper(const Image8bpp *pIn,
                                   float percentage)
{
  if (percentage<0.0 || percentage>1.0)
```

```
      throw runtime_error("Percentage of shot noise must be b/w [0-1.0]");

      srand((unsigned)time(NULL)); // seed generator with current time

      Image8bpp *pOut = new Image8bpp(*pIn); // allocate output image

      // those pixels less than or equal to percentage are to be corrupted
      vector<float> u(pIn->getHeight()*pIn->getWidth());
      AddNoise::UniformDist(&u);

      // corrupt some of pIn's pixels here
      vector<float>::iterator pu = u.begin();
      for (int iRow=0; iRow<pIn->getHeight(); ++iRow) {
        Ipp8u *pOutScanLine = pOut->getPixels() + iRow*pOut->getStride();
        for (int iCol=0; iCol<pIn->getWidth(); ++iCol, pOutScanLine++)
          if (*pu++ <= percentage)
            *pOutScanLine = (AddNoise::RandN()<0.0) ? 0 : 255;
      }

      return pOut;
    }

    void AddNoise::UniformDist(vector<float> *pSamples)
    {
      for (int ii=0; ii<pSamples->size(); ++ii)
        pSamples->at(ii) = (float)rand()/RAND_MAX;
    }

    float AddNoise::RandN()
    {
      vector<float> u(12); // generate normal dist from sum of uniform dist
      AddNoise::UniformDist(&u);
      // Transform uniform dist to normal dist
      // by applying the central limit theorem.
      return accumulate(u.begin(), u.end(), 0.f) - 6;
    }
```

This method for generating shot noise first relies on a uniform distribution, one where the spread of the numbers is evenly distributed throughout a given interval. This distribution determines which pixels are to be corrupted. Prior to generating random numbers, one should *seed* the C random number generator via a call to srand. Failing to seed the random number generator results in always generating the same pseudo-random number sequence, which can be useful in its own right for debugging purposes. By repeatedly calling rand and dividing its output by the macro RAND_MAX (the maximum value that can be returned from rand), a uniform distribution within the interval [0.0,1.0] is generated − such a distribution is generated within the `UniformDist` method and stored within an STL vector.

The next step is to corrupt the condemned pixels by setting their intensity level to either 0 or 255, depending on the sign of a number from a *normal* distribution with zero mean, variance of one, and standard deviation of one, a classic distribution exhibiting the familiar symmetric "bell" shape[17]. There are numerous algorithms for converting a uniform distribution to a normal distribution. The polar method of Box and Muller, is one of the more common and is covered in [18]. `RandN` uses a simpler approximation that exploits the central limit theorem, which states that the means of random samples taken from any distribution tend to be normally distributed. Thus the result of `UniformDist` can be called k times and accumulated. Subtracting the sample mean ($k/2$ in this case because the distribution lies within the interval [0.0,1.0]) and then dividing by the square root of the sample variance:

$$\sqrt{k/12}$$

produces a normal distribution with a variance of one and a standard deviation of one. For the case of $k=12$, all that is required is to sum up 12 uniform samples and subtract 6. The handy STL function `accumulate` is used to implement this technique.

4.5.4.2 Profiling Code in Visual Studio .NET 2003

The purpose of this Visual Studio C/C++ application is two-fold. One, it serves as an interesting example of an image-processing algorithm "test bench". An application like this can be used as a boilerplate for jumpstarting future development. But more importantly, desktop C/C++ code is inherently closer to the eventual DSP implementation than MATLAB code can ever be, and thus because debugging is that much easier on the desktop, it sometimes makes sense to implement certain algorithm tweaks and optimizations at this level. That is not to say that MATLAB or other prototyping/simulation tools do not have their place – they most certainly do. But as we shall see shortly, we are going to be optimizing the median filter algorithm in ways that really are not amenable to a MATLAB-based implementation. However, to measure the efficiency of C/C++ algorithms we need a means of accurately profiling the code in Visual Studio.

There are a number of extremely well polished tools on the market that do a fantastic job of profiling code running on a Wintel machine. Two of the more popular are IBM Rational Quantify and the Intel® VTune™

Performance Analyzer. These tools are very sophisticated and plug into the Visual Studio .NET IDE. The programs in this book use a much simpler profiling mechanism, similar to the usage of the `clock` function seen in previous DSP projects. The Windows high-resolution performance counter API is a useful set of functions that can be used for straightforward code profiling[19].

Essentially this coding technique works just as it did in the DSP case. `QueryPerformanceCounter` obtains the current value of the counter, and the difference between two counter measurements, divided by the frequency of resolution, yields the time taken to execute the code bracketed by the two measurement points. This profiling method is employed in the `MedianFilter` class, which will be discussed in greater detail in the subsequent section. The only caveat is that these counter functions use a `LARGE_INTEGER` data type to store the value of the counter. A `LARGE_INTEGER` is a 64-bit quantity defined as a C union in the Microsoft header files. To access the entire quantity as a single entity for the purposes of performing arithmetic, testing for equality, or casting, the `QuadPart` member of the union should be used, as shown in the following code snippet:

```
LARGE_INTEGER start, stop;
QueryPerformanceCounter(&m_start); // begin timing block

// code that we're profiling ...

QueryPerformanceCounter(&m_stop); // end timing block

// and now convert time to microseconds
LARGE_INTEGER freq;
QueryPerformanceFrequency(&freq);
float delta = stop.QuadPart-start.QuadPart;
float timeMicroSeconds = (delta/ (float)freq.QuadPart) * 1000000.f;
```

Finally, whenever profiling code built using Microsoft Visual Studio, make sure to measure the "Release" configuration build, and not the "Debug" configuration build. The more recent versions of Microsoft's compiler are extremely aggressive, and as is the case with the TI compiler, optimizations are by default completely disabled in the Debug configuration. The **Build|Configuration Manager** menu selection in Visual Studio allows one to change the active build configuration.

4.5.4.3 Various C/C++ Implementations of the Median Filter

The `MedianFilter` class, portions of which are given in Listing 4-12, does the heavy lifting in this application. This class performs median filtering using several different implementations, which are selected from the main GUI dialog via a dropdown list box:

1. Generalized KxK mask, with the neighborhood sorted via the standard C library function `qsort`, and median pixel value extracted from this sorted array (`MedianFilter::processImageQsort`).
2. Same as (1), but use the STL function `sort` instead of `qsort` (`MedianFilter::processImageSTLsort`).
3. Generalized KxK mask, using the Intel IPP function `ippiFilterMedian_8u_C1R` (`MedianFilter::processImageIpp`).
4. Fixed 3x3 mask, using an optimized median finding algorithm specifically targeted for arrays of 9 elements (`MedianFilter::processImage3x3`).
5. Same as (4), but for the 5x5 case (`MedianFilter::processImage5x5`).

The first two implementations are what could be deemed naive implementations in the sense that while they work, they are wholly unoptimized in every sense of the word. A simple means of computing the median pixel intensity in a neighborhood is to first sort the neighborhood, and then pick the central value from this sorted list. This scheme comes directly from the definition of the median (see Figure 4-13).

Listing 4-12: portions of `MedianFilter.cpp`

```
float MedianFilter::processImage(Image8bpp *pNoisyImage,
                                 Image8bpp *pProcessedImage,
                                 Implementation impl)
{
  if (pNoisyImage->getHeight()!=pProcessedImage->getHeight() ||
      pNoisyImage->getWidth()!=pProcessedImage->getWidth())
      throw runtime_error("invalid args: images must be same size!");

  // begin timing block
  m_stop.QuadPart = 0;
  if (!::QueryPerformanceCounter(&m_start))
      m_start.QuadPart = 0;

      if (IPP == impl)
          this->processImageIpp(pNoisyImage, pProcessedImage);
      else if (QSORT == impl)
```

```
      this->processImageQsort(pNoisyImage, pProcessedImage);
    else if (STL_SORT == impl)
      this->processImageSTLsort(pNoisyImage, pProcessedImage);
    else if (FIXED_3x3 == impl)
      this->processImage3x3(pNoisyImage, pProcessedImage);
    else if (FIXED_5x5 == impl)
      this->processImage5x5(pNoisyImage, pProcessedImage);

  // end timing block
  if (m_start.QuadPart)
    if (!::QueryPerformanceCounter(&m_stop))
      m_stop.QuadPart = 0;

  return this->calcMicroSecs(m_start, m_stop);
}

void MedianFilter::processImageIpp(Image8bpp *pNoisyImage,
                                   Image8bpp *pProcessedImage)
{
  int nr = pNoisyImage->getHeight(),
      nc = pNoisyImage->getWidth();
  IppiSize mask = {m_kernelSize, m_kernelSize};
  IppiPoint anchor = {m_khw, m_khw};
  IppiSize dstRoi = {nc-2*m_khw, nr-2*m_khw}; // just interior region

  // offset pointers to start within the interior of the image
  int offset = nc*m_khw + m_khw;
  IppStatus status =
    ippiFilterMedian_8u_C1R(pNoisyImage->getPixels() + offset,
                            pNoisyImage->getStride(),
                            pProcessedImage->getPixels() + offset,
                            pProcessedImage->getStride(),
                            dstRoi, // leaves boundary pixels alone
                            mask,
                            anchor);

  if (ippStsNoErr != status)
    throw runtime_error("median filter operation failed");
}

void MedianFilter::processImageQsort(Image8bpp *pNoisyImage,
                                     Image8bpp *pProcessedImage)
{
  vector<Ipp8u> neighborhood(m_kernelSize*m_kernelSize);
  Ipp8u *pNeighborhood;
  const int nr = pNoisyImage->getHeight(),
            nc = pNoisyImage->getWidth(),
            stride = pNoisyImage->getStride(),
            nks = m_kernelSize*m_kernelSize,
            iMedian = nks/2,
```

```
                    nBackwards = m_kernelSize*stride,
                    noisyNextRowOffset = 2*m_khw + (stride-nc),
                    outputNextRowOffset = m_khw + (stride-nc);
   Ipp8u *pNoisy = pNoisyImage->getPixels(),
            *pOutput = pProcessedImage->getPixels() + m_khw*stride;

   // march through the interior portion of the image
   for (int iRow=m_khw; iRow<nr-m_khw; ++iRow) {
   pOutput += m_khw;
   for (int iCol=m_khw; iCol<nc-m_khw;
            ++iCol, pNoisy++, pOutput++) {

            // splice out the current neighborhood
            pNeighborhood = &neighborhood[0];
            for (int kk=0; kk<m_kernelSize;
                    ++kk, pNoisy+=nc, pNeighborhood+=m_kernelSize) {
                memcpy(pNeighborhood, pNoisy, m_kernelSize*sizeof(Ipp8u));
            }
            pNoisy -= nBackwards;

            qsort(&neighborhood[0], nks, sizeof(Ipp8u),
                    &MedianFilter::compare);
            *pOutput = neighborhood[iMedian]; // output = median value
      }
      pNoisy += noisyNextRowOffset;      // incr pointers in prep
      pOutput += outputNextRowOffset; // for next scan-line
   }
}

int MedianFilter::compare(const void *pElem1, const void *pElem2)
{
  Ipp8u *pPixel1 = (Ipp8u *)pElem1,
          *pPixel2 = (Ipp8u *)pElem2;

  return (*pPixel1<*pPixel2) ? -1 : ( (*pPixel1==*pPixel2) ? 0 : 1 );
}

void MedianFilter::processImageSTLsort(Image8bpp *pNoisyImage,
                                        Image8bpp *pProcessedImage)
{
  vector<Ipp8u> neighborhood(m_kernelSize*m_kernelSize);
  Ipp8u *pNeighborhood;
  const int nr = pNoisyImage->getHeight(),
            nc = pNoisyImage->getWidth(),
            stride = pNoisyImage->getStride(),
            nks = m_kernelSize*m_kernelSize,
            iMedian = nks/2,
            nBackwards = m_kernelSize*stride,
            noisyNextRowOffset = 2*m_khw + (stride-nc),
            outputNextRowOffset = m_khw + (stride-nc);
  Ipp8u *pNoisy = pNoisyImage->getPixels(),
```

```
        *pOutput = pProcessedImage->getPixels() + m_khw*stride;

// march through the interior portion of the image
for (int iRow=m_khw; iRow<nr-m_khw; ++iRow) {
  pOutput += m_khw;
  for (int iCol=m_khw; iCol<nc-m_khw;
        ++iCol, pNoisy++, pOutput++) {

    // splice out the current neighborhood
    pNeighborhood = &neighborhood[0];
    for (int kk=0; kk<m_kernelSize;
          ++kk, pNoisy+=nc, pNeighborhood+=m_kernelSize) {
        memcpy(pNeighborhood,pNoisy,m_kernelSize*sizeof(Ipp8u));
    }
    pNoisy -= nBackwards;

    sort(neighborhood.begin(), neighborhood.end());
    *pOutput = neighborhood[iMedian]; // output = median value
  }
  pNoisy += noisyNextRowOffset;      // incr pointers in prep
  pOutput += outputNextRowOffset; // for next scan-line
  }
}

#define PIX_SWAP(a,b) { temp=(a);(a)=(b);(b)=temp; }
#define PIX_SORT(a,b) { if ((a)>(b)) PIX_SWAP((a),(b)); }

void MedianFilter::processImage3x3(Image8bpp *pNoisyImage,
                                    Image8bpp *pProcessedImage)
{
  const int nr = pNoisyImage->getHeight(),
        nc = pNoisyImage->getWidth(),
        stride = pNoisyImage->getStride(),
        nBackwards = 3*stride,
        noisyNextRowOffset = 2 + (stride-nc),
        outputNextRowOffset = 1 + (stride-nc);
  Ipp8u temp, neighborhood[9], *pNeighborhood;
  Ipp8u *pNoisy = pNoisyImage->getPixels(),
        *pOutput = pProcessedImage->getPixels() + stride;
  Ipp8u *p = neighborhood;

  // march through the interior portion of the image
  for (int iRow=1; iRow<nr-1; ++iRow) {
    pOutput += 1;
    for (int iCol=1; iCol<nc-1; ++iCol, pNoisy++, pOutput++) {
      // splice out the current neighborhood
      pNeighborhood = neighborhood;
      for (int kk=0; kk<3; ++kk, pNoisy+=nc, pNeighborhood+=3)
        memcpy(pNeighborhood, pNoisy, 3*sizeof(Ipp8u));
      pNoisy -= nBackwards;
```

```
    // minimum exchange filter that partially sorts a 9-element
    // array such that the central element contains the median value,
    // in theory it is impossible to get the median using a fewer
    // number of comparisons
    PIX_SORT(p[1],p[2]); PIX_SORT(p[4],p[5]); PIX_SORT(p[7],p[8]) ;
    PIX_SORT(p[0],p[1]); PIX_SORT(p[3],p[4]); PIX_SORT(p[6],p[7]) ;
    PIX_SORT(p[1],p[2]); PIX_SORT(p[4],p[5]); PIX_SORT(p[7],p[8]) ;
    PIX_SORT(p[0],p[3]); PIX_SORT(p[5],p[8]); PIX_SORT(p[4],p[7]) ;
    PIX_SORT(p[3],p[6]); PIX_SORT(p[1],p[4]); PIX_SORT(p[2],p[5]) ;
    PIX_SORT(p[4],p[7]); PIX_SORT(p[4],p[2]); PIX_SORT(p[6],p[4]) ;
    PIX_SORT(p[4],p[2]) ;
    *pOutput = (p[4]) ;
  } // end (for each column)
  pNoisy += noisyNextRowOffset;  // incr pointers in prep
  pOutput += outputNextRowOffset; // for next scan-line
 } // end (for each row)
}

void MedianFilter::processImage5x5(Image8bpp *pNoisyImage,
                                   Image8bpp *pProcessedImage)
{
  const int nr = pNoisyImage->getHeight(),
            nc = pNoisyImage->getWidth(),
            stride = pNoisyImage->getStride(),
            nBackwards = 5*stride,
            noisyNextRowOffset = 4 + (stride-nc),
            outputNextRowOffset - 2 + (stride-nc);
  Ipp8u temp, neighborhood[25], *pNeighborhood;
  Ipp8u *pNoisy = pNoisyImage->getPixels(),
        *pOutput = pProcessedImage->getPixels() + 2*stride;
  Ipp8u *p = neighborhood;

  // march through the interior portion of the image
  for (int iRow=2; iRow<nr-2; ++iRow) {
    pOutput += 2;
    for (int iCol=2; iCol<nc-2; ++iCol, pNoisy++, pOutput++) {
      // splice out the current neighborhood
      pNeighborhood = neighborhood;
      for (int kk=0; kk<5; ++kk, pNoisy+=nc, pNeighborhood+=5)
        memcpy(pNeighborhood, pNoisy, 5*sizeof(Ipp8u));
      pNoisy -= nBackwards;

      // optimized search for the median value in a list of 25 elements,
      // again in theory it is impossible to do this with fewer comparisons

      // PIX_SORT's omitted for brevity's sake, refer to MedianFilter.cpp
      // on CD-ROM for 5x5 minimum comparison sort

      *pOutput = p[12];
    } // end (for each column)
    pNoisy += noisyNextRowOffset;  // incr pointers in prep
    pOutput += outputNextRowOffset; // for next scan-line
```

```
    } // end (for each row)
  }
```

The `qsort` function, which has been part and parcel of the ANSI C library from time immemorial, is an implementation of the "quicksort" sorting algorithm. Quicksort is a well-known algorithm that uses a divide-and-conquer methodology to sort a list of elements stored in an array[20]. The problem we run into when employing `qsort` to perform median filtering is that it is overkill for this purpose. An ANSI C `qsort` function is highly parameterized, so that it can be used to sort arrays of any type of object. This generality is realized through the use of a function pointer, which points to a comparison function that given two objects, returns an integer value based on the ordering between those two elements. In the `MedianFilter::processImageQsort` method, each invocation of `qsort` is passed a pointer to the *static* class method `compare`. When passing a function pointer to a C routine that points to a C++ class method, the function pointer must point to a static class method. The reason for this is with normal class member functions, there is the implicit `this` pointer, where `this` refers to the current object. Passing in a C++ class member function pointer to `qsort` will result in a compilation error.

The incredible overhead associated with using the `qsort` function is apparent when considering the profiling results. Median filtering a 256x256 image with a 3x3 mask using `qsort` takes on average about 64,000 µs on a 2 GHz Pentium 4 workstation. Considering that method 3, which uses the Intel IPP median filtering function, takes on average a mere 431 µs (a better than two orders of magnitude performance increase) is enough evidence that `qsort` is most definitely not the way to go to implement efficient median filtering suitable for use on a DSP.

One red flag is the comparison function pointer. This means that a pointer is dereferenced and a function call is made *for each comparison* during the sort. A complexity analysis of Quicksort shows that on average, about nlog(n) comparisons are required, and in the worst case, n(n-1)/2 comparisons are needed, where n is the total number of pixels in a given neighborhood. Keeping in mind that the neighborhood is shifted about the image for every pixel location, this entails substantial overhead, as variables need to be pushed and popped to and from the stack with each invocation of the comparison function. Moreover, the most common implementation of the quicksort algorithm is recursive, due to its inherent simplicity and elegance – that is, the function calls itself. This function recursion in turn causes even more overhead in the algorithm, for the same reasons as the usage of the comparison function pointer in `qsort` (excessive pushing and popping). In general, due to this overhead a good rule of thumb is to be extremely wary of implementing any recursive algorithm on an embedded real-time system.

Method 2 slightly improves the situation. The implementation for this variant of the algorithm is virtually identical to the first method, except that the call to qsort is replaced with a call to the C++ STL function sort. The beauty of the STL is that it achieves generality with respect to data types through judicious use of C++ templates. The C qsort function achieves equivalent functionality through the use of function pointers in a *type-unsafe* manner (because arrays are treated as void pointers). Not only is the C++ code more aesthetically pleasing, this is also an example where a C++ language feature enables a speed boost. Because the STL consists of small template functions defined in header files, it is highly likely that the compiler will inline the sort function call directly into the processImageSTLsort method. In addition, now that there is no function pointer, the main loop that constitutes the sorting algorithm need not invoke a function to compare two elements. Without going into the specifics of the rules behind the STL, suffice it to say that as long as the objects in your container (i.e. a C array or STL vector) understand the < operator, as all plain old data (POD) types like ints, doubles, and floats do, then sort will work[21]. Finally, the C++ library that comes with Microsoft Visual Studio .NET 2003 uses a non-recursive insertion sort algorithm for small arrays of less than 32 elements, instead of a recursive Quicksort implementation. Yet the performance boost remains quite modest as compared to method 1 – filtering a 256x256 image with a 3x3 kernel still takes on average 55,000 μs.

The real bang for the buck occurs when taking a step back and considering the algorithm itself, instead of focusing on the code. There are really two levels of code optimization: optimization that focuses on low-level details, oftentimes processor or platform specific, and more high-level algorithmic issues. It is the latter where the most significant performance increases are usually attained, and these are best solved first on the desktop, because by the time one is developing on the embedded platform there are so many other issues to deal with. The essence of the performance problem here is that both qsort and sort are shackled by the fact that they are generic algorithms and perform a complete sort of the neighborhood, when in fact to compute a median value from a list of elements all that is needed is a *partial* sort of the data[20]. Just as implementing a minimum or maximum filter through picking either the first or last element from a sorted list obtained via some sorting routine is complete overkill (why not simply iterate through the neighborhood and cache the minimum/maximum intensities?), so is using a bona-fide sorting routine to implement a median filter for a small, fixed number of pixels. Partial sorting in this context means that the central value in the array is guaranteed to contain the median, but the ordering of the other elements in the list is not guaranteed. The partial sort procedure can then be hardwired per kernel size to achieve a blazingly fast implementation, because there will be no looping (and recursion) involved.

Figure 4-20 is a depiction of the exchange network for performing a complete sort of a list of three elements[22], and thus the second element in the list is the median. Given a 3x3 neighborhood then, an initial thought might be to apply the algorithm shown in Figure 4-20 first across the three rows, and then down the middle (second) column. Unfortunately such a procedure only guarantees a pixel intensity of rank 4, 5 (the median), or 6 in the central array index, because the median filter is not separable.

It turns out that the *minimum exchange network* that partially sorts a nine-element list, and guarantees that the central element in that list is the actual median value, requires a total 19 comparisons[23,24,25]. This complete network is shown in Figure 4-21, and is optimal in the sense that without making any assumptions on the input, it is impossible to extract the median in a fewer number of comparison operations. The 3x3 minimum exchange network works by first performing "row" sorts, followed by "column" sorts, and concluding with "diagonal" sorts. This implementation uses the PIX_SWAP and PIX_SORT macros defined just above the processImage3x3 method.

This custom, hardwired 3x3 median filter implementation is substantially faster – filtering a 256x256 image on the same test machine now takes on average about 9,750 µs. While this is still not as fast as the Intel IPP method, from an algorithmic perspective it is optimal – it is impossible to compute the median using a fewer number of comparison operations. The partial sorting optimization can be extended to other filter sizes. A minimum exchange network for a 5x5 median filter requires 99 comparisons, and is implemented in MedianFilter::processImage5x5. Table 4-2 summarizes the profiling results for the median filtering implementations we implemented in this section.

Table 4-3. Performance results for various median filter implementations. Timing measurements taken using Visual Studio .NET 2003 release build (default options), using a 256x256 image, on a 2 GHz Pentium 4 workstation. The quoted times are averages of five runs.

Implementation	MedianFilter class method	Time (µs)
3x3 with qsort	processImageQsort	64,000
3x3 with STL sort	processImageSTLsort	55,000
3x3 with ippiFilterMedian_8u_C1R	processImageIpp	431
3x3 with 9-element minimum exchange network	processImage3x3	9,750

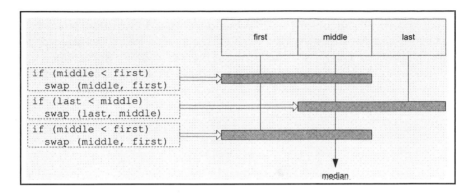

Figure 4-20. Exchange network for performing a complete sort of a list of 3 elements. The thin shaded boxes represent single-element sorters.

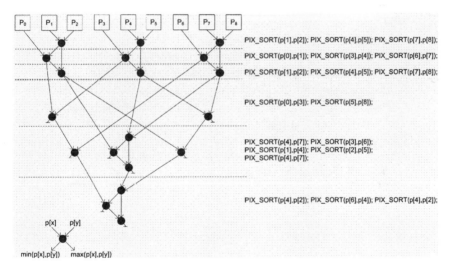

Figure 4-21. Minimum exchange network for computing the median of 9 elements. The C statements on the right come directly from the code in `MedianFilter::processImage3x3` that implements this procedure.

4.5.5 Median Filtering on the TI C6416 DSK

Two median filtering C64x programs can be found on the CD in the `Chap4\median_filter` directory. Both use the infrastructure developed in the previous chapter, where DMA is used to page blocks of image data into internal RAM, those scan-lines are filtered, and then the processed pixels are paged out to external RAM, again via DMA. These programs share two common files, the median_filter.cmd linker command file (the contents of which are identical to the C64x DSK linear filtering programs of 4.4) and image.h, which initializes the in_img buffer with a noisy "Barbara" image (corrupted with 15% shot noise).

The first program, located in the MedianFilterCCS_3x3 subdirectory, uses the IMGLIB function IMG_median_3x3 to implement a 3x3 median filter. Apart from removing a few C pre-processor macros and definitions that are no longer needed, median_filter_3x3.c is very similar to blur3x3_imglib_paging_dma.c (see Listing 4-7). The only function that changes in any significant fashion is filter_block. This function, shown in Listing 4-13, calls IMG_median_3x3 to pass nrows number of image rows (starting at pin), through a 3x3 median filter.

Listing 4-13: filter_block function in median_filter3x3.c.

```
/* Filter one block of an image, returns row index for next block */
int filter_block(int irow, int nrows,
                 const unsigned char *restrict pin,
                 unsigned char *restrict pout)
{
  int jj = irow;
  for (; jj<irow+nrows; ++jj) {
    IMG_median_3x3(pin, Y_SIZE, pout);
    pin += Y_SIZE; /* incr scan-line in preparation for next iteration */
    pout += Y_SIZE;
  }
  return jj-1;
}
```

IMG_median_3x3 works in a row-major fashion[6,11]. This is best illustrated by example, as shown in Figure 4-22. For the first two iterations, default data is used for computation of the median, until the filter slides over enough such that it completely fits over a portion of the image data (iteration 2). As a result, in contrast to the earlier filtering programs where the first and

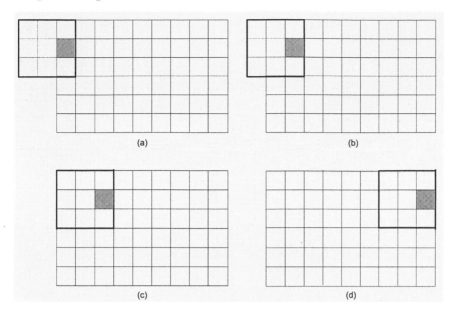

Figure 4-22. IMG_median_3x3 across a single row (shaded box denotes output pixel). (a) Iteration 0, where the 1st two columns of the median neighborhood are filled with default values. (b) Iteration 1, where the 1st column of the neighborhood consists of default values. (c) By iteration 2, the mask has slid over enough such that the entire neighborhood consists of pixels from the input image. This is the first valid output pixel in this row. (d) Final iteration for the current row. With the next iteration, the entire mask slides down a single row and then proceeds to march across the subsequent row.

last columns of the output image are ignored, in this implementation the first two columns of the processed image are meaningless and should be ignored.

IMG_median_3x3 is not as general as the IPP function used in the preceding section, and there are times when a median filter larger than 3x3 is desired. Figure 4-22 shows the result of processing the noisy "Barbara" image using IMG_median_3x3 (the image displays are created using the procedure described in 4.3.1). There remain a few speckles in the processed image, due to the filter size being too small to handle the noise density of the input image. Passing this same image through a 5x5 median filter will further attenuate the effects of the shot noise, albeit at greater cost to the overall "sharpness" of the output.

A C64x program that uses the minimum exchange network for a 5x5 median filter can be found in the directory Chap4\median_filter \MedianFilterCCS_5x5. The overall structure of the source code for this program is very similar to that of the 3x3 case, with the major difference being the inclusion of the partial sorting C code for a 25-element list within

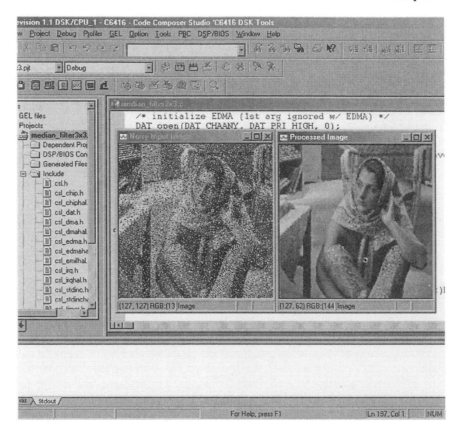

Figure 4-23. Using `IMG_median_3x3` to process clean up an image with 15% shot noise.

`filter_block`, which is more or less taken verbatim from the implementation of `MedianFilter::processImage5x5`.

It should be mentioned that compiling `median_filter_5x5.c` with release configuration optimizations enabled on a 2 GHz Pentium 4 workstation takes approximately 15 minutes. As discussed in 2.1.1, the TI C6x line of DSPs are based on a VLIW architecture. VLIW differs from traditional CISC and in some respects RISC architectures in that it places the burden on the software (compiler) more so than the hardware to provide optimal performance. With VLIW architectures, it is up to the compiler to schedule instructions so that complete fetch packets are generated. This means that there are no NOP instructions within the VLIW instruction and in this way the DSP's functional units are fully utilized. The loop kernel in `filter_block` for the 5x5 median filter has a large number of comparisons – the compiler has to perform substantial dependency analysis in order to generate instruction packets that make use of the DSP's functional units to the fullest extent.

4.6 ADAPTIVE FILTERING

Section 4.5.2 focused on using the median filter to enhance images corrupted by shot noise. It was shown that low-pass filters do not perform as well as median filters in removing impulse noise, because the spatial smoothing mixes any image-independent noisy outliers with the signal content. The median filter, on the other hand, was found to be well-suited for removing such signal-independent statistical outliers, while at the same time preserving the edges of the image. But what if the noise is not completely independent of the signal content, for example additive or multiplicative noise corrupting an image such that the noisy pixels are a mixture of original pixel intensities and noise levels? Might it be possible to construct an intelligent filter that somehow adapts itself to this behavior?

Adaptive filters are designed to address this question, and are the third and final class of spatial image filters we shall explore. The premise behind adaptive image filtering is that by varying the filtering method as the kernel slides across the image (in the same manner as the convolution operation), they are able to tailor themselves to the local properties and structures of an image. In essence, they can be thought of as self-adjusting digital filters. While certain types of adaptive filters may perform better than median filters at removing impulse noise (these are mostly variations on the basic median filtering scheme), they are most often used for denoising *non-stationary* images, which tend to exhibit abrupt intensity changes. Because the filtering operation is no longer purely uniform and instead modulated based on the local characteristics of the image, these filters can be employed effectively when there is little a priori knowledge of the signal being processed.

Adaptive filters find widespread use in countering the effects of so-called "speckle" noise, which afflicts coherent imaging systems like SAR and ultrasound. With these imaging techniques, scattered waves interfere with one another to contaminate an acquired image with multiplicative speckle noise. Various statistical models of speckle noise exist, with one of the more common being

$$g(x, y) = f(x, y) + f(x, y)n(x, y)$$

where g is the corrupted image and $n(x,y)$ is drawn from a zero-mean Gaussian distribution with a given standard deviation. It is clear from the

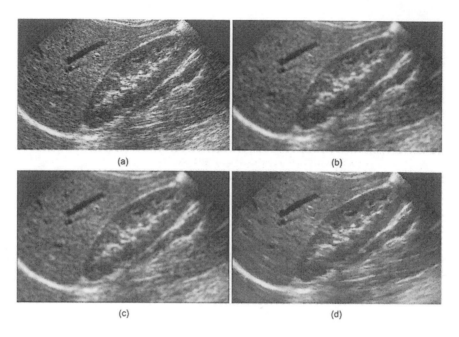

(a)   (b)

(c)   (d)

Figure 4-24. Ultrasound speckle noise removal. (a) Unprocessed ultrasound image of kidney anatomy. (b) Smoothed image (5x5 averaging filter), which while reducing speckle seriously degrades spatial resolution. (c) Result of adaptive filtering, using MATLAB `wiener2` Image-Processing Toolkit function. (d) Commercial quality speckle noise reduction, where image is despeckled using an adaptive geometric filter and then edge sharpened. Images (a) and (d) courtesy of Tetrad Corp., 1995.

above model that speckle noise is dependent on the magnitude of the signal f, and in fact this type of noise is a serious impediment on the interpretability of image data because it degrades both spatial and contrast resolution. This situation is shown with a real-world example in Figure 4-24, where an ultrasound image exhibiting a fairly large amount of speckle noise is enhanced, illustrating the utility of adaptive filtering.

Adaptive filters are effective in reducing the deleterious effects of speckle noise because they are capable of adjusting themselves based on the signal content of the image. Thus it follows that they must use some measure of the local characteristics of the image in order to perform their job. Many adaptive filters are predicated on the use of local pixel statistics, primarily the mean and variance of the pixels within the current neighborhood. The local mean is simply the average pixel intensity of the neighborhood. The local variance is calculated in two stages from the pixels contained within the current neighborhood. First, the mean of the sum of the squares is computed, and then the square of the local mean is subtracted from this number yielding a statistical quantity known as the variance, very often referred to as σ^2. In mathematical terms, these image statistics can be expressed as

$$local\ mean(i,j) = \mu = \frac{1}{NH^2} \sum_{i'=-\frac{NH}{2}}^{\frac{NH}{2}} \sum_{j'=-\frac{NH}{2}}^{\frac{NH}{2}} f(i+i', j+j')$$

$$local\ variance(i,j) = \sigma^2 = \left[\frac{1}{NH^2} \sum_{i'=-\frac{NH}{2}}^{\frac{NH}{2}} \sum_{j'=-\frac{NH}{2}}^{\frac{NH}{2}} f(i+i', j+j')^2 \right] - \mu^2$$

where f is the image and each neighborhood is of size NHxNH pixels. The standard deviation of the neighborhood is the square root of the variance, or σ.

4.6.1 The Minimal Mean Square Error Filter

The adaptive filter we spend the majority of time on in this section is the Minimal Mean Square Error (MMSE) filter. This filter can be used to remove both additive white noise and speckle noise. Consider an observed image $f(i,j)$ and a neighborhood L of size NHxNH. Let σ_n^2 be the noise variance, μ_L be the local mean, and σ_L^2 be the local variance. The σ_n^2 parameter is the variance of a representative background area of the image containing nothing but noise (a technique for estimating this parameter is given later). The linear MMSE filter output is then given by[26]

$$g(i, j) = \left(1 - \frac{\sigma_n^2}{\sigma_L^2}\right) f(i, j) + \frac{\sigma_n^2}{\sigma_L^2} \mu_L$$

This equation describes a linear interpolation between the observed image f and a smoothed version of f. Care should be taken to handle the case where

$$\sigma_L^2 \ll \sigma_n^2$$

in which case a negative output pixel may result (the MMSE filter implemented later in this section clamps the ratio σ_n^2/σ_L^2 to 1). The MMSE filter works as follows:

1. If the local variance is much greater than the noise variance, or if σ_n^2 is small or zero, it produces a value close to the input $f(i,j)$. If $\sigma_L^2 \gg \sigma_n^2$, this part of the image most likely contains an edge and this filter makes the assumption that it is best to leave that portion of the image alone.
2. If the noise variance dominates over the local variance, return the local mean.
3. If the two variance measures are more or less equal, the filter returns a mixture between the input and local mean.

This filter is edge-preserving because of (1), and consequently should tend to retain overall image sharpness, although noise will not be filtered from those portions of the image containing edges. Algorithm 4-2 describes the MMSE filter in pseudo-code form.

Algorithm 4-2: MMSE Adaptive Filter
    INPUT: MxN image I, odd neighborhood size NH, noise variance nv
    OUTPUT: filtered image J

    b = floor(NH/2)   (*half-width of kernel*)

    (*filter interior of image*)
    for (r = b...M-2b+1) (*1ˢᵗ b and last b rows not filtered*)
      for (c = b...N-2b+1) (*1ˢᵗ b and last b cols not filtered*)

        (*compute neighborhood statistics*)
        sum = 0
        sumsq = 0 (*sum of the squares*)
        for (ih = -b...+b)
          for (jh = -b...+b)

            sum += I[r+ih][c+jh]
            sumsq += (I[r+ih][c+jh])²

        end
        end (*for each pixel in current neighborhood*)

Algorithm 4-2 (continued)
    mean = sum / NH^2
    variance = (sumsq / NH^2) – $mean^2$
    α = nv/variance
    if (α > 1.0) then α = 1.0

    (output pixel intensity)
    J[r][c] = (1-α)(I[r][c]) + (α)(mean)

  end (*for each column*)
end (*for each row*)

Figure 4-25 illustrates Algorithm 4-2 in action, comparing its performance with a median filter on an image corrupted with a fair amount of speckle noise. The MMSE filter does a slightly better job of attenuating multiplicative noise while retaining image sharpness, resulting in a performance increase of about 1/3 dB in PSNR. While the MMSE filter is well-suited for dealing with speckle noise, it is not a panacea. A key facet to image enhancement is that the right tool should be used for the job, as clearly illustrated in Figure 4-26. The MMSE filter fails miserably with shot noise and the median filter is the more appropriate filter to use in this case. If the image is contaminated with both shot and speckle noise, a different filter should be used that can handle both types of noise, some of which are discussed in the next section.

4.6.2 Other Adaptive Filters

There exists a large number of adaptive spatial filters, many of them variations on a similar theme and tuned for specific applications. In this section, we take a brief look at some of the more common adaptive image filters, before going to back to the MMSE filter and getting that algorithm to run on the DSP.

The *Double Window-Modified Trimmed Mean* (DW-MTM) filter performs well in the presence of both shot noise and signal-dependent noise like additive Gaussian and multiplicative speckle noise[26]. This filter's namesake arises from the fact it utilizes two neighborhoods, a small one centered about the current pixel from which a median pixel intensity is calculated, and a larger one about the same center pixel from which a modified local mean is computed. This modified local mean is a modified average intensity in that any pixel whose gray-level differs by more than $k\sigma_n m_L$, where k is a constant, σ_n is the standard deviation of the noise, and m_L is the median of the small neighborhood, is excluded from the mean computation. The output intensity is then the average of the remaining pixels. In this way the DW-MTM filter operates as a quasi-smoothing filter but one where statistical outliers do not throw off the mean calculation. Of course, what exactly constitutes a statistical outlier can be difficult to predict. In this case, an outlier is parameterized by k and the magnitude of the noise (both of which remain constant throughout the filtering operation), as well as the median of the small neighborhood, which of course varies throughout the image and provides the adaptive aspect to this filter.

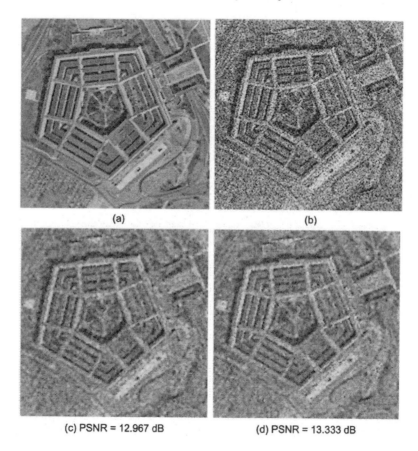

(a)   (b)

(c) PSNR = 12.967 dB   (d) PSNR = 13.333 dB

Figure 4-25. Speckle noise removal using a 3x3 MMSE adaptive filter. (a) Original Pentagon satellite image. (b) Image corrupted with multiplicative speckle noise (modeled using zero-mean Gaussian distribution with variance 0.025). (c) Processed image using 3x3 median filter. (d) Processed image using 3x3 MMSE filter.

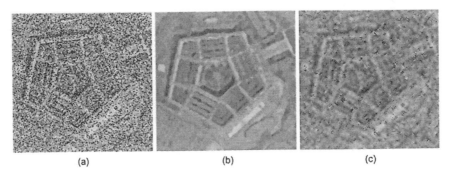

(a)   (b)   (c)

Figure 4-26. MMSE adaptive filtering and shot noise. (a) Original Pentagon satellite image corrupted with 20% salt & pepper noise. (b) Result of 5x5 median filter. (c) Result of 5x5 MMSE filter.

This idea of trimming away outlier pixels finds its way into another adaptive filter, the *alpha-trimmed mean* filter. Given an *m* x *n* neighborhood, where N_{ij} is the current neighborhood, an output image *g* is formed according to the relation

$$g(i,j) = \frac{1}{mn - \alpha} \sum_{(x,y) \in N_{ij}} f'(x,y)$$

where the modified neighborhood $f'(x,y)$ is created from $f(x,y)$ by excluding the maximum $\alpha/2$ brightest and $\alpha/2$ darkest pixels. This operation is best illustrated by example, as in Figure 4-27. The *adaptive local noise reduction* filter is similar in nature to the MMSE filter, in that it uses both the noise variance and local variance to form the output pixel. This filter is given by

$$g(i,j) = f(i,j) - \frac{\sigma_n^2}{\sigma_L^2}\left[f(i,j) - \mu_L\right]$$

where μ_L is the local mean. With this filter, if σ_n^2 is close to zero, it produces an output very close to the input $f(i,j)$. Likewise, if $\sigma_L^2 \gg \sigma_n^2$ it also produces an output pixel close to $f(i,j)$. Otherwise, this filter outputs a pixel close to the local average.

The *signal adaptive median* or SAM filter[26] is interesting in that it uses low-pass and high-pass filtered versions of the input image, in conjunction with an adaptive step involving the noise and local variances to yield a filter that exhibits excellent performance with respect to noise reduction while also preserving edges and overall image "crispness", however at a steep

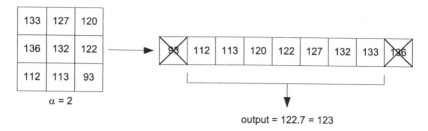

Figure 4-27. Alpha-trimmed mean filter, for the case of an example 3x3 neighborhood and α=2.

computational cost. The output of the SAM filter is a combination of low-pass and high-pass filtered versions of the input image, according to the relation

g = *low-pass image* + *(k)high-pass image*

The factor k is once again based on the ratio σ_n^2/σ_L^2 that we encountered in the adaptive local noise reduction and MMSE filters, and is given by

$$k = \begin{cases} 0, & \dfrac{c\sigma_n^2}{\sigma_L^2} \geq 1 \\[2ex] 1 - \dfrac{c\sigma_n^2}{\sigma_L^2}, & otherwise \end{cases}$$

where c is a positive constant.

4.6.3 Adaptive Image Filtering in MATLAB

The `imnoise` function in the Image Processing Toolbox can be used to add speckle noise to an image by passing in the string `'speckle'` and the noise variance as input arguments to the function. Note that the noise variance argument in this sense is *not* the same as σ_n^2 – in this context it is used to scale a zero-mean uniform distribution of variance 1 to produce noise of the desired magnitude. Obtaining the σ_n^2 parameter used in the MMSE, adaptive local noise reduction, and SAM filters requires looking at the variance of pixel intensities of a noisy background portion of the image, or estimating it through means that shall be discussed shortly.

The MATLAB function uses a slightly different model for corrupting an image with multiplicative speckle noise than what was previously described.

Other models, some of which simulate the noise using a Rayleigh distribution, are also common for modeling speckle noise. Listing 4-14 is a replacement for `imnoise` (when used with the 'speckle' argument), which can be used if the Image Processing Toolbox is not available or if a different model for speckle noise is desired.

Listing 4-14: MATLAB function for corrupting an image with multiplicative speckle noise.

```
function J = add_speckle_noise(I, variance)

J = double(I) + sqrt(variance).*randn(size(I)).*double(I);
if isa(I, 'uint8')
    J = uint8(J+0.5);
elseif isa(I, 'uint16')
    J = uint16(I+0.5);
end
```

The Image Processing Toolbox provides a function `wiener2` (see Figure 4.24c) that implements an adaptive filter of the form

$$g(i,j) = \mu_L + \frac{\sigma_L^2 - \sigma_n^2}{\sigma_L^2}\left[f(i,j) - \mu_L\right]$$

This particular filter is very well-suited to the removal of additive noise. Listing 4-15 is an M-file that implements Algorithm 4-2, the MMSE filter. MMSE uses the built-in MATLAB function `filter2` to compute the local means and local variances. After these temporary matrices have been constructed, the code proceeds to implement the core MMSE algorithm in classic vectorized MATLAB form.

Listing 4-15: MATLAB function for passing an image through the MMSE filter.

```
function J = MMSE(I, sz)

dI = double(I);

% compute local statistics
ncoeffs = sz*sz;
u = filter2(ones(sz)./ncoeffs, dI); % local mean
sigmasq = filter2(ones(sz)./ncoeffs, dI.^2);
sigmasq = sigmasq - u.^2; % local variance
```

% from the local variances estimate the noise variance
noise_variance = mean2(sigmasq);

% the guts of the MMSE adaptive filter
% (linearly interpolate b/w input pixel and local mean,
% depending on the magnitude of the local variance)
alpha = noise_variance ./ sigmasq;
% must be careful to handle the case where noise variance
% much greater than local variance - in this case the output
% pixel should be the local mean (alpha = 1)
ii = find(alpha > 1);
alpha(ii) = 1;
% linearly interpolate
J = (1-alpha).*dI + alpha.*u;

% cast to uint8 or uint16, but for those pixels that are
% out-of-range replace with the local mean
if isa(I, 'uint8')
    J = clamp_pixel_values(J, u, 255);
    J = uint8(J + 0.5);
elseif isa(I, 'uint16')
    J = clamp_pixel_values(J, u, 65535);
    J = uint16(J + 0.5);
end

function J = clamp_pixel_values(J, u, pmax)

% those that are too low receive the local mean pixel value
ilow = find(J < 0);
J(ilow) = u(ilow);

% those that are too high receive the local mean pixel value
ihi = find(J > pmax);
J(ihi) = u(ihi);

One of the drawbacks to adaptive image filtering, aside from increased computational complexity, is the presence of various parameters for tuning the image processing. While these algorithm constants are certainly useful for tweaking the processing, in the case of the ubiquitous noise variance parameter it is generally difficult to obtain this measurement without resorting to less-than-ideal techniques. One course of action is to simply disregard the question entirely, and leave it up to clients to provide a meaningful number, perhaps by extracting a representative background area containing nothing but noise, and computing the variance of this training data beforehand. The MMSE implementation of Listing 4-15 does not expect clients to provide the noise variance – rather, the noise variance is estimated from the input data by calculating the noise "power" from the mean of all the local variances. Strictly speaking, while this estimation is meant for additive noise, it does work adequately in the presence of multiplicative noise, and therefore can be used to despeckle images. This technique is used for all of the adaptive filtering code in this chapter whenever an algorithm calls for σ_n^2.

Listing 4-16 is a MATLAB implementation of the SAM filter. This function again uses `filter2` to compute the μ_L and σ_L^2 matrices (u and sigmasq, respectively) as before, but also uses `filter2` to pass the input image through low-pass and high-pass filters, whose kernels are function arguments to SAM. The `find` function is then used to demarcate those pixels that meet the criterion $k\sigma_n^2/\sigma_L^2 > 1$, and these pixels, whose indices are given by the `ii` vector, are then replaced by a linear weighting of the low-pass and high-pass pixels. A final processing step is to clamp the pixel intensities to a range appropriate for the data type of the input image.

Listing 4-16: Signal Adaptive Mean (SAM) MATLAB function.

```
function J = sam(I, LP, HP, sz, k)

dI = double(I);
[M N] = size(I);
Ilpf = filter2(LP, dI);
Ihpf = filter2(HP, dI);

% compute local statistics
u = filter2(ones(sz)./prod(sz), dI);
sigmasq = filter2(ones(sz)./prod(sz), dI.^2);
sigmasq = sigmasq - u.^2;

% from the local variances estimate the noise variance
noise_variance = mean2(sigmasq);

% find those pixels that meet the SAM criterion
ii = find(k*noise_variance < sigmasq);
Kweights = zeros(M, N);
Kweights(ii) = 1 - (k .* (noise_variance ./ sigmasq(ii)));
```

```
J = Ilpf + Kweights.*Ihpf;

% clip to valid range
if isa(I, 'uint8')
    pmax = 255;
elseif isa(I, 'uint16')
    pmax = 65535;
else
    pmax = 1;
end

ii = find(J < 0);
J(ii) = 0;
ii = find(J > pmax);
J(ii) = pmax;

% return J back in original form
if isa(I, 'uint8')
    J = uint8(J);
elseif isa(I, 'uint16')
    J = uint16(J);
end
```

4.6.4 An MMSE Adaptive Filter Using the Intel IPP Library

The Intel Integrated Performance Primitives library actually consists of three APIs: the image processing component we have dealt with exclusively up until now, a component for one-dimensional signal processing, and a suite of basic mathematical functions for matrix math. The application described here is the first program that uses functionality from the signal processing component, and can be found in the Chap4\ AdaptiveFilter\MMSE_VS directory. The application's interface is very similar to that of the program discussed in 4.5.4, but instead of querying the user to enter in the percentage of shot noise, there is an edit box for the user to enter the noise variance of the multiplicative speckle noise. The project links to the Intel IPP Signal Processing static library ipps20.lib and includes the signal processing library header file ipps.h, in addition to the usual IPP Image Processing dependencies. The AddNoise class has been updated with a new method SpeckleNoise (given in Listing 4-17) that uses ippsRandGauss_Direct_32f and ippsMalloc_32f from the Intel Signal Processing Library to generate Gaussian random variables which are subsequently used to contaminate input pixels with multiplicative speckle noise. The `ippsRandGauss_Direct_32f` function requires a mean and standard deviation to generate the Gaussian distribution. Because the user is prompted to enter in the noise variance on the main application window, the square root of this input parameter is then passed into `ippsRandGauss_Direct_32f`, and then this distribution is multiplied by the image data in a pixel-wise form, thereby simulating a speckled image.

Listing 4-17: `AddNoise::SpeckleNoise`

```
Image8bpp *AddNoise::SpeckleNoise(const Image8bpp *pIn,
                                  float variance)
{
  if (variance<0.0 || variance>1.0)
    throw runtime_error("Noise variance must be b/w [0-1.0]");

  unsigned int N = pIn->getHeight()*pIn->getWidth(),
               seed = (unsigned)time(NULL);

  // allocate memory to store gaussian random #'s
  Ipp32f *pRandNums = (Ipp32f *)ippsMalloc_32f(N*sizeof(Ipp32f));
  if (ippStsNoErr != ippsRandGauss_Direct_32f(pRandNums, N, 0.f,
                                   sqrt(variance), &seed))
    throw runtime_error("Failed to generate gaussian prng sequence!");

  Image8bpp *pOut = new Image8bpp(*pIn); // allocate output image

  // add multiplicative speckle noise here
  Ipp32f *pNoise = pRandNums;
  for (int iRow=0; iRow<pIn->getHeight(); ++iRow) {
    Ipp8u *pOutScanLine = pOut->getPixels()+iRow*pOut->getStride();
    for (int iCol=0; iCol < pIn->getWidth();
         ++iCol, pOutScanLine++, pNoise++) {
      float corrupted = *pOutScanLine + (*pOutScanLine) * (*pNoise);
      *pOutScanLine = (corrupted>255.f) ? 255 : (Ipp8u)corrupted;
      *pOutScanLine = (corrupted<0.f) ? 0.f : (Ipp8u)corrupted;
    }
  }

  ippsFree(pRandNums);
  return pOut;
}
```

The `MMSEFilter` class is at the heart of this application, and its implementation is given in Listing 4-18. This code is somewhat unoptimized (for example, there are Intel IPP functions to compute pixel statistics) but the goal here is provide a starting point, as there are quite a few issues that need to be addressed when porting this algorithm on to the DSP. Essentially, this implementation of the MMSE algorithm proceeds in two phases and this general structure will be carried over in the DSP implementation. If the noise variance is known in advance, then the two phases could be collapsed into one, which should result in a performance boost. However, MMSEFilter does not require clients to know the noise variance, and as a consequence the first phase, which consists of a quadruple loop, entails computing the localized statistics of the image for each neighborhood. In between the two phases, the noise variance is estimated using the same technique as the MATLAB implementation, and then during the second phase we march through the image, replacing pixels if they meet the MMSE criterion.

Listing 4-18: The MMSEFilter class.

```
    void MMSEFilter::processImage(Image8bpp *pNoisyImage,
                                  Image8bpp *pProcessedImage)
    {
     // sanity check
     if (pNoisyImage->getHeight()!=pProcessedImage->getHeight() ||
        pNoisyImage->getWidth()!=pProcessedImage->getWidth())
       throw runtime_error("invalid args: images must be same size!");

     int nr=pNoisyImage->getHeight(),
         nc=pNoisyImage->getWidth(),
         stride=pNoisyImage->getStride();
     float N = (float)m_kernelSize*m_kernelSize;

     // adaptive filtering occurs in 2 steps, first compute local stats
     // and then filter the image based on those states: so here we setup
     // buffers to store these descriptive statistics.
     int nStats = (nr-2*m_khw)*(nc-2*m_khw);
     vector<float> localMeans(nStats);
     // localVariances allocated via Intel library function because later
     // on we're going to use ipps functions to estimate noise power.
     Ipp32f *localVariances = (Ipp32f *)ippsMalloc_32f(nStats);

     // compute localized statistics within the interior of the image
     vector<float>::iterator pLocalMeans = localMeans.begin();
     Ipp32f *pLocalVariances = localVariances;
     for (int iRow=m_khw; iRow<nr-m_khw; ++iRow) {
       Ipp8u *pIn = pNoisyImage->getPixels() + (iRow-m_khw)*stride;
       for (int iCol=m_khw; iCol<nc-m_khw; ++iCol, pIn++) {

         // compute local mean and local variance
         unsigned int sum = 0, sumSquared = 0;
         for (int iw=0; iw<m_kernelSize; ++iw) {
           for (int ih=0; ih<m_kernelSize; ++ih) {
             sum += pIn[iw+ih*stride];
             sumSquared += pIn[iw+ih*stride]*pIn[iw+ih*stride];
```

```
        }
      }
      float mean = (float)sum/N,
          avgSumSquared = (float)sumSquared/N;
      Ipp32f variance = avgSumSquared - mean*mean;
      *pLocalMeans++ = mean;
      *pLocalVariances++ = variance;

    } // end (for each column)
  } // end (for each row)


  // estimate noise power based on local variances
  Ipp32f noiseVariance;
  if (ippStsNoErr != ippsMean_32f(localVariances, nStats,
                                &noiseVariance, ippAlgHintAccurate))
    throw runtime_error("Failed to estimate noise variance!");

  // and now filter the image
  pLocalMeans = localMeans.begin();
  pLocalVariances = localVariances;
  for (int iRow=m_khw; iRow<nr-m_khw; ++iRow) {
    Ipp8u *pIn = pNoisyImage->getPixels() + (iRow-m_khw)*stride,
        *pOut = pProcessedImage->getPixels()+(iRow)*stride+m_khw;
    for (int iCol=m_khw; iCol<nc-m_khw; ++iCol, pIn++, pOut++) {
      float localMean = *pLocalMeans++,
          localVar  = *pLocalVariances++;
      // This is the heart of the MMSE adaptive filter:
      // Linearly interpolate between the local mean and the input pixel.
      // If the local variance is roughly the same as the noise variance,
      // give greater weight to the mean else tilt towards the input
      float alpha = noiseVariance/localVar;
      // pin alpha such that if the noise variance dominates output
      // pixel contribution comes entirely from the local mean
      if (alpha > 1.0)
        alpha = 1.0;
      float outPixel = (1-alpha)*(*pIn) + alpha*localMean;

      // handle out-of-range pixels
      if (outPixel < 0.f)
        outPixel = localMean;
      else if (outPixel > 255.f)
        outPixel = localMean;
      *pOut = (Ipp8u)outPixel+0.5; // clamp to 8 bit range
    } // end (for each column)
  } // end (for each row)

  ippsFree(localVariances);
}
```

4.6.5 MMSE Filtering on the C6416

Unfortunately there are no IMGLIB functions we can leverage to implement adaptive image filtering on the C6x platform. However, since so many adaptive image filtering algorithms are a twist on a single theme that necessitates calculation of local pixel statistics, an MMSE implementation is very useful because once the core algorithm has been implemented and debugged, the basic concepts can be used to implement similar filters. The directory `Chap4\AdaptiveFilter\MMSE_C6416` contains a C64x project that performs adaptive image filtering on a 256x256 image. The algorithmic optimizations, as compared to the Windows application, fall into two categories:

1. The inner loops that comprise the first phase of the algorithm, where the sum and the sum of the squares of the current neighborhood are computed, have been removed.
2. This is a fixed-point implementation. Most of the complications arise from this design decision.

With regards to (1), this program is less general than the Visual Studio version – the kernel size is set to 3x3. The optimization that flows out of this decision is that the inner two loops are unrolled and the number of addition and multiplication operations reduced by 1/3, per each neighborhood. With the quadruple looping structure in Listing 4-18, 9 additions are needed to form the neighborhood sum, and 9 additions and 9 multiplications are needed to calculate the neighborhood sum of the squares, which is used in the computation of the neighborhood variance. This implementation operates in a row-major form (with the neighborhood sliding across the image in a left-to-right manner), and the number of operations is reduced by shifting out the left-most column of the previous neighborhood, and subtracting their contribution to `accum` and `sumsq`, then replacing them with the next 3 pixels corresponding to the incoming column. The end result of this transformation is that the total number of additions and multiplications (where subtraction is treated the same as addition) per neighborhood is now 18, versus 27 in the previous implementation. A reduction in arithmetic operations by a factor of 1/3 is quite significant in neighborhood processing, because these savings are gained for each pixel location within the image.

For the sake of brevity, this program does not employ the DMA paging operation that reduces the EMIF penalty – image buffers are all placed in external memory. Applying this particular optimization is rather straightforward and implementing the technique described in 3.4.3 is left as an exercise for the reader. The two stages of the MMSE filter lead to a natural functional decomposition. In this implementation, shown in Listing 4-19, the first stage (where neighborhood pixel statistics are computed), is found within `collect_local_pixel_stats`. The local pixel statistics are stored in the `local_mean` and `local_variance` buffers, which are then used within `mmse_filter` to actually filter the image.

Listing 4-19: `mmse.c`.

```
#define KERNEL_SIZE 3 /* adaptive filter kernel is of size 3x3 */
#define MARGIN (KERNEL_SIZE/2)

unsigned char out_img[N_PIXELS],
               local_mean[N_PIXELS];
unsigned short local_variance[N_PIXELS];

/*
 * place all buffers in external RAM aligned
 * on double-word boundaries
 */
#pragma DATA_SECTION (in_img, "SDRAM");
#pragma DATA_ALIGN (in_img, 8);
#pragma DATA_SECTION (out_img, "SDRAM");
#pragma DATA_ALIGN (out_img, 8);
#pragma DATA_SECTION (local_mean, "SDRAM");
#pragma DATA_ALIGN (local_mean, 8);
#pragma DATA_SECTION (local_variance, "SDRAM");
#pragma DATA_ALIGN (local_variance, 8);

#define Q 12  /* right-shift used in collect_local_pixel_stats */
#define S 455 /* 1/9 * 2^12, box filter divisor in Q12 format */
#define Q15_ONE (1L<<15)

unsigned short collect_local_pixel_stats()
{
  int ir=MARGIN, ic;
  unsigned char *pin1=in_img, *pin2=in_img+Y_SIZE,
               *pin3= in_img+2*Y_SIZE;
  unsigned char *plocal_mean = local_mean+Y_SIZE+MARGIN;
  unsigned short *plocal_var = local_variance+Y_SIZE+MARGIN;
  unsigned long accum, sumsq; /* sumsq = "sum of squares" */
  unsigned long sumvar = 0, /* sum of the local variances */
               m; /* temporary local mean storage */
  float avg_variance;
```

```
    for (; ir<X_SIZE-MARGIN; ++ir) {
      accum = pin1[0] + pin1[1] + pin1[2] +
              pin2[0] + pin2[1] + pin2[2] +
              pin3[0] + pin3[1] + pin3[2];
      sumsq = pin1[0]*pin1[0] + pin1[1]*pin1[1] + pin1[2]*pin1[2] +
              pin2[0]*pin2[0] + pin2[1]*pin2[1] + pin2[2]*pin2[2] +
              pin3[0]*pin3[0] + pin3[1]*pin3[1] + pin3[2]*pin3[2];
      for (ic=MARGIN; ic<Y_SIZE-MARGIN;
           ++ic, plocal_mean++, plocal_var++) {
        m = (accum * S);
        accum -= pin1[0] + pin2[0] + pin3[0];
        accum += pin1[3] + pin2[3] + pin3[3];

        *plocal_var = ( (sumsq * S) - ((m*m)>>Q) ) >> Q;
        *plocal_mean = m>>Q;
        sumsq -= pin1[0]*pin1[0] + pin2[0]*pin2[0] + pin3[0]*pin3[0];
        sumsq += pin1[3]*pin1[3] + pin2[3]*pin2[3] + pin3[3]*pin3[3];
        sumvar += *plocal_var;

        pin1++; pin2++; pin3++;
      } /* end (for each column) */
      pin1 += 2*MARGIN;
      pin2 += 2*MARGIN;
      pin3 += 2*MARGIN;
      plocal_mean += 2*MARGIN;
      plocal_var += 2*MARGIN;
    } /* end (for each row) */

  avg_variance =
          (float)sumvar / ((X_SIZE-2*MARGIN)*(Y_SIZE-2*MARGIN));
  return (unsigned short)(avg_variance + 0.5f);
}

inline unsigned short newton_raphson_Q15(unsigned short x,
                                         unsigned short seed)
{
  unsigned short r = seed, rprev = r;
  int ii=0;

  for (; ii<6; ++ii) {
    r = (r*(65536 - x*r)) >> 15;
```

```
      if (rprev==r)
        return r;
      else
        rprev = r;
    }

  return r;
}


/* Newton-Raphson seed lookup tables */
const unsigned short seed_1_to_256[] = {
  2048,1024,682,512,409,341,292,256,
  227,204,186,170,157,146,136,128
};

const unsigned short seed_128_to_9362[] = {
  256,85,51,36,28,23,19,17,15,13,12,11,10,9,8,8,
  7,7,6,6,6,5,5,5,5,5,4,4,4,4,4,4,3,3,3,3,3,3
};

inline unsigned short recip_Q15(unsigned short x)
{
  static const unsigned short reciprocals_1_to_5[] =
                                      {32768,16384,10923,8192};
  if (x<5)
    return reciprocals_1_to_5[x];
  else if (x > 21845)
    return 1;
  else if (13107<x && x<21845)
    return 2;
  else if (9326<x && x<13107)
    return 3;
  else if (x<256)
    return newton_raphson_Q15(x, seed_1_to_256[x>>4]);
  else
    return newton_raphson_Q15(x, seed_128_to_9362[x>>8]);
}


void mmse(unsigned short noise_var)
{
  int ir=MARGIN, ic;
  unsigned char *pin = in_img+Y_SIZE+MARGIN,
```

```
        *plocal_mean = local_mean+Y_SIZE+MARGIN,
        *pout = out_img+Y_SIZE+MARGIN;
unsigned short *plocal_var = local_variance+Y_SIZE+MARGIN;
unsigned short alpha; /* (noise variance) / (local variance) */

for (; ir<X_SIZE-MARGIN; ++ir) {
  for (ic=MARGIN; ic<Y_SIZE-MARGIN;
      ++ic, plocal_mean++, plocal_var++, pin++) {
    if (noise_var > *plocal_var)
      *pout++ = *plocal_mean;
    else {
      alpha = noise_var * recip_Q15(*plocal_var);
      *pout++ =
              (Q15_ONE-alpha)*(*pin) + (alpha)*(*plocal_mean) >> 15;
    }
  }
  plocal_mean += 2*MARGIN;
  plocal_var += 2*MARGIN;
  pin += 2*MARGIN;
  pout += 2*MARGIN;
  }
}

void main()
{
  unsigned short noise_variance_est;

  DSK6416_init(); /* initialize the DSK board support library */

  noise_variance_est = collect_local_pixel_stats();
  mmse(noise_variance_est);
}
```

The fixed point algorithm in `collect_local_pixel_stats` assumes a radix point between the third and fourth binary digit, meaning that bit-shifts of 12 are called for. After the local 3x3 sum is computed and stored in `accum`, this sum is then divided by 9. Instead of an extremely expensive division operation (which on the fixed-point C6416 DSP results in a function being called) the equivalent result is obtained by multiplying the sum by 1/9 in Q12 format, or 455. Right-shifting this result by 12 bits then yields the local mean. Proper calculation of the local variance in fixed-point is more involved, because one must be cognizant of the order of operations, or else significance loss degrades the end result. Consider the code snippets in Listing 4-20. The code on the top is taken from mmse.c, whereas the bottom-most snippet is incorrect, producing inaccurate variances even though at first glance it appears nothing may be awry.

Listing 4-20: Correct fixed-point computation of the local variance (top) versus an incorrect version (bottom). The constants S=455, and Q=12 are from mmse.c (see Listing 4-19). Both versions of code produce an identical local mean, but the version on the left propagates rounding-error into the computation of the local variance.

```
/* correct sequence of statements */
*plocal_mean = (accum * S) >> Q;
*plocal_var = ((sumsq * S)>>Q) - ((*plocal_mean)*(*plocal_mean));

/* incorrect sequence */
m = (accum * S);
*plocal_var = ((sumsq*S)-((m*m)>>Q)) >> Q;
*plocal_mean = m>>Q;
```

The reason why the code at the bottom of Listing 4-20 should not be used is because of the limited number of bits allocated for the fractional portion of the Q12 numbers used in the calculations in mmse.c. Any round-off error introduced by this fixed-point format is magnified when intermediate results are subsequently multiplied by a large number (in this case, 455). Incidentally, this type of fixed-point arithmetic error was the cause of the failure of the American Patriot Missile battery in Dharan, Saudi Arabia, during the First Gulf War in 1991[27]. The situation can be simulated in MATLAB, as shown in Listing 4-21.

Listing 4-21: MATLAB simulation of the fixed-point operations used in collect_local_pixel_stats from Listing 4-19.

```
accum = 1311; % example sum of pixel intensities for 3x3 neighborhood
sumsq = 194199; % example sum of squares for same neighborhood
Q = 12; % Q factor for fixed-point arithmetic

% floating-point
variance_floating_point = (sumsq/9) - (accum/9)*(accum/9);
disp(sprintf('Floating point: %.3f - %.3f^2 = %.3f', ...
    (sumsq/9), (accum/9), variance_floating_point));

% fixed-point (1)
% u = (accum*S) >> Q;
```

```
% v = (sumsq*S) >> Q;
% variance_fixed_point1 = v - u*u;
S = round(1/9 * 2^Q);
term1 = bitshift(sumsq*S,-Q); % sumsq*455 >> 12
term2 = bitshift(accum*S,-Q); % accum*455 >> 12
variance_fixed_point1 = term1 - term2*term2;
disp(sprintf('Fixed point #1 = %d - %d^2 = %d', ...
    term1, term2, variance_fixed_point1));
disp(sprintf('Fixed point #1 |delta| = %f', ...
    abs(variance_floating_point-variance_fixed_point1)));

% fixed-point (2): different order of operations
% usq = (accum*S)*(accum*S)
% variance_fixed_point1 = (sumsq*S - usq>>Q) >> Q
term1 = sumsq*S;
term2 = accum*S * accum*S;
term2 = bitshift(term2, -Q);
variance_fixed_point2 = bitshift(term1 - term2, -Q);
disp(sprintf('Fixed point #2 = %d - %d^2 = %d', ...
    term1, term2, variance_fixed_point2));
disp(sprintf('Fixed point #2 |delta| = %f', ...
    abs(variance_floating_point-variance_fixed_point2)));
```

Through judicious use of the built-in MATLAB functions `bitshift`, `floor`, and `round`, fixed-point math can be simulated fairly easily in MATLAB, which is by nature a floating-point language. Using a Q value of 12, this simulation's output is the following:

```
Floating point: 21577.667 - 145.667^2 = 358.889
Fixed point #1 = 21572 - 145^2 = 547
Fixed point #1 |delta| = 188.111111
Fixed point #2 = 88360545 - 86869681^2 = 363
Fixed point #2 |delta| = 4.111111
```

Clearly the second order of operations results in a far more accurate calculation of the local variance. When running these types of simulations, care must be taken to remember that the `unsigned long` type on the C64x is 40 bits long, and unfortunately MATLAB does not have an internal 40-bit integer type (it does provide `uint8`, `uint16`, `uint32`, and `uint64`). The MATLAB simulation suggests that even greater accuracy may be obtained by increasing Q. However, this tactic does not translate over to the C64x architecture precisely because of the 40-bit restriction, as a

Q value greater than 12 will result in potential integer overflow. In an ANSI-compliant C, compiler the maximum value for all of the supported data types, as well as a variety of other useful constants, can always be found in the limits.h header file.

The second stage of the algorithm takes place in the mmse function, and the main issue here is the calculation of the weighting factor α for linear interpolation between the input pixel and local mean, should the current pixel meet the MMSE criterion. The crux of the problem is that the C64x, unlike the floating-point C67x architecture, does not offer a division instruction, and thus the operation a/b is extremely expensive, as the compiler will be forced to inject a slow software routine in the middle of this tight loop to perform the division. Note that the division operation is only needed if the noise variance is less than the local variance, because otherwise the output pixel is simply the local mean. Yet if the weighting factor needs to be computed, there needs to be some means of calculating the ratio of the noise variance to the local variance.

In general, in fixed-point systems the way to get around this lack of division conundrum is to implement what is known as *division by reciprocal approximation* - that is replacing a/b with a(1/b). Assuming it is possible to efficiently approximate the reciprocal of b, this reformulation of the division operation should be more efficient than a direct division. The DSPLIB library provides a function DSP_recip16 which may be used to calculate the reciprocals of a vector of fixed point numbers in Q15 format; however it is unsuitable in this program for two reasons. First, we do not necessarily need to calculate the reciprocal of every local variance (for some of the pixels we simply use the local mean as the output); while one could conceivably call DSP_recip16 to calculate the reciprocals of an entire vector, and then simply discard those values that are not needed, such a strategy is quite wasteful. Second, DSP_recip16 returns the reciprocals as separate fractional and exponent parts[28]. Recall that a prerequisite of Q15 arithmetic are that the numbers must be scaled so that they are less than 1 in absolute value (see 2.1.2), which is typically accomplished by dividing by a power of 2 so that divisions can be efficiently performed using a right bit-shift. Due to the way DSP_recip16 is implemented, the maximum scaling factor that can be used is $2^{11} = 2048$ (because the exponent overflows with powers of 2 greater than 11 and less than 15). Since the local variance of 3x3 neighborhoods can easily exceed 2048, this potential input to DSP_recip16 cannot be scaled properly to a Q15 format.

There are a multitude of potential schemes for getting around the problem of efficiently computing the reciprocal 1/b:

1. The reciprocal, which in Q15 format is actually 32767/b, may be obtained directly by repeated subtractions. An optimized implementation of such a scheme would use the conditional integer subtract and shift C intrinsic _subc[29].
2. The FastRTS library[30], which consists of optimized floating-point software functions, can be used. This library contains highly optimized assembly functions that can serve as a useful stop-gap measure when porting a floating-point algorithm onto a fixed-point architecture.
3. Lookup tables, where reciprocals are precomputed, may be used. If the input range is known beforehand, and is reasonably limited, this can be a very effective approach as the reciprocal operation then translates into simply indexing into an array. The downside to this approach is that it can be prohibitively memory consumptive, especially because the lookup table would most likely need to be placed entirely within internal on-chip RAM.
4. Newton-Raphson iteration can be used to estimate the reciprocal.

Using Newton-Raphson iteration, or "Newton's method", to evaluate reciprocals is a venerable technique that dates back to the age of early digital computers[18]. Newton's method is predicated on the existence of the inverse for the function being evaluated, which for the case of finding 1/b is met so long as b is not zero. An initial guess, or seed value, is made for $f(x) = $ *reciprocal* and the iteration proceeds to successively refine the solution y according to the following equation:

$$y_{n+1} = y_n - \frac{g(y_n) - x}{g'(y_n)}$$

where

- x = input value
- y_n = current solution to $f(x)$
- $g(y_n)$ = $f^{-1}(x)$ evaluated at y_n
- $g'(y_n)$ = derivative of $f^{-1}(x)$ evaluated at y_n
- y_{n+1} = refined solution

The iteration proceeds until either the desired accuracy is achieved or until a certain tolerance is met. For the case of finding the reciprocal of x, a Newton-Raphson iteration can be derived as so:

$$f(x) = y = 1/x$$
$$g(y) = y^{-1} = x$$
$$g'(y) = -y^{-2}$$

And then the iteration follows from the definition of Newton's method:

$$y_{n+1} = y_n - \frac{y_n^{-1} - x}{-y_n^{-2}}$$

$$= y_n + y_n^2 \left(y_n^{-1} - x \right)$$

$$= y_n + y_n - xy_n^2$$

$$= 2y_n - xy_n^2$$

$$= y_n \left(2 - xy_n \right)$$

Newton-Raphson iteration converges quadratically to $1/x$. Translating the above scheme so as to converge to the Q15 reciprocal $2^{15}/x$ using 32-bit arithmetic leads to the following C statement:

```
y = (y * (65536 - x*y)) >> 15;
```

A variety of methods exist for seeding the iterative scheme. The C67x instruction set includes instructions for calculating single-precision (RCPSP) and double-precision (RCPDP) floating-point reciprocal approximations, which can then be further refined by employing Newton-Raphson iteration[14]. In fact, reciprocal approximation via Newton-Raphson iteration is sometimes used in floating-point architectures when full accuracy is not required, simply due to the speed benefits the technique offers when compared to IEEE floating-point math. The approach used in this program is a combination of Newton-Raphson iteration and a lookup table approach. A lookup table is used to seed the Newton-Raphson iteration, or even return the actual reciprocal, should we be so lucky that the function's input argument be found within the lookup table. Newton's method is then used to find the "exact" reciprocal in Q15 form by iterating until either $y_{n+1}=y_n$, or six iterations have been performed - at which point the iteration terminates. A MATLAB simulation showed that at most six iterations are required to obtain the reciprocal for the range of values needed for this program (0-32767). This simulation performed fixed-point Newton-Raphson iteration, seeded by the lookup tables used in the C64x program, and then compared the result to the known reciprocal value.

The reciprocal finding algorithm is implemented in Listing 4-19 in the functions `recip_Q15` and `newton_rhapson_Q15`. After the Q15 reciprocal of the local variance has been calculated, the weighting factor α is computed, and finally linear interpolation between the input pixel and the local mean is accomplished again using Q15 arithmetic via the following C code:

```
out_pixel = ((32767-alpha)*in_pixel+alpha*local_mean)>>15;
```

REFERENCES

1. Claus, H., and Meusel, H., *Filter Practice* (Focal Press, 1964).
2. Gonzalez, R., and Woods, R., *Digital Image Processing* (Addison-Wesley, 1992).
3. Russ, J., *The Image Processing Handbook* (CRC Press, 1999).
4. *MATLAB Image Processing Toolbox User's Guide*, fspecial Function.
5. *MATLAB Function Reference Vol. 1: A-E*, conv2 Function.
6. Texas Instruments, *TMS320C62x Image/Video Processing Library Programmer's Reference* (SPRU400a.pdf), Chapter 5, section 5.3, *IMG_corr_gen*.
7. Texas Instruments, *TMS320C62x DSP Library Programmer's Reference* (SPRU402b.pdf).
8. Texas Instruments, *TMS320C6000 Peripherals Reference Guide* (SPRU190d.pdf).
9. Texas Instruments, *TMS320C6000 Chip Support Library API User's Guide* (SPRU401f.pdf).
10. Texas Instruments, *TMS320C6000 DMA Example Applications* (SPRA529A.pdf).
11. Texas Instruments, *TMS320C64x Image/Video Processing Library Programmer's Reference* (SPRU023A.pdf).
12. Texas Instruments, *Image Processing Examples Using the TMS320C64x Image/Video Processing Library* (SPRA887.pdf).
13. Texas Instruments, *TMS320C6000 Optimizing Compiler User's Guide* (SPRU187k.pdf).
14. Texas Instruments, *TMS320C6000 Programmer's Guide* (SPRU198G.pdf).
15. *MATLAB Image Processing Toolbox User's Guide*, medfilt2 Function.
16. *MATLAB Image Processing Toolbox User's Guide*, imnoise Function.
17. Ross, S., *A First Course in Probability, 3rd Edition* (MacMillan Publishing Co., 1988).
18. Knuth, Donald, *The Art of Computer Programming Volume 2: Seminumerical Algorithms, 3rd Edition* (Addison-Wesley, 1998).
19. Microsoft Developers Network (MSDN), see "How to Use QueryPerformanceCounter to Time Code" Retrieved October 2004 from:
    http://support.microsoft.com/default.aspx?scid=kb;en-us;172338

20. Knuth, Donald, *The Art of Computer Programming Volume 3: Sorting and Searching, 2nd Edition* (Addison-Wesley, 1998).
21. Musser, D. et al., *STL Tutorial and Reference Guide: C++ Programming with the Standard Template Library, 2nd Edition* (Addison-Wesley, 2001).
22. Waltz, F.M., Hack, R., Batchelor, B.G., "Fast, efficient algorithms for 3x3 ranked filters using finite-state machines," *Proc. SPIE Conf. on Machine Vision Systems for Inspection and Metrology VII, Vol. 3521, Paper No. 31,* Boston, Nov. 1998.
23. Paeth, A. "Median Finding on a 3x3 Grid." In *Graphics Gems*, Andrew Glassner (Academic Press, 1990).
24. Smith, J., "Implementing Median Filters in XC4000E FPGAs," Retrieved October 2004 from: http://www.xilinx.com/xcell/xl23/xl23_16.pdf
25. Devillard, N. "Fast median search: an ANSI C implementation," Retrieved October 2004 from: http://ndevilla.free.fr/median/median.pdf
26. Phillips, D., "Removing Image Noise with Adaptive Filters," *C/C++ User's Journal*, Dec. 2000; pp. 10-23.
27. United States General Accounting Office, *Patriot Missile Defense: Software Problem Led to System Failure at Dhahran, Saudi Arabia.* Retrieved October 2004 from: http://www.fas.org/spp/starwars/gao/im92026.htm
28. Texas Instruments, *TMS320C64x DSP Library Programmer's Reference* (SPRU565a.pdf).
29. Texas Instruments, *TMS320C6000 Integer Division* (SPRA707.pdf).
30. Texas Instruments, *TMS320C62x/64x FastRTS Library Programmer's Reference* (SPRU653.pdf).
31. Mack, B., Govoni, P., "Direct Memory Access," *C/C++ User's Journal*, Jan. 2005, pp. 40-45.

Chapter 5

EDGE DETECTION AND SEGMENTATION

Chapters 3 and 4 dealt with image enhancement. In this chapter, the focus shifts to image *analysis*, which in some respects is a more difficult problem. Image analysis algorithms draw on information present within an image, or group of images, to extract properties about the scene being imaged. For example, in machine vision applications, images are analyzed to oftentimes determine the orientation of objects within the scene in order to send feedback to a robotic system. In medical imaging registration systems, radiographs, or slices from 3D volumes, are analyzed or fused together by a computerized system to glean information that a clinician or medical device may be interested in, such as the presence of tumors or the location and orientation of anatomical landmarks. In military tracking systems, interest lies in identifying targets, such landmarks, enemy combatants, etc. There are many other real-world applications where image analysis plays a large role.

Generally speaking, image analysis boils down to image *segmentation*, where the goal is to isolate those parts of the image that constitute objects or areas of interest. Once such objects or regions have been separated from the image, various characteristics (e.g. center-of-mass or area) can be computed and used towards a particular application. Many of these algorithms utilize various heuristics and localized domain knowledge to help steer the image processing algorithms to a desired solution, and therefore it is essential for the practitioner to be well-grounded in the basics. Universally applicable image segmentation techniques general enough to be used "straight out of the box" do not exist – it is almost always the case that the basic algorithms must be tweaked so as to get them to work robustly within a particular environment. In this chapter we discuss some of the basic algorithms that aid in the segmentation problem. First, the topic of edge detection is introduced, and an interactive embedded implementation featuring real-time data transfer between a host PC and embedded target is provided. After an initial exploration of classical edge-detection techniques (we come back to this topic in Chapter 6), we move on to image segmentation proper, and in turn will use some of these algorithms to develop a more flexible and generalized classical edge detector.

5.1 EDGE DETECTION

In a monochrome image, an edge is defined as the border between two regions where large changes in intensity occur, as shown in Figure 5-1. An input image may be edge *enhanced*, meaning which the image is processed so that non-edges are culled out of the input, and edge contours made more pronounced at the expense of more constant areas of the image. A more interesting problem is that of edge detection, whereby an image processing system autonomously demarcates those portions of the image representing edges in the scene by first edge enhancing the input and subsequently utilizing a thresholding scheme to binarize the output. Edge detection typically requires the services of one or more spatial filters, so these algorithms naturally build upon the techniques developed in Chapter 4 to spatially filter an image.

From Figure 5-1 it is apparent that the key to detecting edges is robustly locating abrupt changes in such intensity profiles. Thus it is not terribly surprising that edge enhancement filters are predicated on the idea of utilizing local derivative or gradient operators. The derivative of the intensity profile from a scan-line of an image is non-zero wherever a change in intensity level occurs, and zero wherever the image is of constant intensity. The gradient of an image ∇f at location *(x,y)* is

$$\nabla f(x, y) = \begin{bmatrix} G_x \\ G_y \end{bmatrix} = \begin{bmatrix} \partial f / \partial x \\ \partial f / \partial y \end{bmatrix}$$

where G_x and G_y are the partial derivatives in the x and y directions, respectively. Discrete gradients are approximated with finite differences, i.e.

$$\frac{\partial f(t)}{\partial t} \approx \frac{f(t+1) - f(t-1)}{2}$$

Given this approximation, the convolution kernel that produces the above central first difference is then [1 0 -1]. This kernel yields a response for

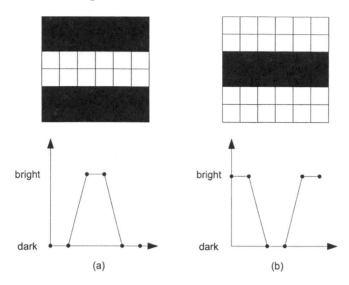

Figure 5-1. Vertical edges in simulated digital image. (a) Intensity profile of a vertical line through a white stripe on a dark background. (b) Intensity profile of a vertical line through a dark stripe on a white background.

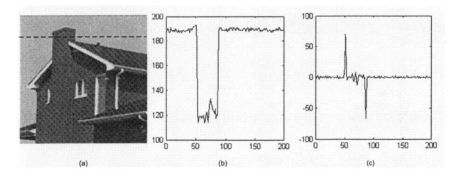

Figure 5-2. Derivative of scan-line through house image. (a) Dashed line indicates the horizontal scan-line plotted in this figure. (b) Intensity profile of the scan-line, plotted as a function of horizontal pixel location. The abrupt change in intensity levels at around pixel locations 50 and 90 are the sharp edges from both sides of the chimney. (c) Numerical derivative of this intensity profile, obtained using central first differences. This signal can be obtained by convolving the intensity profile with [1 0 -1], and the sharp peaks used to gauge the location of the edges.

vertical edges. The transposition of this kernel gives an approximation to the gradient in the orthogonal direction, and produces a response to horizontal edges:

$$\begin{bmatrix} 1 \\ 0 \\ -1 \end{bmatrix}$$

An example of how these gradient kernels operate is shown in Figure 5-2. In this example, a scan-line from a row of an image is plotted in 5-2b, and the gradient of this signal (which in one dimension is simply the first derivative) is shown in Figure 5-2c. Note that the gradient exhibits distinct peaks where the edges in this particular scan-line occur. Edge detection could then be achieved by thresholding – that is, if $|\partial f(t)/\partial t| > T$, mark that location as an edge pixel. Extending the central first difference gradient kernel to a 3x3 (or larger) size has the desirable result of reducing the effects of noise (due to single outlier pixels), and leads to the Prewitt edge enhancement kernel:

$$G_x = \begin{bmatrix} 1 & 0 & -1 \\ 1 & 0 & -1 \\ 1 & 0 & -1 \end{bmatrix} \quad G_y = \begin{bmatrix} 1 & 1 & 1 \\ 0 & 0 & 0 \\ -1 & -1 & -1 \end{bmatrix}$$

G_x approximates a horizontal gradient and emphasizes vertical edges, while G_y approximates a vertical gradient and emphasizes horizontal edges. The more widely used Sobel kernels are weighted versions of the Prewitt kernels:

$$G_x = \begin{bmatrix} 1 & 0 & -1 \\ 2 & 0 & -2 \\ 1 & 0 & -1 \end{bmatrix} \quad G_y = \begin{bmatrix} 1 & 2 & 1 \\ 0 & 0 & 0 \\ -1 & -2 & -1 \end{bmatrix}$$

The Sobel edge enhancement filter has the advantage of providing differencing (which gives the edge response) and smoothing (which reduces noise) concurrently. This is a major benefit, because image differentiation acts as a high-pass filter that amplifies high-frequency noise. The Sobel kernel limits this undesirable behavior by summing three spatially separate discrete approximations to the derivative, and weighting the central one more than the other two. Convolving these spatial masks with an image, in the manner described in 4.1, provides finite-differences approximations of the orthogonal gradient vectors G_x and G_y. The magnitude of the gradient, denoted by ∇f, is then calculated by

$$\nabla f = \sqrt{(\partial f / \partial x)^2 + (\partial f / \partial y)^2} = \sqrt{G_x^2 + G_y^2}$$

A commonly used approximation to the above expression uses the absolute value of the gradients G_x and G_y:

$$\nabla f = \sqrt{G_x^2 + G_y^2} \approx |G_x| + |G_y|$$

This approximation, although less accurate, finds extensive use in embedded implementations because the square root operation is extremely expensive. Many DSPs, the TI C6x included, have specialized instructions for obtaining the absolute value of a number, and thus this gradient magnitude approximation becomes that much more appealing for a DSP-based implementation. Figure 5-3 shows how to apply the Sobel filter to an image. The process is exactly the same for the Prewitt filter, except that one substitutes different convolution kernels. The input image is passed through both Sobel kernels, producing two images consisting of the vertical and horizontal edge responses, as shown in 5-3b and 5-3c. The gradient image ∇f is then formed by combining the two responses using either of the two aforementioned formulas (in this case the square root of the sum of the squares of both responses is used). Edge detection is then accomplished by extending the one-dimension thresholding scheme to two dimensions:

$$\sqrt{G_x^2 + G_y^2} > T \quad \text{or} \quad \left(|G_x| + |G_y| \right) > T$$

where T may be user-specified or determined through some other means (see 5.2.2 for techniques for automatically determining the value of T). Algorithm 5-1 describes a Sobel-based edge detection algorithm, which we shall port to the DSP at the end of 5.1.2.2.

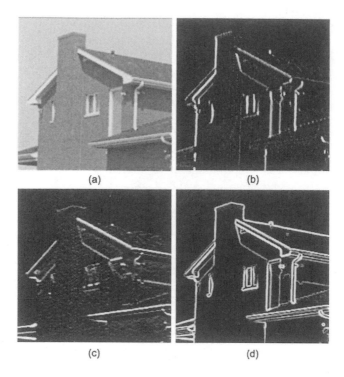

(a)   (b)

(c)   (d)

Figure 5-3. 2D edge enhancement using the Sobel filters. Edge response images are scaled and contrast-stretched for 8-bit dynamic range. (a) House input image. (b) G_x image, the vertical edge response. (c) G_y image, the horizontal edge response. (d) Combined edge image, sqrt(G_x* G_x + G_y* G_y).

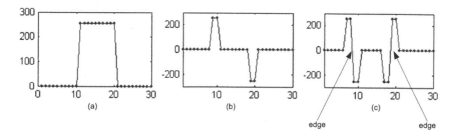

Figure 5-4. Using the second-order derivative as an edge detector. (a) A fictionalized profile, with edges at locations 10 and 20. (b) Numerical 1st-order derivative, obtained by convolving the profile with [1 0 -1]. (c) The second-order derivative, obtained by convolving the 1st-order derivative with [1 0 -1]. The zero-crossings, highlighted by the arrows, give the location of both edges.

Algorithm 5-1: 3x3 Sobel Edge Detector
INPUT: MxN image I, gradient threshold T
OUTPUT: binary edge image J

(*initialize output image to zero*)
for (r = 0...M-1)
   for (c = 0...N-1)
      J[r][c] = 0
   end
end

(*calculate gradient pixel, and then threshold*)
for (r = 1...M-2) (*1st and last rows not filtered*)
   for (c = 1...N-2) (*1st and last cols not filtered*)

   p_{00}=I[r-1][c-1]; p_{01}=I[r-1][c]; p_{02}=I[r-1][c+1];
   p_{10}=I[r][c-1]; p_{12}=I[r][c+1];
   p_{20}=I[r+1][c-1]; p_{21}=I[r+1][c]; p_{22}=I[r+1][c+1];

   G_x = p_{00}-p_{02}+2p_{10}-2p_{12}+p_{20}-p_{22} (*horizontal gradient*)
   G_y = p_{00}+2p_{01}+p_{02}-p_{20}-2p_{21}-p_{22} (*vertical gradient*)

   if (($|G_x|$+$|G_y|$) > T   (*then this pixel is an edge*)
      J[r][c] = 1;
   end

   end (*for each col*)
end (*for each row*)

There are a multitude of other edge enhancement filters. The Prewitt and Sobel gradient operators work well on images with sharp edges, such as those one might encounter in industrial applications where the lighting and overall imaging environment is tightly controlled, as opposed to, say, military tracking or space imaging applications. However, such filters tend not to work as well with "natural" images that are more continuous in nature and where edge transitions are more nuanced and gradual. In these cases the second-order derivative can *sometimes* be used as a more robust edge detector. Figure 5-4 illustrates the situation graphically – here the zero-crossings in the second derivative provide the edge locations. One distinct advantage to these types of edge detection schemes is that they do not rely on a threshold to separate the edge pixels from the rest of the image.

To implement a zero-crossing based edge detector, we need the second-order derivative of a 2D function, which is given by the Laplachian:

$$\nabla^2 f = \partial^2 f / \partial x^2 + \partial^2 f / \partial y^2$$

Several discrete approximations to the 2D Laplachian, in the form of spatial filter kernels, exist. One of the more common is

$$\begin{bmatrix} 0 & -1 & 0 \\ -1 & 4 & -1 \\ 0 & -1 & 0 \end{bmatrix}$$

This particular filter is appealing because of the presence of the zeroes within the kernel, as the output pixel $g(i,j)$ is

$$g(i,j) = 4f(i,j) - [f(i-1,j) + f(i,j-1) + f(i,j+1) + f(i+1,j)]$$

In contrast, some of the other Laplachian approximations include more dense spatial kernels, such as

$$\begin{bmatrix} 1 & 4 & 1 \\ 4 & -20 & 4 \\ 1 & 4 & 1 \end{bmatrix}$$

and

$$\begin{bmatrix} 1 & -2 & 1 \\ -2 & 4 & -2 \\ 1 & -2 & 1 \end{bmatrix}$$

Even though the Laplachian method has the advantage of not requiring a threshold in order to produce a binary edge image, it suffers from a glaring susceptibility to image noise. The presence of any noise within the image is the cause of a fair amount of "streaking", a situation where extraneous edges consisting of a few pixels obfuscate the output. This type of noisy output can be countered somewhat by smoothing the image prior to differentiating it with the 2D Laplachian (a pre-processing step also recommended with the

Sobel and Prewitt filters), and then applying sophisticated morphological image processing operations to filter out the edge response (for example, using a connected-components algorithm to throw away single points or small bunches of pixels not belonging to a contour). Yet at the end of the day, all of this processing usually does not lead to a robust solution, especially if the imaging environment is not tightly controlled. The Canny edge detector was designed to circumvent these problems and in many respects is the optimal classical edge detector. Compared to the edge detectors described thus far, it is a more complex and sophisticated algorithm that heavily processes the image and then dynamically selects a threshold based on the local properties of the image in an attempt to return only the true edges. The Canny edge detector proceeds in a series of stages[1]:

1. The input image is low-pass filtered with a Gaussian kernel. This smoothing stage serves to suppress noise, and in fact most edge detection systems utilize some form of smoothing as a pre-processing step.
2. The gradient images G_x and G_y are obtained using the Sobel operators as defined previously. The magnitude of the gradient is then calculated, perhaps using the absolute value approximation ($|G| = |G_x| + |G_y|$).
3. The direction of the individual gradient vectors, with respect to the x-axis, is found using the formula $\theta = \tan^{-1}(G_y/G_x)$. In the event that $G_x = 0$, a special case must be taken into account and θ is then either set to 0 or 90, as determined by inspecting the value of G_y at that particular pixel location. These direction vectors are then typically quantized into a set of eight possible directions (e.g. 0°, 45°, 90°, 135°, 180°, 225°, 315°).
4. Using the edge directions from (3), the resultant edges are "thinned" via a process known as non-maximal suppression. Those gradient values that are not local maxima are flagged as non-edge pixels and set to zero.
5. Spurious edges are removed through a process known as hysteresis thresholding. Two thresholds are used to track edges in the non-maxima suppressed gradient. In essence, edge contours are created by traversing along an edge until the gradient dips below a threshold.

The Intel IPP library includes a function `ippiCanny_16s8u_C1R` in its computer vision component that performs Canny edge detection. In addition, the library includes a variety of other derivative-based edge detectors, such as the 3x3 Sobel and Laplachian[2].

5.1.1   Edge Detection in MATLAB

The Image Processing Toolbox includes a suite of edge detectors through its `edge` function. This functionality allows one to specify any of the derivative filters discussed in the preceding section, as well as a few other variants on the general theme. The edge function accepts an intensity image and returns a MATLAB binary image, where pixel values of 1 indicate where the detector located an edge and 0 otherwise. An interactive demonstration of The MathWorks Image Processing Toolbox edge detection algorithms can be accessed via the edgedemo application.

Listing 5-1 is a port of Algorithm 5-1 that does not rely on any Image Processing Toolbox functions. The built-in MATLAB function filter2 is used to form the vertical and horizontal edge responses by convolving the input with the two Sobel kernels, and the gradient image is then computed using the absolute value approximation. The MATLAB statement

J = logical(zeros(size(G)));

creates a MATLAB binary matrix and initializes all matrix elements to false or zero. A simple thresholding scheme is then used to toggle (i.e. set to 1) those pixel locations found to be edges.

Listing 5-1: MATLAB Sobel edge detection function.

```
function J = sobel_edge_detect_fixed(I, T)

I = double(I);

% compute gradient image
Gx = filter2([1 0 -1;2 0 -2;1 0 -1], I, 'valid');
Gy = filter2([-1 -2 -1;0 0 0;1 2 1], I, 'valid');
G = abs(Gx) + abs(Gy);

% edge detection via thresholding on gradient image
J = logical(zeros(size(G)));
indices = find(G > T);
J(indices) = true;
```

The call to sobel_edge_detect_fixed in an industrial-strength system would more than likely be preceded by a low-pass filtering step. However, in actuality Listing 5-1 is not suitable for a real application, as it is incumbent upon the client to provide the threshold value that obviously plays a large role in determining the output (see Figure 5-5). Unless the image is known to be high-contrast and the lighting conditions are known and stable, it is difficult for any client of this function to provide a meaningful threshold value. The Image Processing Toolbox edge function circumvents this problem by employing a technique for estimating a

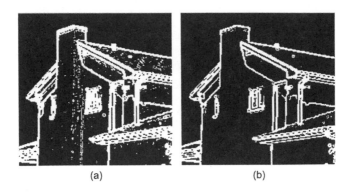

(a)   (b)

Figure 5-5. The effect of choosing different thresholds for use with
`sobel_edge_detect_fixed` on the house image. (a) Binary image returned using a
threshold of 50, clearly exhibiting the presence of noisy edges within the interior of the house.
(b) A much cleaner binary edge image, obtained with a threshold value of 100.

threshold value described by Pratt[4]. This technique uses an RMS estimate of
the noise in the input to generate a threshold value. In 5.2.5.1, we enhance
the Sobel edge detector in Listing 5-1 with additional functionality for
automatically determining an appropriate threshold.

Until now, all of the MATLAB code presented has been in the form of
functions residing in M-files. A MATLAB GUI application, modeled after
`edgedemo`, can be found in the `Chap5\edge\SobelDemo` directory on
the CD-ROM. This application was built with the MATLAB R14 GUIDE
user-interface designer tool[5]. The motivation behind developing and
explaining this application is not to wax poetic about developing GUIs in
MATLAB (although this functionality is extremely well done and quite
powerful), but to introduce a starting point for an image processing front-end
that will eventually be augmented with the capability of communicating with
a TI DSP using Real-Time Data Exchange (RTDX) technology. Figure 5-6
is a screen-shot of the application. The user imports an image into the
application, enters the threshold value used to binarize the processed image,
and finally calls `sobel_edge_detect_fixed` to run Sobel edge
detection on the image and display the output. If the input image is a color
image, it is converted to gray-scale prior to edge-detection. All code that
implements the described functionality does not rely on anything from the
Image Processing Toolbox, which is important for two reasons:

1. If the toolbox is not available, obviously an application using such
   functionality would not be very useful!
2. For Release 13 users, a MATLAB application that uses toolbox functions
   cannot be compiled via `mcc` (the MATLAB compiler[6]), to a native

application that can be run on a machine without MATLAB installed. With Release 14 of MATLAB, this restriction has been lifted.

Since this application has no dependencies on the Image Processing Toolbox, it does not use `imshow`, which for all intents and purposes can be replaced with the built-in MATLAB `image` function. The application uses a MEX-file, `import_grayscale_image.dll`, in lieu of the MATLAB `imread` function. This DLL serves as a gentle introduction to the topic MEX-file creation, which is covered extensively in Appendix A. Like most interpreted languages, MATLAB supports the idea of a plug-in architecture, where the developer can implement their own custom commands and hook them into the language. In the MATLAB environment, this facility is provided via such MEX-files. The MATLAB statement

    I = import_grayscale_image('some_file.bmp');

is roughly equivalent to

    I = imread('some_file.bmp');

except that `import_grayscale_image` automatically converts RGB color images to grayscale (8 bpp) and only supports a few color formats.

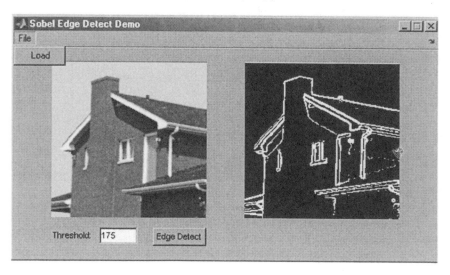

Figure 5-6. The MATLAB `SobelEdge` demo application.

This MEX-file uses the GDI+ library to parse and read in image files. However its usage is so similar to that of imread that if desired, imread can be used instead.

GUIDE creates an M-file and associated MATLAB figure (.fig) file for each user-defined dialog. This application, since it consists of a single dialog, is made up of three files: SobelDemo.fig, SobelDemo.m, and sobel_edge_detect_fixed.m. The figure file SobelDemo.fig contains information about the layout of the user interface components. The M-file SobelDemo.m has the actual code that provides the functionality for the GUI. To run the application, call the GUI entry-point function by typing SobelDemo at the MATLAB command-line prompt. Most of the implementation is created automatically by GUIDE, similar to how the Microsoft wizards in Visual Studio generate most of the glue code that give a GUI its innate functionality. The pertinent portions of SobelDemo.m are two callback functions, shown in listing 5-2.

Listing 5-2: SobelDemo.m callback functions.

```
function MenuFileLoad_Callback(hObject, eventdata, handles)

try
    % prompt user for input image filename
    [filename, pathname] = uigetfile( ...
        { '*.bmp;*.jpg;*.gif;*.tif', 'Images (*.bmp,*.jpg,*.gif, *.tif)';
          '*.bmp', 'Windows Bitmap Files (*.bmp)'; ...
          '*.jpg', 'JPEG Files (*.jpg)'; ...
          '*.gif', 'GIF Files (*.gif)'; ...
          '*.tif', 'TIFF (*.tif)'; ...
          '*.*', 'All Files (*.*)' }, ...
        'Select an image');

    if ~isequal(filename,0) & ~isequal(pathname,0)
        % Remember in C/C++, format strings use the \ character for
        % specifications (\t, \n, and so on) - so replace all
        % '\' with '\\' so as not to confuse the MEX function.
        imgpath = strrep(fullfile(pathname,filename), '\', '\\');
        I = import_grayscale_image(imgpath);
        % display input image
        subplot(handles.input_image);
        imagesc(I); axis off;
        set(handles.edge_detect_button, 'Enable', 'on');
    end
```

```
catch
   errordlg(lasterr, 'File I/O Error');
end


function edge_detect_button_Callback(hObject, eventdata, handles)

T = str2num(get(handles.threshold_edit, 'String'));

% get pointer to image data - could just have easily created another
% field in the handles struct, but since it is held by a child
% of the axes object, in the interest of reducing clutter I access
% it this way.
H = get(handles.input_image, 'Children');
I = get(H, 'CData');

% now pass I through the edge detector
try
   J = sobel_edge_detect_fixed(I, T);
   subplot(handles.edge_image);
   imagesc(double(J).*255);
   axis off;
catch
   disp(lasterr, 'Edge Detection Error');
end
```

The first MATLAB subfunction in Listing 5-2 gets executed when the user selects the **File|Load** menu item, and the second subfunction gets called when the user clicks the button labeled "Edge Detect". `MenuFileLoad_Callback` uses the built-in MATLAB function `uigetfile` to bring up a dialog prompting the user for the filename, and then uses `import_grayscale_image` to read the user-specified image file. After reading the image file, it then uses `image` to display the input in the left display. In `edge_detect_button_Callback`, the input pixel data is extracted from the left image display, and subsequently handed off for processing to `sobel_edge_detect_fixed`. Lastly, the processed binary output is shown in the right display.

The vast majority of the MATLAB code in this book should run just fine in Release 14 and Release 13 of MATLAB, and quite possibly earlier versions as well, although this has not been verified. However, this statement no longer holds true with applications built using GUIDE. Attempting to execute `SobelDemo` from MATLAB R13 results in an error emanating from the `SobelDemo.fig` figure file, therefore readers using older versions of MATLAB should recreate the dialog shown in Figure 5-6 using the older version of GUIDE.

If the MATLAB compiler (mcc) is available, this GUI application can be compiled to a native executable which can be run like any other program, outside of the MATLAB environment and even on computers that do not have MATLAB installed. These PCs will still need the MATLAB runtime redistributables, however – consult the MATLAB documentation for further details. The Release 14 command to compile SobelDemo.m and create a Windows executable SobelDemo.exe is

```
mcc -m SobelDemo
```

Under MATLAB R13 the correct command is

```
mcc -h -B sgl SobelDemo
```

5.1.2 An Interactive Edge Detection Application with MATLAB, Link for Code Composer Studio, and RTDX

Previous embedded DSP applications in this book either used CCStudio file I/O facilities (accessed via various menu options within the IDE) to feed data to and from the DSP, or image data was hard-coded into an array at compile-time by placing the actual pixel values in C header files. Obviously, both of these methodologies do not lend themselves to enabling any sort of interactive prototyping or proof-of-concept type applications. To that end, TI provides a technology called Real-Time Data Exchange (RTDX)[8,9]. With RTDX, developers gain continuous and real-time visibility into applications running on the DSP. Data can be transferred from applications running on a *host* PC down to an embedded *target* DSP, and vice-versa.

Figure 5-7 illustrates just how RTDX fits into the loop. In our application, the host client is a MATLAB application that hooks up to CCStudio, also running on the host. CCStudio in turn links up with the target, where the DSP is executing a program. This figure brings out a key point with respect to RTDX, namely that the CCStudio application must be running on the host in order for RTDX to function. In Chapter 6, another TI technology, HPI, is used to accomplish much the same thing except that the services of the CCStudio IDE will no longer be required. Also note that RTDX data transfer pipes are half-duplex in that they are unidirectional in nature. Two-way (or full-duplex) data transfer can be implemented using two separate RTDX pipes and configuring them appropriately.

Figure 5-7 also shows a Component Object Model (COM) interface between the host client application and CCStudio. COM is a Microsoft Windows technology for building applications from binary software components enabling a "plug-in" architecture. It is COM that serves as the underpinnings of the Internet Explorer (IE) web browser, so that various software vendors can develop components that seamlessly integrate into the browser. As such, utilizing COM on the host for RTDX is a natural choice as it enables developers to use RTDX services from within a myriad of contexts: custom C/C++ applications (a route taken in 5.2.5.2), Visual Basic, or even Excel (by using Visual Basic for Application, or VBA, to bring data

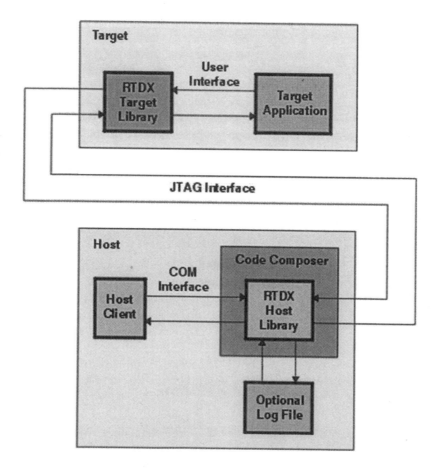

Figure 5-7. Interplay between RTDX and various components on the host and target devices. Figure reprinted courtesy of Texas Instruments.

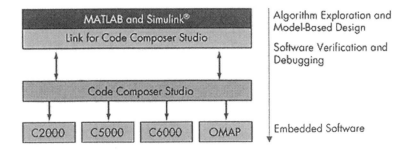

Figure 5-8. Using Link for Code Composer Studio to communicate with an embedded DSP target from within MATLAB. The Link for Code Composer Studio Development Tools includes a set of functions and structures that call upon various TI host utility libraries to shuttle data between CCStudio and MATLAB. Figure reprinted courtesy of The MathWorks.

in from an RTDX pipe into an Excel worksheet). In this case the host client application takes the form of a MATLAB program, but how does MATLAB initiate the connection to RTDX over COM?

MATLAB offers built-in support so that it can act as a COM client, or alternatively one could develop a MEX-file in C/C++ that uses the TI RTDX COM object directly and provides a MATLAB interface to it. However, The MathWorks has already developed the Link for Code Composer Studio[10] to facilitate precisely this sort of communication with a TI DSP development platform, and which encapsulates the TI COM object in a fashion amenable to MATLAB developers. Figure 5-8 shows where this toolbox resides within the hierarchy of a host-target MATLAB application.

This product provides a suite of MATLAB functions for communicating with CCStudio, transferring information back and forth via RTDX, and extracting information about data and functions residing on the target. The product is at the core of the host application presented in this section, for without it communication between the host and target becomes a painful process. Unfortunately we will get a feel for this pain in 5.2.5.2 when we implement a subset of this functionality, without the services of MATLAB in a C/C++ Visual Studio application. While it is theoretically possible to implement a MEX-file that provides a MATLAB wrapper around the TI COM objects, implementing and debugging such a beast is not for the faint of heart. In addition to the RTDX communication functionality featured here, the Link for Code Composer Studio also provides a wealth of additional tools for instrumenting DSP code and running hardware-in-the-loop or processor-in-the-loop simulations.

The Sobel edge detection application featured in this section consists of two parts, the MATLAB front-end host that provides user input and data visualization, and the DSP back-end target that runs the image processing algorithm. The MATLAB GUI front-end looks the same as its stand-alone predecessor and its implementation can be found on the CD-ROM in the directory Chap5\edge\SobelFixedRTDX\Host. Two target DSP projects are provided, one for the C6416 DSK and another for the C6701 EVM. While the code for the MATLAB host remains the same no matter which target is used (except for the code that loads the target project), the implementation for each of the two targets differs slightly.

To run the application one first needs to modify the MATLAB host (Chap5\edge\SobelFixedRTDX\Host\SobelDemo.m) to point to the location of the desired target COFF executable file. Starting the GUI by issuing the command SobelDemo initializes the DSP by invoking the CCStudio IDE, connecting to it, loading the project file and associated binary executable, and finally running the program on the DSP. All of this is accomplished using various commands provided by the Link for Code Composer Studio. If the CCStudio IDE is already running, there is no need to close it down prior to starting SobelDemo, simply ignore any warnings that appear in the MATLAB command window. After initialization, from the point-of-view of the end-user, the entire application operates in the same manner as before. Of course, the major internal difference is that a DSP is doing the processing instead of sobel_edge_detect_fixed.m. The embedded DSP program meanwhile sits in a continual loop, eagerly awaiting image data from the host PC. When it receives a threshold and image via RTDX, the program proceeds to use IMGLIB functions to locate the edges in the image.

The protocol between the host and target is shown in Figure 5-9. The receipt of a single integer value (the threshold) is a signal to the target that it should expect the image data to immediately follow. After this data is received and the image processed, the target sends the processed pixels back to the host. As shown in Figure 5-9, the manner in which data is sent from the host and is received from the target differs. From the target's vantage point, reading large blocks of data is less problematic than writing blocks data. An entire image can be sent from the host to the target in one fell swoop, but unfortunately that is not the case in the opposite direction. In fact, the processed image has to be sent one row at a time, and for the C6701 EVM a pause in the writing must be inserted to give the host a chance to catch its breath, so to speak.

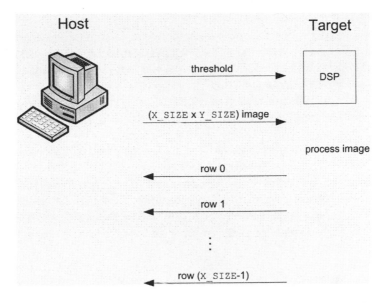

Figure 5-9. Data flow protocol between the Sobel edge detection host and embedded DSP target doing the actual image processing.

This particular data transfer protocol assumes a fixed image size, because the image dimensions are never transferred between the host and target. In this application, the image dimensions are X_SIZE (rows) and Y_SIZE (columns), both of which are set to 256 by default in the image.h header file. As a consequence, the host uses the imresize Image Processing Toolbox function to resize the input image to be of dimensions 256x256, prior to sending it to the target[11].

5.1.2.1 DSP/BIOS

Until this point, all CCStudio projects have been fairly bare-bones, eschewing a formal operating system, using manually crafted linker command (.cmd) files, and configuring peripherals manually through the CSL API (e.g., DMA and EDMA in 4.3.4). The CCStudio RTDX tutorial and sample projects suggest it is possible to continue along this route – however it is the author's experience that when used in the context of transferring large amounts of data, RTDX works far more reliably when used in conjunction with the TI real-time operating system, DSP/BIOS[12]. In fact, some of the very simple C6701 EVM tutorial samples do not work at all without recreating them to incorporate DSP/BIOS. DSP/BIOS is a fully preemptive real-time operating system (RTOS), featuring:

- Real-time scheduling based on a preemptive thread management kernel.
- Real-time I/O facilities, like RTDX, that engender two-way communication between different threads or tasks running on the target, or between the target and host PC.
- Real-time analysis tools, such as Software Logic Analyzer, Message Log, CPU Load Graph and Real-Time Profiler.

The DSP/BIOS Real-Time OS is scalable, meaning that the developer has the ability to reduce its overall footprint by using only what is required for a particular application. This type of scalability is enormously important in embedded systems where resources come at a premium. DSP/BIOS is configured using a graphical interface from within CCStudio, and the steps to add DSP/BIOS support to an existing or new project are enumerated below:

1. Select **File|New|DSP/BIOS Configuration**, and choose the appropriate base seed (e.g. C6701.cdb for the C6701 EVM or C64xx.cdb/dsk6416.cdb for the C6416 DSK).
2. One normally then proceeds to configure the chip's memory map (which we have previously done using linker command files), although this is not necessary if external RAM is not going to be utilized. We do this by adding one or more memory blocks that point to external RAM. The "System" DSP/BIOS configuration tree contains four items: the Memory Section Manager, Buffer Pool Manager, System Settings, and Module Hook Manager. After expanding the System tree, right-clicking on the Memory Section Manager shows an option to insert a "MEM" module. Right-clicking on this new module allows one to rename it and also set its properties, which should normally mirror what goes into the linker command file. For example, in the case of the C6416 DSK, one would add a 16 MB segment conforming to the addresses specified in [13], as shown in Figure 5-10. The base address and length may be garnered from working linker command files. Data buffers can now be mapped in the source code to this segment using the familiar DATA_SECTION pragma, where the identifier specified in the pragma declaration would refer to the name specified in the DSP/BIOS MEM configuration. If 6416dsk.cdb was chosen as the base seed configuration, then this step is not required as the DSP/BIOS configuration already includes an SDRAM MEM module.
3. After the memory configuration has been set, save the DSP/BIOS configuration to a .cdb file, and then add this file to the CCStudio project.

4. The act of saving the DSP/BIOS configuration in step 3 also results in the creation of a linker command file, which should never be hand-modified by the user, as any such changes will be overwritten by the DSP/BIOS configuration tool should the configuration change in the future. Anything that can be done by manually altering the linker command file can be accomplished in one form or another through the DSP/BIOS configuration tool. The newly generated linker command file must also be added to the project.

There is no longer any need to explicitly link to the runtime-support library (`rts6400.lib` for the C6416 DSK or `rts6701.lib` for the C6701 EVM) for DSP/BIOS projects. In addition, it is no longer required to call `DSK6416_init` (for C6416 DSK projects) or `evm_init` for C6x EVM projects), as the OS now takes care of properly initializing the development board.

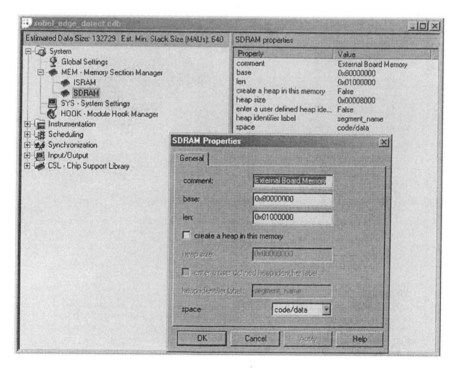

Figure 5-10. Using the DSP/BIOS configuration utility to set up the memory map for the C6416 DSK. In this example the SDRAM external memory section is configured.

5.1.2.2 C6416 DSK Target

The C6416 DSK Sobel project is located in the Chap5\edge\SobelFixedRTDX\ Target\C6416DSK directory. In general, RTDX works more reliably on this development platform, as compared to the C6701 EVM. Prior to using RTDX in a project, it must be correctly configured, with one of the main settings being the maximum size of the buffers used internally by the RTDX system. The online tutorial states that to change the buffer size a copy of the rtdx_buf.c file should be modified and explicitly linked into the project. When using DSP/BIOS, this inconvenience is avoided and the DSP/BIOS configuration tool should be used instead. There is a separate RTDX section from within DSP/BIOS, under the "Input/Output" tree. Expanding this tree, and right-clicking on the RTDX icon, exposes a dialog where various RTDX settings are made. This dialog is shown in Figure 5-11.

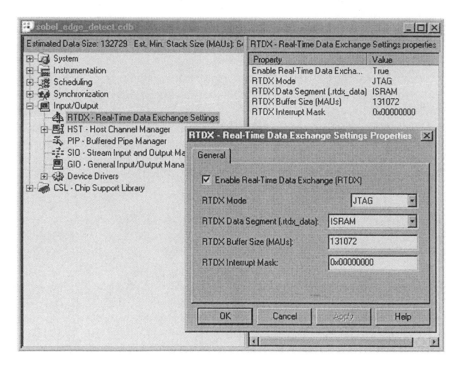

Figure 5-11. Configuring RTDX for the C6416 DSK target project, using the DSP/BIOS configuration tool.

Since this project defines both X_SIZE and Y_SIZE as 256, and the pixels are stored in an 8 bpp format, the RTDX buffer size should really only need to be 65536 (64K) bytes; however, given the vagaries of RTDX, the bigger the better, so long as the buffer can be stored in internal on-chip RAM (for performance reasons). Experiments show that doubling the RTDX buffer size increases overall system reliability, hence for the C6416 target the buffer size is set to 128K, which still fits into the C6416's internal chip RAM because of that processor's expanded memory (as compared to the older C62x and C67x DSPs).

For the sake of simplicity, this project, as well as its close C6701 cousin, does not utilize the paging and DMA optimizations described in 4.3.5 and 4.4.2, and all image buffers are placed in external RAM. Listing 5-3 is the full contents of the C6416 target source file sobel_edge_detect.c, which implements the protocol described in Figure 5-9 and of course the Sobel edge detection image processing functionality.

Listing 5-3: sobel_edge_detect.c

```c
#include <rtdx.h>    /* target API */
#include <stdio.h>
#include <IMG_sobel.h>
#include <IMG_thr_le2min.h>
#include "target.h" /* defines TARGET_INITIALIZE() */
#include "image.h"  /* dimensions */

RTDX_CreateInputChannel(ichan); /* input,T come down this pipe */
RTDX_CreateOutputChannel(ochan); /* processed data via this pipe */
#pragma DATA_SECTION(img_buf1, "SDRAM");
#pragma DATA_ALIGN (img_buf1, 8);
unsigned char img_buf1[N_PIXELS];
#pragma DATA_SECTION(img_buf2, "SDRAM");
#pragma DATA_ALIGN (img_buf2, 8);
unsigned char img_buf2[N_PIXELS];

static void sobel_edge_detect(unsigned char T)
{
  /* input & output buffers for IMG_thr_le2min may NOT
   * alias (see IMG_thr_le2min docs), thus operation
   * cannot be done "in-place". */
  IMG_sobel(img_buf1, img_buf2, Y_SIZE, X_SIZE);
  IMG_thr_le2min(img_buf2, img_buf1, Y_SIZE, X_SIZE, T);
}
```

```c
void main()
{
  int status, T, ii;

  TARGET_INITIALIZE();
  while (!RTDX_isInputEnabled(&ichan))
    ;/* wait for channel enable from MATLAB */
  while (!RTDX_isOutputEnabled(&ochan))
    ;/* wait for channel enable from MATLAB */
  printf("Input & Output channels enabled ...\n");

  while (1) {
    /* wait for the host to send us a threshold */
    if (sizeof(T) != (status = RTDX_read(&ichan, &T, sizeof(T))))
      printf("ERROR: RTDX_read of threshold failed!\n");
    else
      printf("Received threshold (%d) from host\n", T);

    /* now we're expecting X_SIZE x Y_SIZE worth of image data */
    if (N_PIXELS != (status=RTDX_read(&ichan,
                                      img_buf1,
                                      N_PIXELS))) {
      printf("ERROR: RTDX_read of image failed (%d)!\n", status);
      exit(-1);
    }
    printf("Received %dx%d image\n", X_SIZE, Y_SIZE);

    /* process the image */
    sobel_edge_detect((unsigned char)T);

    /* send it back to host */
    printf("Sending processed image data back to host ...\n");
    for (ii=0; ii<X_SIZE; ++ii) {
      /* write one row's worth of data */
      if (!RTDX_write(&ochan, img_buf1+ii*Y_SIZE, Y_SIZE)) {
        printf("ERROR: RTDX_write of row %d failed!\n", ii);
        exit(-1);
      }
    } /* end (for each row) */
  } /* end (while forever) */
}
```

This program sits in a continual loop, waiting for data to jump-start the processing. The receipt of a single integer value, containing the threshold used during segmentation of the edge enhanced image, is a signal to the target that the host will be sending down pixel data for processing. The blocking function (meaning it does not return until the operation has completed) `RTDX_read` is used to read data from an RTDX "channel", or pipe. There is a non-blocking version of `RTDX_read`, called `RTDX_readNB` which returns immediately. When using this function, it is then the responsibility of the application to poll RTDX via the `RTDX_channelBusy` macro to wait for the read operation to complete. In this fashion, the DSP would be free to perform other useful work while the RTDX system does the grunt work of transferring data to or from the host.

After receiving the threshold value from the host, the target reads the entire image in one fell swoop with another call to `RTDX_read`. Once this invocation of `RTDX_read` returns, the target has all the data it needs to process the image. The actual image processing occurs within `sobel_edge_detect`, where the image is passed through 3x3 Sobel vertical and horizontal kernels using the IMGLIB function `IMG_sobel` and the edge enhanced image is formed by combining the intermediate results using the absolute value approximation to the gradient. `IMG_sobel` is implemented in such a manner that the first and last columns of the output are invalid, and the output is two rows shorter than the input. This implementation detail is utilized by the host application, which zeros out those portions of the returned image prior to display.

Once the gradient image is available, the actual edge detection is carried out by thresholding the gradient image using another IMGLIB function, `IMG_thr_le2min`. This function is similar to the MATLAB Image Processing Toolbox function `im2bw`[14], except that it zeros out those pixels that lie below a given threshold – the remaining pixels are left unmodified, whereas with `im2bw` all non-zero pixels are set to 1. IMGLIB provides four variants of the basic threshold operation (`IMG_thr_gt2max`, `IMG_thr_gt2thr`, `IMG_thr_le2min`, `IMG_thr_le2thr`) that allow one to set pixels above or below a given threshold value to either an extreme value (0 or 255) or a specified intensity. All four of these IMGLIB thresholding functions leave the pixels that do not meet the threshold criterion untouched, which has ramifications on our host application because it displays the processed output as a binary image.

A significant difference between this code and previous TI implementations in this book is in that the input image buffer is actually overwritten during the image processing. As is evident from a close inspection of `sobel_edge_detect`, the output of `IMG_sobel` is placed into `img_buf2`; `img_buf2` is then the input to `IMG_thr_le2min`, which outputs the thresholded image back into `img_buf1`. The reason for this is that `IMG_thr_le2min`, for performance reasons, dictates that its two input buffers not alias – that is, point to the same memory location. One way of circumventing this restriction would be to use three image buffers, which of course comes at the cost of a larger memory footprint. Yet if this edge detection were part of a larger system requiring access to the original pixel data further downstream the processing chain, then such a strategy would have to be used.

After the edge detection runs its course, all that remains is to send the pixel data back to the host. In contrast to reading image data from an RTDX input channel, writing the same amount of data in a single block is more problematic. Using a single RTDX call to write the processed data to the output channel in a single contiguous block should look like `RTDX_write(&ochan, out_img, N_PIXELS)`. Unfortunately, that function call does not work as advertised. In fact, splitting the target-to-host image transfer into two operations (sending the first half, followed by the second half) also fails to work reliably. It was found that the most robust and reliable means of performing the data transfer is to send the data on a row-by-row basis, which is the reason for the final loop in `sobel_edge_detect.c`.

5.1.2.3 C6701 EVM Target

A target project meant to be used with C62xx or C67xx EVM boards can be found in `Chap5\edge\SobelFixedRTDX\Target\C6701EVM`. This project has been tested on the C6701 EVM, and is quite similar to the C6416 DSK project. Since the C6701 EVM's memory map differs from that of the C6416 DSK, the DSP/BIOS MEM module configuration reflects these differences, which can be discerned by comparing two linker command files for each development platform. The RTDX configuration also differs, in that the `.rtdx_data` segment (see Figure 5-11) is mapped to external RAM due to the smaller amount of space available on the C62xx/C67xx DSPs, as compared to the more powerful C6416. Such a configuration affects RTDX throughput, because of the latencies involved in accessing the external memory. Due to these timing changes, the target application does need to change to account for this difference. Moreover, on the C62xx/C76xx EVM platform setting the RTDX buffer size to values greater than 64K results in RTDX completely failing to function correctly, and thus in this project the buffer size is set to 65536 (versus 128K as was done in the C6416 version).

The timing changes manifest themselves in a change to how the processed pixel data is written back to the host client application. Basically, a failure to insert a pause when writing the image rows to the RTDX output channel `ochan` results in `RTDX_write` returning an error just prior to sending the last few rows of image data. Therefore, an artificial pause is inserted after having sent half of the image rows, in order to give the host client application time to read some of the image data and flush out the RTDX buffers. A better means would be to somehow query the RTDX state programmatically, and temporarily stop writing data to the output channel until it has been determined that the client has had a chance to read enough of the data so that subsequent calls to `RTDX_write` succeed. Unfortunately, there appears to be no easy way of implementing such a strategy.

The question now becomes how exactly to implement this pause? On POSIX and Windows systems there is a function `sleep` (UNIX) or `Sleep` (Win32) that can be used to suspend a running process. These operating systems are not real-time, and thus one is guaranteed only that the suspended process shall be halted for *at least* the amount of time specified in the `sleep` or `Sleep` call. It is entirely possible that the suspended process will be halted for longer than that amount of time. DSP/BIOS provides a function `TSK_sleep` which fits the bill, suspending the currently running *task* for "exactly" the specified amount of time, where "exactly" depends on the clock resolution and whether there are higher priority tasks waiting to run after the task in question is ready to resume. None of our embedded DSP code thus far has utilized separate tasks, rather the programs have been of a fairly simple variety and called a few functions from `main`, or as in the previous program's case, executed in an infinite loop within `main`. In the C6701 EVM Sobel project, the DSP/BIOS configuration tool is used to create a task which hosts this infinite loop. This task can then be suspended at the right time in order to allow the host client application time to catch up.

Figure 5-12 is a screen-shot from the DSP/BIOS configuration tool illustrating how to create a new task and incorporate it into a project. The entry-point function, where a DSP/BIOS task begins its life, is specified in the "Task function" field, and the function name (if it is implemented in C) must be preceded by an underscore, otherwise the build fails at link time because the linker decorates all C function names with a preceding underscore.

Listing 5-4 is a portion of the C6701 `sobel_edge_detect.c` source code (the remainder is identical to the C6416 version), showing that `main` is now mostly empty. The DSP/BIOS kernel starts the new `processingTsk` task at startup, which sends the program flow into the `processing_task` function. Most tasks follow a simple motif: an infinite loop that performs some processing or executes a series of commands when asked to from some outside entity. This task is no different, and in fact its internals are quite similar to that of the `main` function in Listing 5-3, except for the portion of code that sends the processed data back to the host client. After the program

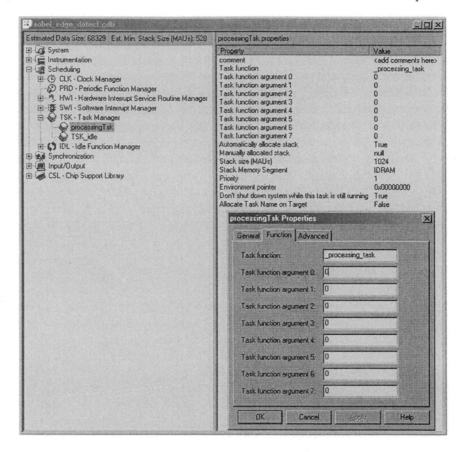

Figure 5-12. Using DSP/BIOS to add a new task to a project. The other tabs in the task properties dialog have settings for a variety of other options, such as the location of the task's stack and its size and the task priority.

has transferred half of the processed edge image rows, `TSK_sleep` is called to suspend the task for 1000 clock ticks, thereby giving the client the opportunity to read the data the target has already written and flush the RTDX buffers. If there was another lower-priority task waiting to run, the DSP/BIOS scheduler would allow this task to execute during `processingTsk`'s suspension. Such determinism is in stark contrast to general-purpose non-real-time operating systems like Windows or Linux. In those operating systems, the scheduler is typically of a "round-robin" nature, and the scheduler's charter is to guarantee fairness so that no process is starved of CPU time, as opposed to a hard real-time system like DSP/BIOS where the overriding goal is to ensure deterministic behavior.

Listing 5-4: The C6701 EVM version of `sobel_edge_detect.c`. This version differs from that of Listing 5-3 in that it uses a DSP/BIOS task, and during the write operation this task is suspended so that the RTDX operation succeeds.

```
void main()
{
  printf("Sobel Edge Detector started.\n");
}


void processing_task()
{
  int status, T, ii;

  TARGET_INITIALIZE();
  while (!RTDX_isInputEnabled(&ichan))
    ;/* wait for channel enable from MATLAB */
  while (!RTDX_isOutputEnabled(&ochan))
    ;/* wait for channel enable from MATLAB */
  printf("Input & Output channels enabled ...\n");

  for (;;) {
    /* wait for the host to send us a threshold */
    if (sizeof(T) != (status = RTDX_read(&ichan, &T, sizeof(T))))
      printf("ERROR: RTDX_read of threshold failed!\n");
    else
      printf("Received threshold (%d) from host\n", T);

    /* now we're expecting X_SIZE x Y_SIZE worth of image data */
    if (N_PIXELS != (status = RTDX_read(&ichan,
                                        img_buf1,
                                        N_PIXELS))) {
      printf("ERROR: RTDX_read of image failed (%d)!\n", status);
      exit(-1);
    }
    printf("Received %dx%d image\n", X_SIZE, Y_SIZE);

    /* process the image */
    sobel_edge_detect((unsigned char)T);

    /* send it back to host */
    printf("Sending processed image data back to host ...\n");
```

```
for (ii=0; ii<X_SIZE; ++ii) {
  /*
   * halfway through the xfer, pause a bit to give
   * host a chance to read out some of the buffer.
   */
  if ((X_SIZE>>1)==ii) TSK_sleep(1000);

  /* write one row's worth of data */
  if (!RTDX_write(&ochan, img_buf1+ii*Y_SIZE, Y_SIZE)) {
    printf("ERROR: RTDX_write of row %d failed!\n", ii);
    exit(-1);
  }
} /* end (for each row) */
} /* end (for ;;) */
}
```

5.1.2.4 Host MATLAB Application

The host application that provides a front-end user interface for both targets takes the form of a modified version of the MATLAB Sobel demo application, and is located in `Chap5\edge\SobelFixedRTDX\Host`. As before, the MATLAB source code for the application resides in `SobelDemo.m`. The GUI behavior remains the same, but underneath MATLAB makes a connection to Code Composer Studio via the Link for Code Composer Studio, and subsequently uses this connection to feed data to the target and read processed data from the target over RTDX channels.

Depending on the target, the only change that needs to be made to the target source code is to modify a single string variable that points the application to a directory on the host from which the COFF binary file is loaded onto the DSP. This type of initialization code is usually placed within a startup subfunction that the GUIDE tool automatically creates whenever one builds a MATLAB GUI application. Listing 5-5 shows the contents of this subfunction, `SobelDemo_OpeningFcn`.

Listing 5-5: MATLAB GUI initialization code, which establishes a connection to CCStudio and configures RTDX for image data transfer.

```
% --- Executes just before SobelDemo is made visible.
function SobelDemo_OpeningFcn(hObject, eventdata, handles, varargin)
% This function has no output args, see OutputFcn.
% hObject    handle to figure
% eventdata  reserved - to be defined in a future version of MATLAB
% handles    structure with handles and user data (see GUIDATA)
```

```
% varargin   command line arguments to SobelDemo (see VARARGIN)

% Choose default command line output for SobelDemo
handles.output = hObject;

% initialize main window
subplot(handles.input_image);
colormap(gray(255)); image(zeros(200)); axis off;
subplot(handles.edge_image);
colormap(gray(255)); image(zeros(200)); axis off;

% Code Composer Link initialization (load & run project)
try
  handles.cc = ccsdsp('boardnum', 0); % connect to target
  set(handles.cc,'timeout', 3); % set the CCSDSP timeout to 3s
  visible(handles.cc,1); % Code Composer Studio available to user
  target = 'C6416DSK'; % either 'C6416DSK' or 'C6701EVM'
  projfile = [pwd '\..\Target\' target '\sobel_edge_detect.pjt'];
  projpath = fileparts(projfile);
  open(handles.cc, projfile);
  cd(handles.cc, projpath);
  build = 'Debug'; % Debug or Release
  load(handles.cc, [build '\sobel_edge_detect.out']);
  goto(handles.cc, 'main');

  % Initialize RTDX
  handles.image_size = [256 256]; % target expects image of this size
  buflen = prod(handles.image_size);
  if buflen < 1024
    buflen = 1024; % min RTDX buflen is 1024 bytes
  end
  handles.cc.rtdx.configure(buflen, 2);
  handles.cc.rtdx.open('ichan', 'w'); % open write channel
  handles.cc.rtdx.open('ochan', 'r'); % open read channel

  % Enable RTDX
  enable(handles.cc.rtdx);
  if 0 == isenabled(handles.cc.rtdx)
    error('Failed to enable RTDX.');
  end
  set(handles.cc.rtdx,'timeout', 3); % set RTDX timeout value to 3s
  run(handles.cc); % run target
```

```
% ENABLE channels
enable(handles.cc.rtdx, 'ichan');
if 0 == isenabled(handles.cc.rtdx, 'ichan')
  error('Failed to enable target write channel.');
end
enable(handles.cc.rtdx, 'ochan');
  if 0 == isenabled(handles.cc.rtdx, 'ochan')
     error('Failed to enable target read channel.');
  end
catch
  errordlg(lasterr, 'CCS Link Error');
  delete(hObject);
end


% Update handles structure
guidata(hObject, handles);
```

The handles data structure is where common data, used throughout the GUI, should be placed so that all callback subfunctions have access to it. Most importantly for this discussion, handles.cc is a Link for Code Composer Studio object through which a connection is made to CCStudio and RTDX data transfers take place. After the binary executable is loaded onto the target DSP by invoking various toolbox functions, the host side of RTDX is configured and enabled. After the size of the host RTDX buffer is set, the host RTDX channels are then configured. The open method creates a read-only or write-only RTDX channel and associates it with the name given as the first argument. The naming conventions used here are based on the target's point-of-view; that is, the channel over which MATLAB sends data to the DSP is given the name 'ichan', because with respect to the target this is the input channel. Likewise, the reverse direction (MATLAB reading processed data emanating from the target) is given the name 'ochan'.

The other area in the host application where the Link for Code Composer Studio is utilized is in the updated callback for the "Edge Detect" button, shown in Listing 5-6. Because the target is expecting a 256x256 image, the image the user initially loaded in from the File menu is resized to those dimensions using imresize prior to sending it down to the target.

Listing 5-6: MATLAB callback subfunction for the "Edge Detect" button, which sends an image down to the target and then reads and displays the processed image.

```
% --- Executes on button press in edge_detect_button.
function edge_detect_button_Callback(hObject, eventdata, handles)
% hObject    handle to edge_detect_button (see GCBO)
% eventdata  reserved - to be defined in a future version of MATLAB
% handles    structure with handles and user data (see GUIDATA)

T = str2num(get(handles.threshold_edit, 'String'));

% get pointer to image data - I could just have easily created
% another field in the handles struct, but since it is held by a
% child of the axes object, in the interest of reducing clutter I
% access it this way.
H = get(handles.input_image, 'Children');
I = get(H, 'CData');

% now pass image I through the edge detector on the DSP
try
  set(gcf, 'Pointer', 'watch');
  % send threshold T over to the target DSP
  if iswritable(handles.cc.rtdx,'ichan'),
    writemsg(handles.cc.rtdx,'ichan', int32(T));
  else
    error('ichan not writable!');
  end

  % if the image isn't of size 256x256, resize it
  I = imresize(I, handles.image_size, 'bicubic');

  % target is now waiting for the image data
  % remember MATLAB is column-major but C is row-major
  J = I';
  writemsg(handles.cc.rtdx, 'ichan', uint8(J(:)));

  % let the DSP process it, and read the data back
  % enable read channel, and verify
  enable(handles.cc.rtdx,'ochan');
  if 0 == isenabled(handles.cc.rtdx,'ochan')
    error('Failed to enable read channel.');
  end
```

```
% Query for number of available messages
num_of_msgs = msgcount(handles.cc.rtdx,'ochan');

% Read messages
J = zeros(handles.image_size(1), handles.image_size(2));
n_recvd = 1;
while (handles.image_size(1)-n_recvd)>1
  for msg=1:num_of_msgs
    % read a single row and place it in the image matrix
    J(n_recvd,:) = readmsg(handles.cc.rtdx, 'ochan', 'uint8');
    n_recvd = n_recvd + 1;
  end
  num_of_msgs = msgcount(handles.cc.rtdx, 'ochan');
end
% 1st and columns, and last 2 rows contain meaningless data
J(:,1) = 0; J(:,handles.image_size(2)) = 0;
J(handles.image_size(1)-1,:) = 0;
J(handles.image_size(1),:) = 0;
J = logical(J); % target doesn't return a binary image

subplot(handles.edge_image);
imagesc(double(J).*255);
axis off;

catch
  set(gcf, 'Pointer', 'arrow');
  errordlg(lasterr, 'Edge Detection Error');
end

set(gcf, 'Pointer', 'arrow');
```

Because RTDX can accommodate a 256x256 byte host-to-target transfer, a single call to the writemsg Link for Code Composer function suffices to send pixel data to the target. However, care must be taken to transpose the matrix prior to flattening the two-dimensional array due to the differences in data layout between MATLAB and the target. Recall that in MATLAB, multi-dimensional arrays are stored in column-major format while the C language is row-major in nature. In addition, an explicit cast to the uint8 data type is necessary to guard against the case where J is not of type uint8, in which case the result would be extraneous data sent to the target.

For reasons discussed previously, reading the processed image data from the target is more involved, since it cannot be performed with a single function call. Once the edge detection completes, the target sends back 256 rows, or more strictly speaking 256 RTDX *messages*. Reading all of the processed data from the target is accomplished via a while loop that queries RTDX via `msgcount` for how many messages are sitting in a FIFO (first-in, first-out), and then comparing this count to how many image rows are needed to completely fill an image matrix. There is some post-processing of the data required due to the implementation details behind the IMGLIB functions used to implement Sobel edge detection on the target. The first and last columns are meaningless, and are therefore set to zero. Similarly, the valid image data is actually two rows less than the original image, and so those rows are also set to zero. Finally, the target does not send back a binary image per se, but rather the way the target is implemented any non-zero pixel should be interpreted as an edge, with a binary value of 1 or true. The built-in MATLAB function `logical` is used to transform the edge image into a binary image consisting of just ones and zeros.

5.1.2.5 Ideas for Further Improvement

The motivation behind this project was two-fold: first, to show how edge detection can be easily implemented on TI DSPs using IMGLIB functions, and second, to introduce RTDX and present a MATLAB host application that can be used as a template for a complete interactive image processing application. That being said, there remains much room for improvement with regards to this particular application, in terms of functionality as well as performance.

With regards to functionality, the restriction of a fixed image size is rather limiting. A more useful application would allow the user to read in any image and feed it to the DSP to be processed, as opposed to forcing a 256x256 size on the user. Of course, if the edge detection is part of a larger system (say for example a video surveillance product) the image (or individual video frames, as the case may be) is always going to be a certain size and this point is moot. But for testing purposes, a useful addition would be to change the protocol between the host and target such that the host sends the target the image dimensions, and then the target adjusts its internal algorithms appropriately. There are some limitations that must be always be accounted for, such as a maximum image size and power-of-2 considerations for the various IMGLIB functions.

The target performance can be dramatically improved by utilizing paging and DMA to send image blocks into internal chip memory. This technique has been described in full detail in Chapter 4, and here the DSP/BIOS DMA module would prove useful – a DMA Application Programming Interface (API) within DSP/BIOS exists that roughly mirrors that of the CSL library.

Finally, collapsing the Sobel edge detection and image thresholding into a single loop will further increase performance. Currently, the edge image is formed by first calling IMG_sobel and following that with a call to IMG_thr_le2min. By performing the filtering and thresholding within a single nested loop, a la Algorithm 5-1, the looping overhead is reduced. To garner maximal performance out of the algorithm, the loop kernel would need to be fully optimized, which has already been done separately for us in the two IMGLIB functions. More information on how to implement an efficient threshold loop kernel on the C64x platform can be found in [15].

5.2 SEGMENTATION

The goal of image segmentation is to divide an image into constituent parts that correlate to objects within the image. Once these objects have been extracted from the scene, information about them, such as their location, orientation, area, and so on may be gleaned and used towards a specific application. Typically this involves passing such information back to some higher-level processing code that then uses this data to make decisions pertaining to the application at hand.

Image segmentation is a vast topic, however the classes of algorithms fall roughly under two categories: those that are based on abrupt changes in intensity and those where the image is divided into regions that are similar in accordance with a predefined criterion. Edge detection, the topic of 5.1, falls under the first category. Here we focus on the second category of segmentation algorithms, and in particular histogram thresholding. Histogram thresholding was introduced in 3.2 within the context of the window/level technique, and thresholding of the gradient was used in 5.1 to turn edge-enhanced images into binary images. In 3.2, a human operator was assumed to be in the loop to alter the window and level parameters. In this section, the focus is on autonomous methods of inspecting the histogram to choose an appropriate threshold value for use in image segmentation. In addition, the techniques developed here will also be used to improve the Sobel edge detection algorithm (see Algorithm 5-1).

Image segmentation applications are too numerous to list exhaustively. Some noted examples include:

- X-ray luggage inspection: Inspection of carry-on luggage is an essential airport security measure, and various image processing techniques are used to aid the screener in interpretation of these images. Automatic or semi-automatic image processing systems are in use today to help combat concealment of weapons and dangerous items. Obviously, the images are quite complex but sophisticated segmentation algorithms are used to help separate certain dangerous materials from innocuous items.

- Microscopy: Segmentation is often employed to separate specimens and various structures from the background of images acquired from a microscope.
- Medical imaging: In computer-aided diagnosis, segmentation algorithms operating on radiographs or x-ray projection images are used for automatic identification of pathological lesions in the breast or lung.
- Industrial applications: Automated defect inspections of silicon wafers or electronic assemblies via image segmentation have found widespread use in the semiconductor and other industries.
- Retinal image processing: Interestingly enough, retinal image processing is an active topic of research, and segmentation plays an important role in two rather different camps. In biometric applications, various structures in the retina form a distinct signature along the lines of one's fingerprint. Thus, extraction of this signature aids immensely in automated identification of individuals. Along different lines, object extraction of the retinal blood vessel tree, the optic nerve, and various other structures in the human retina, is performed as part of image-guided laser treatments and computer-aided diagnosis. See Figure 5-13 for an example.

It should be noted that universal and foolproof segmentation algorithms, guaranteed to work on all images, is not a feasible design goal. It is almost always the case that some set of heuristics or assumptions, based on specific knowledge of the problem domain, is by necessity incorporated into the overall process. Complex scenes are extraordinarily difficult to segment, and the goal should be to tune and tweak segmentation algorithms so that they work robustly and reliably for images one expects to encounter within a specific application.

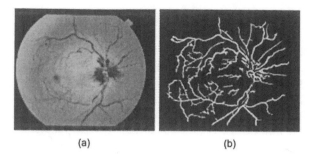

(a)   (b)

Figure 5-13. Blood-vessel segmentation of retinal image (images courtesy Adam Hoover of Clemson University and Michael Goldbaum of UC San Diego). (a) Green channel of an RGB retina image. (b) Segmented image, produced using the method reported in [16], that shows just the blood vessels.

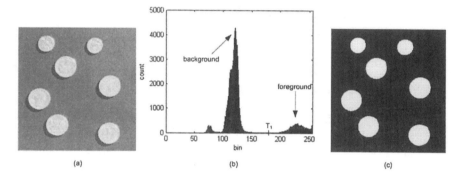

(a)   (b)   (c)

Figure 5-14. Segmentation of an image with a strong bimodal distribution. (a) Unprocessed coins image. (b) Bimodal histogram, with the two strong peaks representing the background gray-level values and foreground gray-level values, respectively. The threshold value T_1 serves to separate the two portions of the image. (c) Result of thresholding the image using a value of T_1=180.

5.2.1 Thresholding

Image thresholding, the process whereby all pixels in an image less than some value T are set to zero, or alternatively all pixels greater than some value T are set to zero, has previously been encountered numerous times in this book. Gray-level thresholding is the simplest possible segmentation process, and formed an important part of the Sobel edge-detector that was the focus of 4.1. The general idea behind using a threshold to separate objects from a background is conceptually simple – consider the histogram of an image with light objects on a dark background, as shown in Figure 5-

14. The image in 5-14a consists of multiple coins with approximately the same pixel intensity differing from the intensity level of the background, which results in a *bimodal* histogram exhibiting two distinct peaks, or modes. One of these modes corresponds to the background area and the other to the objects-of-interest (in this case, the coins). Hence, it follows that a single threshold value divides the image into separate regions that are homogenous with respect to brightness, as shown in 5-14c where a value of T_1 serves to discriminate the coins from the grayish background. While a single threshold value is sufficient to segment an image whose histogram is bimodal, as one would expect it is also quite possible to have histograms that are multi-modal in nature. Figure 5-15 illustrates the scenario for the case of a histogram exhibiting three modes, with two types of light objects on a dark background. In this case, the basic thresholding approach is extended to use two thresholds, so that for an image with a dark background, the two-level threshold segmentation scheme classifies a point in the image $f(i,j)$ as either background, object class 1, or object class 2 according to the following criteria:

- if $f(i,j) \leq T_1$ then pixel is background
- if $T_1 < f(i,j) < T_2$ pixel belongs to object class 1
- if $f(i,j) \geq T_2$ then pixel belongs to object class 2

This scheme is known as multi-level thresholding and in general is far less reliable than using a single threshold value.

5.2.2 Autonomous Threshold Detection Algorithms

The central problem afflicting the Sobel edge detection routine of Algorithm 5-1 is still unresolved – namely, how to have a computer (or DSP) find a good threshold? If the image is guaranteed to be of high contrast, then simply selecting a brightness threshold somewhere within the middle of the dynamic range may be sufficient (i.e. 128 for 8-bit images). Obviously this is not always going to be the case and in this section we explore algorithms that derive a "good" threshold from the histogram of the image, where the goodness criterion is one where the number of falsely classified pixels is kept to a minimum. As was alluded to in the introduction to this section, using properties of the image that are known a priori can greatly aid in the selection of a good threshold value. For example, in the case of optical character recognition (OCR) applications, it may be known that text covers 1/p of the total canvas area. Thus it follows that the optimal algorithm for OCR is to select a threshold value such that 1/p of the image area has pixel intensities less than some threshold T (assuming the text is dark and the sheet is white), which is easily determined through inspection of the histogram. This method is known as p-tile-thresholding.

Alternative techniques relying on histogram shape analysis are used when such knowledge is not available. One method that has been shown to work well under a large variety of image contrast conditions is the iterative isodata clustering algorithm of Ridler and Calvard[17]. The histogram is initially segmented into two sections starting with an initial threshold value T^0 such as 2^{bpp-1}, or half the dynamic range. The algorithm then proceeds as follows:

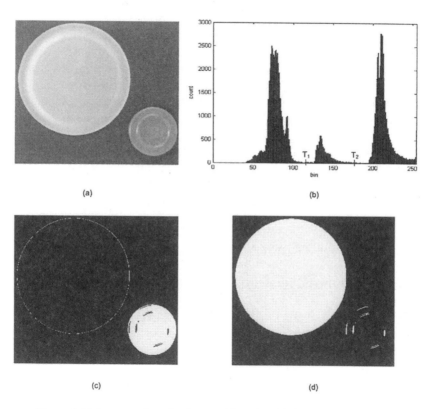

(a)   (b)

(c)   (d)

Figure 5-15. Segmentation of an image with a trimodal distribution. (a) Unprocessed bottlecaps image. (b) Trimodal histogram, with the two strong peaks representing both objects and a third corresponding to the background. The two threshold values T_1 and T_2 can be used to separate out both object types from the background. (c) Binary image resulting from thresholding the image using a value of T_1=125 and T_2=175, that is all pixels outside the range $[T_1$-$T_2]$ are set to zero and the remaining non-zero pixels set to logical value of 1. (d) Result of thresholding the image using a value of T_2=175, all pixels less than this gray-level intensity are set to zero and the remaining pixels set to logical value of 1.

1. Compute sample mean of the pixel intensities of the foreground m_f
2. Compute sample mean of the pixel intensities of the background m_b
3. Set $T^{i+1} = (m_f + m_b) / 2$
4. Terminate iteration if $T^{i+1} = T^i$, else go to step 1

Algorithm 5-2 is a complete image segmentation algorithm that estimates the background/object distinction using the isodata method to automatically determine the threshold value

Algorithm 5-2: Threshold detection using the isodata method.
    INPUT: q-bit image I, length K of smoothing filter, max iterations N
    OUTPUT: threshold T_i

```
MP = 2^q-1
hist[] = int(MP+1)    (discrete PDF or histogram of I)
H[] = int(MP+1)       (smoothed histogram)
NH = floor(K/2)       (half-width of 1D smoothing filter)

foreach (b = 0...MP) (compute histogram of input image)
  hist[b] = 0
end
foreach (pixel p in I)
  hist[p] += 1
end

(zero-phase smoothing of histogram, so peaks don't shift)
for (ii = NH ... MP-NH)
  sum = 0
  for (jj = -NH...NH)
    sum += hist[ii+jj]
  end
  H[ii] = sum / K
end

(copy over unsmoothed margins of histogram array)
for (ii = 0 ... NH-1)
  H[ii] = hist[ii]
  H[MP-ii] = hist[MP-ii]
end
```

Algorithm 5-2 (continued)
   (*isodata algorithm*)
   $T_{i-1} = -1$
   $T_i = 2^{(q-1)}$
   ii = 1
   while (ii≤N) and ($T_{i-1} \neq T_i$)
     (*center of mass above current T*)
     sumhi = prod = 0
     for (jj = T_i ... MP)
       sumhi += H[jj]
       prod += (jj)(H[jj])
     end
     m_f = prod / sumhi (*sample mean of foreground gray-levels*)
     (*center of mass below current T*)
     sumlo = prod = 0
     for (jj = 0 ... T_i-1)
       sumlo += H[jj]
       prod += (jj)(H[jj])
     end
     m_b = prod / sumhi (*sample mean of background gray-levels*)
     $T_{i-1} = T_i$
     T_i = round((m_f+m_b)/2);
     ii += 1
   end

    Most threshold detection algorithms based on histogram shape analysis benefit from applying a smoothing filter to the histogram. Smoothing the histogram removes small fluctuations from the signal that tend to inject noise into the algorithm, and Algorithm 5-2 accomplishes this by computing a running average of the raw histogram. If H is a histogram with $b = 2^{bpp}$ bins then

$$\hat{H}[i] = \frac{1}{K} \sum_{j=-\lfloor K/2 \rfloor}^{j=\lfloor k/2 \rfloor} H[i+j] \qquad K \text{ odd and } i = 0 \ldots 2^{bpp-1}$$

smoothes the histogram and in effect, low-pass filters the one-dimensional signal H. In fact, the above expression describes a convolution in one-dimension of H and a box filter of length K, or in other words a weighted average or FIR filter. Figure 5-16 illustrates this algorithm's effectiveness on a few different images, and in 5.2.5.1 we shall incorporate the isodata threshold detection routine into the Sobel Edge Detector to increase its practicality.

The triangle algorithm, attributed to Zack[18], is particularly effective when the image histogram exhibits a long tail with the foreground objects forming a rather weak peak, in comparison to a larger peak consisting of the background pixels. Conceptually, the triangle algorithm works in the following fashion, and is illustrated in diagrammatic form in Figure 5-17:

1. Fit a line between the peak of the histogram at bin b_{max} and the end of the longer tail of the histogram at bin b_{min}
2. Calculate the perpendicular distance d_{b_i} between this line and every point in the histogram between b_{min} and b_{max}
3. Set as the threshold the maximum d_{bi} found

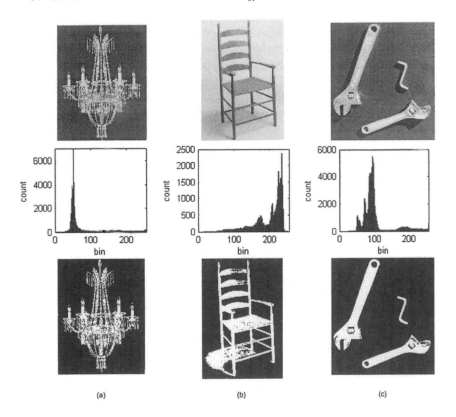

(a)　　　　　　　　(b)　　　　　　　　(c)

Figure 5-16. Examples of image segmentation using the isodata threshold detection algorithm. Top image is the raw, unprocessed image; middle plot is the histogram of this image; and bottom image is the binary segmented image. Images a and b courtesy of Professor Perona, Computational Vision at Caltech, http://www.vision.caltech.edu. (a) T = 120. (b) T = 177. (c) T = 143.

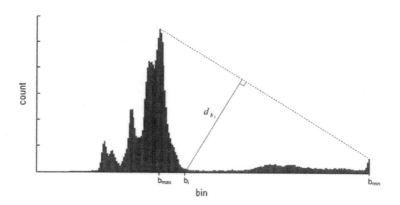

Figure 5-17. Single iteration of the triangle threshold detection algorithm.

5.2.3 Adaptive Thresholding

In general, a major problem with global threshold segmentation techniques is that they rely on objects in the image being roughly of the same brightness level. In most real-world imaging environments gray-level variations are to be expected, and such variations tend to alter the histogram shape in a manner not amenable to global threshold techniques. One rather common source of trouble, especially in retinal image processing and face recognition systems, is that of non-uniform illumination, of which the effects on image thresholding are illustrated in Figure 5-18. There are means of compensating for this effect, and such techniques should be attempted prior to applying more sophisticated threshold detection algorithms that will be described shortly. One means of compensating for nonuniformity of illumination is to calibrate the system by taking a flat-field image and then using this image to normalize the captured image. This method of course assumes that the illumination pattern remains constant. If, on the other hand, the nonuniformity changes from image to image, it may be possible to measure or estimate the background illumination, which can then be subtracted from the image. There are a variety of other techniques, including gamma intensity correction, modified histogram equalization, and others specific to various applications that also find common use. Figure 5-19 is an example of such a method from the field of retinal image processing.

When none of these methods are available, adaptive thresholding can be used to handle variations in the histogram shape. The general approach is to divide the image into a set of non-overlapping subimages, and then determine a suitable threshold independently for each subimage. The segmentation occurs by processing each subimage separately, with respect to its local threshold. An example is shown in Figure 5-20.

While conceptually simple, adaptive thresholding in reality is not as straightforward as one might expect. For starters, great care must be taken to choose subimages of sufficient size, so that each histogram contains enough information to make a decent estimate of a meaningful threshold value. It still may be the case that for some of the subimages a threshold cannot be found (e.g. background portions of the image that are more or less constant) – in this event, one strategy may be to interpolate from surrounding sub-

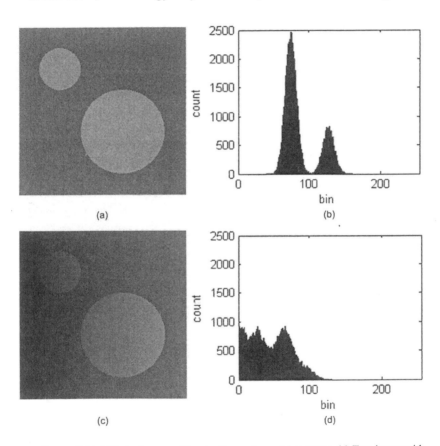

Figure 5-18. Effect of uneven illumination on image histograms. (a) Test image with completely uniform illumination. (b) Histogram of test image, which could be easily segmented using a variety of automatic threshold detectors. (c) Test image with an uneven illumination field. (d) Histogram of test image with uneven illumination. The two peaks corresponding to the background and foreground have now blurred into one another, which would cause problems for image segmentation algorithms relying on a single, global threshold value.

images. Chow and Kaneko describe a variation on the basic adaptive threshold algorithm whereby a thresholding "surface" is formed by interpolating from the various thresholds found within each subimage[19]. With this method, it is conceivable that each pixel in the original image takes on a different threshold value, and it has found widespread use in medical imaging.

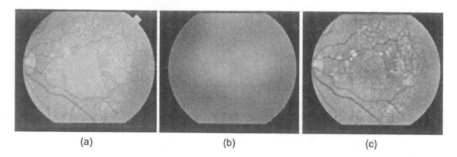

(a)   (b)   (c)

Figure 5-19. Compensating for uneven illumination in retinal image processing (images courtesy Adam Hoover of Clemson University). (a) Due to the imaging process, the illumination across the image is uneven, obscuring some of the structures. (b) Intensity adjustment field, obtained by segmenting the blood vessels using the method reported in [16], and computing the average local blood vessel pixel intensity. (c) Corrected image.

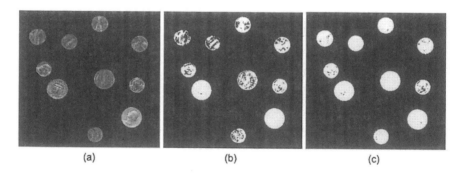

(a)   (b)   (c)

Figure 5-20. Adaptive versus global thresholding. (a) Original coins image, with a fair amount of shadow contributing to uneven lighting. (b) Result of thresholding using a single threshold determined via application of the isodata algorithm (Algorithm 5-2). (c) Binary segmented image obtained using adaptive thresholding, with each subimage processed using a separate invocation of the isodata algorithm. While the segmentation is not perfect, compared to the image on its left it is far better.

5.2.4 MATLAB Implementation

The Image Processing Toolbox includes two functions that are very useful for performing image segmentation, im2bw[14] and graythresh[20]. The im2bw function converts a grayscale or color image to binary using a normalized threshold value that lies between 0 and 1. Thus to segment an 8-bit image I using a threshold value of 127, the proper function call is

```
im2bw(I, 127/255) ;
```

The graythresh function can be used to automatically select a threshold value, using Otsu's method[21]. The saving of binary images in MATLAB in the Windows BMP format can be troublesome. The returned image from im2bw is of type logical, and the following MATLAB code snippet does not work as expected:

```
J = im2bw(I, graythresh(I)); % I is a uint8 matrix, J is a logical matrix
imwrite(J, 'filename.bmp' );
```

In fact, casting the logical image to a uint8 image also fails:

```
imwrite((uint8(J), 'filename.bmp' ); % does not work
```

The binary image needs to be scaled correctly, and the correct function call is as follows:

```
imwrite(uint8(double(J).*255), 'filename.bmp' );
```

The built-in MATLAB function times may be used in lieu of the cast to double followed by the .* operator; that operator only works on double matrices:

```
imwrite(times(J,255), 'filename.bmp' );
```

Listing 5-7, segment_isodata.m, is a MATLAB port of Algorithm 5-2 that does not utilize any functions from the Image Processing Toolbox. This function first smoothes the image histogram by convolving it with a five-sample moving average filter, and then falls into a while loop where the main isodata algorithm takes place. The iterative process terminates after 10 trips through the while loop, and if the stopping condition is never reached the last computed pixel intensity is chosen as the threshold. After a threshold value has been found, the actual segmentation occurs by using find to set all pixels falling above the threshold to logical value one, and zero otherwise.

Listing 5-7: MATLAB function for determining a threshold using the isodata algorithm.

```
function J = segment_isodata(I)

I = double(I);
J = logical(zeros(size(I))); % return matrix

% image histogram
max_pixel = max(I(:));
bins = 0:max_pixel;
H = histc(I(:),bins)';
% zero-phase smoothing of the histogram (so peaks don't shift)
H_smoothed = conv(H, [1 1 1 1 1]./5);
H(2:length(H)) = H_smoothed(5:length(H_smoothed)-1);

% isodata algorithm
nh = length(H);
T_prev = -1;
T = round(max_pixel/2);
if T % guard against the pathological case where I=0
    kk = 1;
    while (T_prev ~= T) & (kk <= 10)
        % mean (center-of-mass) above T
        sumhi = sum(H(T:nh));
        if sumhi
            mhi = sum(bins(T:nh).*H(T:nh)) ./ sumhi;
        else % handle case where upper portion of hist all zeros
            mhi = T + length(T:nh)/2;
        end
        % mean (center-of-mass) below T
        sumlo = sum(H(1:T-1));
        if sumlo
            mlo = sum(bins(1:T-1).*H(1:T-1)) ./ sumlo;
        else % handle case where lower portion of hist all zeros
            mlo = (T-1)/2;
        end
        T_prev=T;
        T=round((mhi+mlo)/2);
        kk=kk+1;
```

```
    end % end (while no convergence)
  end % if image isn't entirely all zeros

% segmentation
indices = find(I > T);
J(indices) = true;
```

We can enhance the previous `sobel_edge_detect_fixed` function by incorporating the threshold detection algorithm implemented in Listing 5-7. Listing 5-8 is an improved Sobel edge detector, `sobel_edge_detect`, that either calls upon `segment_isodata` to pick a threshold value for segmenting the gradient image, or otherwise acts identically to `sobel_edge_detect_fixed` – that is, it relies on a user-specified value.

Listing 5-8: Improved Sobel edge detector that does not necessarily require the callee to pass in a threshold for segmenting the gradient image.

```
function J = sobel_edge_detect(I, T)

I = double(I);

% compute gradient image
Gx = filter2([1 0 -1;2 0 -2;1 0 -1], I, 'valid');
Gy = filter2([-1 -2 -1;0 0 0;1 2 1], I, 'valid');
G = abs(Gx) + abs(Gy);

if 1 == nargin % using the histogram, select a good threshold
    J = segment_isodata(G);
else % binarize image using user-provided threshold value
    J = logical(zeros(size(G))); % return matrix
    indices = find(G > T);
    J(indices) = true;
end
```

Another segmentation function, `segment_triangle`, is located in the `Chap5` directory on the CD-ROM, and as its name suggests it segments an image using the triangle method for determining the threshold value. Since this function has a functionally equivalent signature to that of `segment_isodata`, it can easily be used as a replacement for that function within `sobel_edge_detect` if so desired. The final image segmentation M-file provided on the CD-ROM implements a simple adaptive thresholding algorithm that calls upon the services of segment_isodata. This function, segment_adaptive_isodata, is given in Listing 5-9.

Listing 5-9: MATLAB function for adaptive thresholding using the isodata algorithm.

```
function J = segment_adaptive_isodata(I)

if isa(I, 'uint8'), t = 256/4;
elseif isa(I, 'uint16'), t = 65536/4;
else error('Only support 8 or 16 bit images.');
end

I = double(I);
J = logical(zeros(size(I))); % return matrix

% partition image into 10 contiguous blocks
NB = 5;
MN = size(I);
R = [[1:round(MN(1)/NB):MN(1)] MN(1)+1];
C = [[1:round(MN(2)/NB):MN(2)] MN(2)+1];
skipped_subimages = [];
for irow = 1:NB
  for icol = 1:NB

    sub_image = I(R(irow):R(irow+1)-1, C(icol):C(icol+1)-1);
    if max(sub_image(:)) - min(sub_image(:)) > t
      subimage_binary = segment_isodata(sub_image);
    else
      % could be used to do something more intelligent for
      % those subimages that had to be skipped.
      skipped_subimages = [skipped_subimages [irow; icol]];
      subimage_binary = logical(zeros(size(sub_image)));
    end
    J(R(irow):R(irow+1)-1, C(icol):C(icol+1)-1) = subimage_binary;

  end % (for each block col index)
end % (for each block row index)
```

This function splits the input image into a set of 5x5 non-overlapping regions and applies the isodata algorithm using the block processing paradigm depicted in Figure 4-2. Recall that it is quite likely that a suitable threshold may not be found for some of the blocks in the image, and the criterion employed within this function is based on a sense of the spread in the y dimension of the local histogram function. That is, if the delta between the maximum frequency count and minimum frequency count is greater than 1/4 of the total dynamic range of the image, then the local histogram is assumed to contain enough information for estimation of a suitable threshold. The premise behind this choice is that in nearly constant background (or foreground, for that matter) regions of the image, the local histogram will be more or less flat. Therefore, selecting a threshold based on that histogram shape will result in a noisy segmented image. More sophisticated criterions could also be used, for example shape analysis that searches for a bimodal or multi-modal distribution would probably be preferable.

Another shortcut taken in `segment_adaptive_isodata` has to do with what is done in the case when a threshold could not be calculated for a subimage of `I`. This function does nothing in this event and merely sets all pixels in those regions to logical zero, which is a viable strategy only for those images where the background is dark and the foreground is light. This logic would have to be negated for images of the opposite polarity, with skipped blocks set to logical value one. A more robust implementation could potentially utilize one of the following strategies:

1. Use the average threshold value from all of the subimages where a viable threshold was found, or
2. Interpolate the threshold value from the surrounding blocks.

The second strategy is most likely preferred, although it would be harder to implement as one would need to be cognizant of the fact that some of the neighboring subimages also may not have a threshold. The function of Listing 5-9 maintains a list – although not actually used – of subimages where a threshold value was not computed. This `skipped_subimages` variable could thus be used to implement either of the aforementioned strategies after the double for loop completes execution.

5.2.5 RTDX Interactive Segmentation Application with Visual Studio and the TI C6416

An interactive application based on RTDX was dissected in 5.1.2, and it demonstrated how MATLAB can be used to drive an embedded edge detection algorithm implemented on the TIC6x EVM or DSK development platform. In this section, we build upon that application and increase its usability by integrating Algorithm 5-2, the isodata threshold selection algorithm, into the C64xx DSK target. After this enhancement, the new target performs either enhanced Sobel edge detection (augmented with automatic selection of a threshold value), or simple segmentation, again using the isodata method to choose the threshold. In addition, in this case the host application is a native Windows application, built using Visual Studio .NET 2003. This host communicates with the DSK over RTDX using the Texas Instruments COM objects that come with the CCStudio IDE.

5.2.5.1 C6416 DSK Implementation

The source code for the target application, found in Chap5\segmentation\SegmentRTDX\Target\C6416DSK, is shown in Listing 5-10. This project is similar to its predecessor (Listing 5-3) in that it utilizes DSP/BIOS and does not page external memory into on-chip RAM. In addition, it also presumes a fixed size 256x256 input image. One significant difference is that the corresponding host program does not enable RTDX remotely, as the MATLAB host in 5.1.2.4 did. Therefore, RTDX must be manually enabled after loading the program onto the DSP, and prior to running it. After the program has been loaded onto the DSP, RTDX can be enabled by bringing up the RTDX configuration control window in CCStudio via **Tools|RTDX|Configuration Control**, and clicking on the "Enable RTDX" checkbox. The MATLAB host also took care of enabling the RTDX read and write channels, while in this program the channels are enabled on the target end by using the RTDX_enableInput and RTDX_enableOutput macros.

Listing 5-10: Enhanced image segmentation C6416 program that communicates over RTDX.

```
#include <rtdx.h>      /* target API */
#include <stdio.h>
#include <IMG_sobel.h>
#include <IMG_thr_le2min.h>
#include <IMG_histogram.h>
#include <fastrts62x64x.h> /* spuint() & divsp() */
#include "target.h" /* defines TARGET_INITIALIZE() */
#include "image.h"   /* dimensions */

RTDX_CreateInputChannel(ichan); /* input, T come down this pipe */
RTDX_CreateOutputChannel(ochan); /* processed data via this pipe */

#pragma DATA_SECTION(img_buf1, "SDRAM");
#pragma DATA_ALIGN (img_buf1, 8);
unsigned char img_buf1[N_PIXELS];

#pragma DATA_SECTION(img_buf2, "SDRAM");
#pragma DATA_ALIGN (img_buf2, 8);
```

```
unsigned char img_buf2[N_PIXELS];

#pragma DATA_ALIGN (hist, 8);
unsigned short hist[256];

#pragma DATA_ALIGN (smoothed_hist, 8);
unsigned short smoothed_hist[256];

#pragma DATA_ALIGN (t_hist, 8);
unsigned short t_hist[1024]; /* temp storage needed for IMG_histogram */

/*
 * Computes histogram, smoothes it, and returns the
 * maximum pixel value
 */
unsigned char calc_hist(const unsigned char * restrict pimg)
{
  short * restrict phist = (short *)hist; /* to remove warnings */
  short * restrict pthist = (short *)t_hist; /* see above */
  unsigned short *pH = hist;
  unsigned int accum;
  int ii;
  const unsigned int DIVISOR = 13107; /* (1/5)*2^16 */
  unsigned char max_pixel = 0;

  /* set histogram buffers to 0 */
  memclear(t_hist, 1024*sizeof(unsigned short));
  memclear(hist, 256*sizeof(unsigned short));

  IMG_histogram(pimg, N_PIXELS, 1, pthist, phist);

  /* now smooth histogram using 5-sample running average */
  accum = pH[0]+pH[1]+pH[2]+pH[3]+pH[4];
  for (ii=2; ii<254; ++ii, ++pH) {
    smoothed_hist[ii] = (accum*DIVISOR)>>16; /* accum/5 (Q.16) */
    accum -= pH[0];
    accum += pH[5];
    /*
     * We read one element beyond the end of the array,
     * so that's why we look for pH[4] instead of pH[5]
     * when updating the max pixel variable
     */
    if (pH[4])
      max_pixel = ii+2;
  }

  smoothed_hist[0] = hist[0];
  smoothed_hist[1] = hist[1];
  smoothed_hist[254] = hist[254];
  smoothed_hist[255] = hist[255];
  return max_pixel;
}
```

```
/*
 * Used by isodata() to compute center-of-mass under histogram
 */
unsigned long dotproduct(int lo, int hi)
{
  /* 0, 1, 2, ..., 255 */
  static const unsigned short pixval[]={0,1,2, /* 3,5,...,252 */ ,253,254,255};
  unsigned long sum1 = 0, sum2 = 0, sum3 = 0, sum4 = 0, sum;
  const int N = hi-lo;
  int ii=0, jj=lo, remaining;
  double h1_h2_h3_h4, b1_b2_b3_b4;
  unsigned int h1_h2, h3_h4, b1_b2, b3_b4;

  /* unrolled dot-product loop with non-aligned double word reads */
  for (; ii<N; ii+=4, jj+=4)
  {
    h1_h2_h3_h4 = _memd8_const(&smoothed_hist[ii]);
    h1_h2 = _lo(h1_h2_h3_h4);
    h3_h4 = _hi(h1_h2_h3_h4);

    b1_b2_b3_b4 = _memd8_const(&pixval[ii]);
    b1_b2 = _lo(b1_b2_b3_b4);
    b3_b4 = _hi(b1_b2_b3_b4);

    sum1 += _mpyu(h1_h2, b1_b2);    /* (h1)(b1) */
    sum2 += _mpyhu(h1_h2, b1_b2);   /* (h2)(b2) */
    sum3 += _mpyu(h3_h4, b3_b4);    /* (h3)(b3) */
    sum4 += _mpyhu(h3_h4, b3_b4);   /* (h4)(b4) */
  }
  sum = sum1 + sum2 + sum3 + sum4;
  /*
   * loop epilogue: if # iterations guaranteed to
   * be a multiple of 4, then this would not be required.
   */
  remaining = N - ii;
  jj = N - remaining;
  for (ii=jj; ii<N; ii++)
    sum += smoothed_hist[ii]*pixval[ii];

  return sum;
}

/*
 * Finds a threshold for segmentation using the isodata algorithm
 */
unsigned char isodata(unsigned char max_pixel)
{
  const int max_iterations = 10;
  unsigned char T = max_pixel>>1,
                T_prev,
                kk = 0, mhi, mlo;
```

```
  unsigned long sumhi, sumlo, dotprod;
  int ii;

  if (T) { /* guard against pathological case where image all zero */
    do {

      /* mean (center-of-mass) above current T */
      sumhi = 0;
      for (ii=T; ii<256; ++ii)
        sumhi += smoothed_hist[ii];
      if (sumhi) {
        dotprod = dotproduct(T, 256);
        mhi = spuint( divsp((float)dotprod, (float)sumhi) );
      }
      else /* upper portion of hist all zeros */
        mhi = T + (256-T)>>1;

      /* mean (center-of-mass) below current T */
      sumlo = 0;
      for (ii=0; ii<T; ++ii)
        sumlo += smoothed_hist[ii];
      if (sumlo) {
        dotprod = dotproduct(0, T);
        mlo = spuint( divsp((float)dotprod, (float)sumlo) );
      }
      else /* lower portion of hist all zeros */
        mlo = (T-1)>>1;

      T_prev = T;
      T = (mhi+mlo)>>1;

    } while (T_prev!=T && ++kk<max_iterations);
  }

  printf("threshold = %d\n", T);
  return T;
}

/*
 * Most of the work emanates from here, returns
 * the processed image data pointer
 */
unsigned char *process_image(int run_edge_detection)
{
  unsigned char max_pixel, T;

  if (run_edge_detection) {

    IMG_sobel(img_buf1, img_buf2, Y_SIZE, X_SIZE);
    max_pixel = calc_hist(img_buf2);
    T = isodata(max_pixel); /* calc threshold value */
```

```
    IMG_thr_le2min(img_buf2, img_buf1, Y_SIZE, X_SIZE, T);
    return img_buf1;

  } else { /* no edge detection, just straight up segmentation */

    max_pixel = calc_hist(img_buf1);
    T = isodata(max_pixel); /* calc threshold value */
    IMG_thr_le2min(img_buf1, img_buf2, Y_SIZE, X_SIZE, T);
    return img_buf2;

  }
}

void main()
{
  int status, ii;
  int run_edge_detector;
  unsigned char *pimg = NULL;

  TARGET_INITIALIZE();
  RTDX_enableOutput(&ochan); /* enable output channel */
  RTDX_enableInput(&ichan); /* enable input channel */
  printf("Input & Output channels enabled ...\n");

  while (1) {
    /* wait for the host to send us a threshold */
    if (sizeof(run_edge_detector) !=
                        (status = RTDX_read(&ichan, &run_edge_detector,
                                            sizeof(run_edge_detector)))))
      printf("ERROR: RTDX_read of edge detection flag failed!\n");
    else
      printf("Edge detection = %s\n", (run_edge_detector)?"yes":"no");

    /* now we're expecting X_SIZE x Y_SIZE worth of image data */
    if (N_PIXELS !=
                   (status = RTDX_read(&ichan, img_buf1, N_PIXELS))) {
      printf("ERROR: RTDX_read of image failed (%d)!\n", status);
      exit(-1);
    }
    printf("Received %dx%d image\n", X_SIZE, Y_SIZE);

    pimg = process_image(run_edge_detector);

    /* send processed image back to host */
    printf("Sending processed image data back to host ...\n");
    for (ii=0; ii<X_SIZE; ++ii) {
      /* write one row's worth of data */
      if (!RTDX_write(&ochan, pimg+ii*Y_SIZE, Y_SIZE)) {
        printf("ERROR: RTDX_write of row %d failed!\n", ii);
        exit(-1);
```

```
    }
  } /* end (for each row) */
  printf("Image segmentation completed.\n");
  }
}
```

The host/target protocol has been slightly changed in this application. Recall that in the previous RTDX example, the entire process was kicked off when an integer threshold value was sent from the host and received by the target. An RTDX message consisting of a single integer still initiates the image processing, however now this integer (run_edge_detector) is treated as a boolean value. If run_edge_detector is zero, then no edge detection is performed and the isodata algorithm operates on a smoothed histogram of the input image stored in img_buf1. On the other hand, if run_edge_detector is non-zero then the isodata algorithm operates on a gradient image created using the IMGLIB function IMG_sobel. Regardless of whether edge detection is performed, IMG_thr_le2min is used as before to perform the actual thresholding.

The smoothing of the image histogram is an essential part of the isodata algorithm; calc_hist calls upon the services of IMG_histogram to compute the histogram of either a gradient image or original input image. The third argument to IMG_histogram is an integer boolean that specifies whether the pixel counts should be added or subtracted to the histogram buffer. This argument proves very useful when implementing the paging optimization, because as blocks of the image are paged into internal memory, their corresponding frequency counts are successively added to a single buffer so that after all blocks have been transferred, the buffer contains the histogram for the entire image.

Following the call to IMG_histogram, which places the histogram into the hist array, a low-pass filtered version of hist is computed and stored in smoothed_hist. This step was accomplished in the MATLAB version of the isodata algorithm by convolving the histogram with a 1x5 box filter, i.e. [1 1 1 1 1]/5. While we do have the luxury of using similar canned fixed-point DSPLIB functions to implement the same operation on the DSP, we choose not to in this case in the interest of optimizing the performance. The running average computation is nothing more than a special case of a 1D FIR filter – recall we used the DSPLIB function DSP_fir_gen to implement 2D low-pass filtering on a flattened image matrix in 4.3.2. While we could easily adapt DSP_fir_gen to the current context (in fact, the application of the DSPLIB function is more straight-forward here because we are using it for the exact purpose for which it was originally designed – 1D signal processing), we choose not to because we can further optimize the algorithm's performance by taking advantage of the simplicity of the smoothing filter. Consider the C equivalent of DSP_fir_gen, which looks something like Listing 5-11.

Listing 5-11: "Behavioral" C code for a fixed-point 1D FIR filter, i.e. a function akin to DSP_fir_gen.

```
void DSP_fir_gen(short x[], short h[], short r[], int nh, int nr)
{
  int i, j, sum;
  for (j = 0; j < nr; j++) {
    sum = 0;
    for (i = 0; i < nh; i++)
      sum += x[i + j] * h[i];
    r[j] = sum >> 15; % fixed-point, assume Q15 format
  }
```

Listing 5-11 is implemented directly from the definition of convolution in one dimension, and is general enough that the individual filter coefficients (the h array) could conceivably differ from one another. In the case of the function in Listing 5-10 where the histogram is smoothed, calc_hist, we know that the averaging box filter coefficients are all 1/5. Based on this knowledge, instead of using DSP_fir_gen, the optimized histogram smoothing code maintains a running summation of the five most recent frequency counts in the accumulator variable accum. The average or smoothed value is then computed using Q16 fixed-point arithmetic. Multiplying accum by floor($2^{16}/5$) = 13107 and then right-shifting by 16 bits effectively divides accum by 5, thus producing an averaged frequency count. The oldest frequency count is then shifted out by subtraction, and the subsequent one is shifted in by addition, in preparation for the next sample (histogram bin).

Moving on to the isodata function, this is more or less a C port of Listing 5-7, with a few twists to account for the fixed-point nature of the implementation. Each iteration of the isodata algorithm entails computing the center-of-mass of the current two histogram modes, represented here by mhi and mlo. In the MATLAB version of the algorithm, the two center-of-masses are computed either with the statement

mhi = sum(bins(T:nh).*H(T:nh)) ./ sum(H(T:nh));

or

mlo = sum(bins(1:T-1).*H(1:T-1)) ./ sum(H(1:T-1));

The numerators of the right-hand sides of both of the above expressions are the dot product of two vectors, in particular the frequency bins and the histogram counts. Computation of the dot product is an oft-cited code optimization exercise[22,23,24,25,26], because the optimization techniques are quite effective and the underlying operation simple to understand. An optimized dot product operation takes place within the dotproduct function, with detailed discussion of this function's implementation deferred until Appendix B. Suffice it to say that the C intrinsics employed in this function make use of the LDDW instruction, which is only available on the C67xx and C64xx DSPs, and therefore this program as it stands cannot be ported to run on a C62xx DSP.

Computation of mhi and mlo both involve division, which as we saw in 4.6.5 can be rather involved to implement efficiently on a fixed-point architecture like the C64x. Because these division operations do not occur on a pixel-by-pixel basis, as they did in the case of the adaptive filter implementation of 4.6.5, and only occur a relatively small number of times within the isodata loop, the FastRTS library[27] is used here to compute mhi and mlo. The FastRTS library function divsp performs single-precision floating-point division in software, and then this value is "casted" to an integer value using another FastRTS function, spuint. Incorporating FastRTS into a C64xx or C62xx project is just like using IMGLIB or DSPLIB - add the library include path to the preprocessor directive and link to the static library, in this case fastrts64x.lib. The computation of mhi and mlo could also of course be obtained using the reciprocal approximation described in 4.6.5, where the fixed-point reciprocals of sumhi and sumlo (i.e. 2^Q/sumhi or 2^Q/sumlo) would be calculated, multiplied by the numerator, and finally right-shifted by Q bits. For a 256x256 image, the maximum value the denominator could take on is 65536, and for that reason the recip_Q15 function from Listing 4-19 is not guaranteed to work. A reciprocal approximation function suitable for inclusion into Listing 5-10 would need to account for a larger range of values than did recip_Q15. This implies a larger look-up table for seeding Newton-Raphson iteration, and quite possibly a larger number of iterations to converge to an acceptable approximation of the reciprocal.

5.2.5.2 Visual Studio .NET 2003 Host Application

The directory Chap5\segmentation\SegmentRTDX\Host contains the C/C++ host application, and leverages the Image8bpp class used in all other Visual Studio applications. The GUI operation is self-explanatory, and should be apparent from Figure 5-21.

The MATLAB host application from the previous section used objects and functions provided by the Link for Code Composer Studio to pass data back and forth between the host and target over RTDX channels. This program does the same thing but uses a Texas Instruments COM component that comes standard with CCStudio to access the various RTDX facilities. COM, short for Component Object Model, is a venerable Microsoft technology that in some respects has been supplanted by the .NET initiative – however due to COM's installed base it is not going away any time soon[28]. So called "pluggable" software components are implemented through COM and its cousin ActiveX. Thus, with the Texas Instruments RTDX COM component, data can be sent over RTDX channels from a variety of languages and environments, such as MATLAB, Visual Basic, C/C++, or even Excel (using Visual Basic for Applications, or VBA).

Before making use of anything COM-related in a C/C++ application, the `CoInitialize` function must be called, and it is also a good idea at the end of the application's lifetime to call `CoUnInitialize`. Whenever building an MFC program using the Visual Studio wizards, an application class is automatically created, and the `InitInstance` method in this object can be thought of as akin to `main` in a standard C program. Here, `CSegmentationHostApp::InitInstance` is therefore a logical place to make both of the aforementioned COM library initialization calls. Incorporating a COM object within an "unmanaged" or native C++ program (so-called "managed C++" is a .NET-aware Microsoft variant of C++ and is not used in this book) into code built using the Visual Studio .NET compiler can be done in a variety of ways. One of the easiest is to use the `#import` directive, and explicitly reference the DLL hosting the COM component. The TI RTDX COM component is hosted within `rtdxint.dll`, which can be found under the directory where CCStudio is installed – e.g. `<TI>\cc\bin`, where `<TI>` is normally `C:\TI`. The following `#import` directive is found in the pre-compiled header file `stdafx.h` for this Visual Studio project, meaning which that any C++ source file using pre-compiled headers will have access to the RTDX COM interfaces:

```
#import "C:\ti\cc\bin\rtdxint.dll"
```

COM shines when used with higher-level languages like Visual Basic – using COM from C/C++ is quite verbose and frankly not very elegant. Developers of RTDX host applications are unfortunately saddled with such COM luminaries like `VARIANT` and `SAFEARRAY`, because such applications typically transmit and receive arrays of data and COM is not built to deal with native C arrays. If the RTDX host library was provided in the form of a suite of C/C++ functions, reading and writing arrays of data would most likely involve passing in "raw" arrays of data to functions that would do the actual work of shuttling the data to CCStudio, which in turn deals directly with the DSP. However, because we are in the land of COM, we are forced to deal with the vagaries of VARIANTs and SAFEARRAYs. The VARIANT type is actually an immense union, structured in such a manner that it can contain a 4-byte integer, 1-byte integer, COM string, or any number of basic data types supported by COM. The VARIANT is the underlying mechanism of how certain higher-level COM-aware languages like Visual Basic or Delphi implement weak type checking, whereby a variable at run time can refer to any number of data types. This is in stark contrast to a strongly-typed language such as C/C++, in which once a variable has been declared, it can only be assigned values that are of the declared type. Of course in C/C++ variables may be implicitly promoted (int to float), implicitly demoted (float to int), or explicitly casted to circumvent the type rules (and in the embedded world such practices should be kept to a bare minimum in the interests of performance), but the fact remains that simulating the same pseudo-polymorphic behavior in C requires cumbersome data structures like VARIANT. Microsoft, to its credit, does provide certain wrapper classes to ease the pain for C++ developers, for example the _variant_t and _bstr_t classes that wrap the COM VARIANT and BSTR ("b-string") types.

In the case of this implementation, our VARIANTs refer to an array type, or SAFEARRAY. The SAFEARRAY is a COM-compliant array of elements whose bounds are not necessarily zero-based as in C/C++ or one-based as in MATLAB, in fact they are programmer-specified. In the interests of good

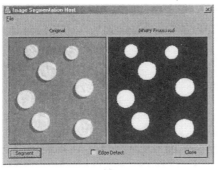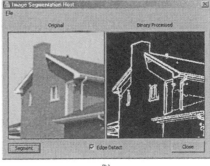

(a)   (b)

Figure 5-21. The two usage examples of the image segmentation host program. (a) If the checkbox labeled "Edge Detect" is not selected, the host informs the target of the fact by initiating the handshaking protocol with a value of 0, and the isodata and segmentation operations operate on the raw input image. (b) If edge detection is enabled, then the gradient image is passed through the isodata/segmentation algorithm running on the DSP.

object-oriented design, all of the RTDX-related functionality implemented through the COM facilities has been encapsulated into a single class, RTDX. This class has been placed within its own namespace (Qureshi) to avoid future name clashes, due to its less-than-unique name. The interface and implementation for this class is given in Listing 5-12.

Listing 5-12: Interface and implementation for C++ class that encapsulates the RTDX host-side communication.

```
//
// Interface for RTDX class (RTDX.h)
//
namespace Qureshi {

  class RTDX {

  public:

    RTDX();
    ~RTDX();

    void sendInteger(int x);
    void sendImage(Image8bpp *pImage);

    Image8bpp *readImage(); // client is responsible for deallocating!

  private:

    // Helper method used by readImage() to copy single scan-line,
    // returns true if read operation completed
    bool readAndCopyRow(int nPixels, Ipp8u *pDst);

    // pointers to RTDX COM interfaces
    RTDXINTLib::IRtdxExpPtr m_rtdxRead, m_rtdxWrite;

  };

} // namespace Qureshi


//
// Implementation for RTDX class (RTDX.cpp)
//
#include "stdafx.h"
#include "RTDX.h"

using namespace Qureshi;
using namespace std;

RTDX::RTDX()
{
  HRESULT hr =
```

```
      m_rtdxWrite.CreateInstance(_uuidof(RTDXINTLib::RtdxExp) );
  if (FAILED(hr))
    throw _com_error(hr);
  hr = m_rtdxRead.CreateInstance(_uuidof(RTDXINTLib::RtdxExp) );
  if (FAILED(hr))
    throw _com_error(hr);

  // open read and write channels
  if (0 != m_rtdxRead->Open("ochan", "R"))
    throw runtime_error("Failed to open RTDX read channel.");

  if (0 != m_rtdxWrite->Open("ichan", "W"))
    throw runtime_error("Failed to open RTDX write channel.");
}

RTDX::~RTDX()
{
  m_rtdxRead->Close();
  m_rtdxRead.Release();
  m_rtdxWrite->Close();
  m_rtdxWrite.Release();
}

void RTDX::sendInteger(int x)
{
  long bufferState;  // holds the state of the host's write buffer
  if (0 != m_rtdxWrite->WriteI4(x, &bufferState))
    throw runtime_error("WriteI4 failed.");
}

void RTDX::sendImage(Image8bpp *pImage)
{
  // send entire image in a single RTDX write message

  unsigned int nPixels = pImage->getHeight()*pImage->getWidth();
  VARIANT sa;  // pointer to a SAFEARRAY
  SAFEARRAYBOUND rgsabound[1];  // SAFEARRAY dimensions

  ::VariantInit(&sa);  // initialize VARIANT data structure
  sa.vt = VT_ARRAY — VT_I4;  // SAFEARRAY of 32-bit integers
  rgsabound[0].lLbound = 0;  // lower bounds of the SAFEARRAY
  rgsabound[0].cElements = nPixels>>2;  // upper bounds
  sa.parray = SafeArrayCreate(VT_I4, 1, rgsabound);

  // fill SAFEARRAY with pixel data
  unsigned char *pDst = NULL;
  HRESULT hr = SafeArrayAccessData(sa.parray, (void **)&pDst);
  if (FAILED(hr))
    throw _com_error(hr);

  if (Y_SIZE != pImage->getStride()) {  // copy on a row-by-row basis
```

```
    for (int iRow=0; iRow<pImage->getHeight(); ++iRow) {
      unsigned char *pSrc =
                          pImage->getPixels() + iRow*pImage->getStride();
      memcpy(pDst+iRow*Y_SIZE, pSrc, Y_SIZE>>2);
    }
  } else // a single copy suffices
      memcpy(pDst, pImage->getPixels(), nPixels);

  hr = SafeArrayUnaccessData(sa.parray);
  if (FAILED(hr))
    throw _com_error(hr);

  // now send data to the target
  long bufferState;
  if (0 != m_rtdxWrite->Write(sa, &bufferState))
    throw runtime_error("RTDX write failed.");

  ::VariantClear(&sa); // cleanup
}

Image8bpp *RTDX::readImage()
{
  static Ipp8u *pReadBuffer = NULL;
  static int stride = -1;
  if (!pReadBuffer) // 1st time around allocate aligned buffer
    pReadBuffer = ippiMalloc_8u_C1(Y_SIZE, X_SIZE, &stride);

  for (int iRow=0; iRow<X_SIZE; ++iRow)
    if (!this->readAndCopyRow(Y_SIZE, pReadBuffer+iRow*stride))
      TRACE("read row %d failed, try again...\n", iRow--);

  return new Image8bpp(X_SIZE, Y_SIZE, stride, pReadBuffer);
}

bool RTDX::readAndCopyRow(int nPixels, Ipp8u *pDst)
{
  VARIANT sa; // pointer to a SAFEARRAY

  ::VariantInit(&sa); // initialize VARIANT
  if (0 != m_rtdxRead->ReadSAI4(&sa)) // read pixel data
    return false;

  unsigned char *pSrc = NULL; // pointer to raw data from RTDX
  HRESULT hr = SafeArrayAccessData(sa.parray, (void **)&pSrc);
  if (FAILED(hr))
    throw _com_error(hr);

  memcpy(pDst, pSrc, nPixels); // could use ippiCopy_8u_C1R here

  // clean up before returning
  hr = SafeArrayUnaccessData(sa.parray);
  if (FAILED(hr))
```

```
    throw _com_error(hr);
::VariantClear(&sa);

return true;
}
```

Two separate COM interface objects are utilized within `Qureshi::RTDX`, one for writing data (`m_rtdxRead`) and another for reading data (`m_rtdxWrite`). A data type for the RTDX COM interface, `RTDXINTLib::IRtdxExpPtr`, is available when `rtdxint.dll` is imported into the project. This COM interface mirrors very closely the functionality provided by the RTDX portion of the MATLAB Link for Code Composer Studio (without having seen the toolbox implementation, one may venture a guess that it is a thin MATLAB wrapper around the COM interface used here). The CCStudio online help enumerates the complete C++ interface to the RTDX COM object, and contains more information about each of the exported interface methods.

The two interesting methods in `Qureshi::RTDX` are `sendImage` and `readImage`. Again, similar to the MATLAB host, the writing of the image data occurs in a single RTDX message, whereas in `readImage` the rows are read one by one until all `X_SIZE` rows have been read. One very convenient aspect about coding the host in C++ versus MATLAB is that with a C/C++ host application, header files can be shared between the target and the host. In the MATLAB source code from 5.1.2.4, the expected image dimensions (256x256) were duplicated by hard-coding the dimensionality into the MATLAB source – in contrast now we simply include `image.h`. Thus if the image dimensions ever change a recompile of both projects is sufficient to propagate the changes to both host and target.

Unfortunately, Microsoft does not provide a COM wrapper class akin to `variant_t` or `_bstr_t` for SAFEARRAY (see [29] for a very complete C++ wrapper over SAFEARRAY with an STL interface). Thus in the `sendImage` and `readAndCopyRow` methods low-level COM C API functions are used to marshal image data to a form the RTDX COM interface can deal with. In `sendImage`, we marshal the pixel data contained in an `Image8bpp` object to a SAFEARRAY of 32-bit integers. The SAFEARRAY upper bound is then the total number of pixels divided by four, since we intend to send packed 8-bit data over the RTDX channel. Any code accessing the buffer encapsulated by the SAFEARRAY must be bracketed by corresponding calls to `SafeArrayAccessData` and `SafeArrayUnaccessData`. Likewise, the VARIANT union must first be initialized with a call to `VariantInit` and then deallocated with a call `VariantClear` to avoid resource leaks. When using a well-designed wrapper class like `_variant_t`, it is not expected that the programmer be

cognizant of such function calls. Rather, we would expect that
VariantInit be called in the wrapper's constructor and
VariantClear be called in the wrapper's destructor - this common C++
idiom is known as "resource acquisition is initialization" or RAII. After the
SAFEARRAY is populated with the packed pixel data, the RTDX COM
method Write is invoked on the m_rtdxWrite COM interface to
transmit the image data over the RTDX channel.

The readImage method proceeds along similar lines, except of course
in the opposite direction. The receiving protocol differs somewhat from the
MATLAB version. In the MATLAB host application, the receiving loop
continually queries the number of outstanding RTDX messages in the host
read channel, and keeps reading until either the number of outstanding
messages is zero or the number of received rows equals the number of
expected rows. This loop simply keeps reading into a local buffer until
X_SIZE rows are received, and if the target is unable to keep up with the
host, the RTDX COM interface method ReadSAI4 fails and the loop
counter in readImage is decremented and the host keeps trying again until
that read operation succeeds. After all of the processed pixel data has been
received from the target and transferred into a local memory buffer
(pReadBuffer), an Image8bpp object is instantiated using a new
constructor that accepts pixel data provided by a client of Image8bpp.
This implementation of this new constructor is shown in Listing 5-13.

Listing 5-13: An updated constructor for the Image8bpp class that either
initializes an nr x nc image from pixels provided by the callee, or sets all
pixel values to zero if the client passes in a NULL pointer.

```
Image8bpp::Image8bpp(int nr, int nc,
                     int srcStride /* = -1 */,
                     Ipp8u *pSrc /* = NULL */)
 : m_pBitmap(NULL), m_pPixels(NULL)
{
  // allocate aligned memory
  if (NULL == (m_pPixels = ippiMalloc_8u_C1(nc, nr, &m_stride)))
    throw runtime_error("Out of memory.");

  IppiSize roi = {nc, nr};
  if (!pSrc) { // set entire image to 0
    if (ippStsNoErr != ippiSet_8u_C1R(0, m_pPixels, m_stride, roi))
    throw runtime_error("ippiSet_8u_C1R() error");
  } else { // copy pixel data
    if (ippStsNoErr != ippiCopy_8u_C1R(pSrc, srcStride,
                              m_pPixels, m_stride, roi))
```

```
      throw runtime_error("ippiCopy_8u_C1R() error");
  }


  // instantiate the GDI+ object that will by used for display
  // purposes (note that Bitmap ctor will not perform a deep copy,
  // as m_pPixels is shared)
  m_pBitmap = new Bitmap(nr, nc, m_stride,
                            PixelFormat8bppIndexed, m_pPixels);
  Status s = m_pBitmap->GetLastStatus();
  if (Ok != s)
    throw runtime_error((LPCSTR)GdiplusUtil::getErrorString(s));


  // without the correct color palette the gray-scale image
  // will not display correctly
  m_pBitmap->SetPalette(GdiplusUtil::get8bppGrayScalePalette());
  }
```

Two additional methods have been added to the `Image8bpp` class for the purposes of this application, `resize` and `binarize`. These methods are shown in Listing 5-14.

Listing 5-14: The two new methods added to the `Image8bpp` class, `resize` and `binarize`.

```
  Image8bpp &Image8bpp::resize(int newWidth, int newHeight)
  {
    if (m_pBitmap && m_pPixels) {
    Ipp8u *pResizedPixels = NULL;
    Bitmap *pNewBitmap = NULL;
    try {
      IppiSize srcSize = { m_pBitmap->GetWidth(),
                      m_pBitmap->GetHeight() };
      IppiRect srcRoi = { 0, 0, srcSize.width, srcSize.height };
      IppiSize dstRoiSize = { newWidth, newHeight };

      // allocate new buffers, that will replace current ones
      int newStride;
      if (NULL==(pResizedPixels=ippiMalloc_8u_C1(newWidth,
                                      newHeight,
                                      &newStride)))
        throw runtime_error("Out of memory.");
```

```
    // resize the image pixel data
    double xFactor = (double)newWidth/(double)srcSize.width,
          yFactor = (double)newHeight/(double)srcSize.height;
    if (ippStsNoErr != ippiResize_8u_C1R(m_pPixels,
                                         srcSize,
                                         m_stride,
                                         srcRoi,
                                         pResizedPixels,
                                         newStride,
                                         dstRoiSize,
                                         xFactor, yFactor,
                                         IPPI_INTER_CUBIC))
      throw runtime_error("Resize operation failed.");

    // instantiate new Bitmap object
    pNewBitmap = new Bitmap(newHeight, newWidth, newStride,
                            PixelFormat8bppIndexed,
                            pResizedPixels);
    Status s = pNewBitmap->GetLastStatus();
    if (Ok != s)
      throw runtime_error((LPCSTR)GdiplusUtil::getErrorString(s));

    // without the correct color palette the gray-scale image will
    // not display correctly
    pNewBitmap->SetPalette(GdiplusUtil::get8bppGrayScalePalette());

    // update this object's internal data structures
    m_stride = newStride;
    ippiFree(m_pPixels); m_pPixels = pResizedPixels;
    delete m_pBitmap; m_pBitmap = pNewBitmap;
  } catch (runtime_error &e) {
   if (pNewBitmap)
     delete pNewBitmap;
   if (pResizedPixels)
     ippiFree(pResizedPixels);
   throw (e);
  }

 } // end (if this image contains pixel data)

 return *this;
}
```

```
void Image8bpp::binarize()
{
  IppiSize size = {m_pBitmap->GetWidth(), m_pBitmap->GetHeight()};
  if (m_pPixels)
    if (ippStsNoErr != ippiThreshold_Val_8u_C1IR(m_pPixels,
                                                  m_stride,
                                                  size, 0, 255,
                                                  ippCmpGreater))
      throw runtime_error("ippiThreshold_Val_8u_C1R failed");
}
```

The target is designed to handle images of exactly X_SIZE x Y_SIZE dimensionality. Therefore, after the user loads in an arbitrarily sized image from disk, resize is invoked to resample the image so that it is of the expected target size. This new Image8bpp image resizing method is a port of the MATLAB Image Processing Toolbox function imresize, and calls up the services of the IPP function ippiResize_8u_C1R to perform bicubic interpolation to fit the pixel data into a X_SIZE x Y_SIZE matrix[30].

The binarize method is used for display purposes. The target uses IMG_thr_le2min to threshold the image, and this IMGLIB function sets only those pixels that lie below the threshold value to zero, leaving the remaining pixels unaltered. Since the host application displays 8-bit binary images in its right-hand display pane (see Figure 5-21), pixels need to be either 0 (black) or 255 (white) in order to visualize the binary image correctly and the IPP ippiThreshold_Val_8u_C1IR function is used in this context to perform a "search-and-replace" operation where all non-zero pixels are set to 255[2].

REFERENCES

1. Canny, J., "A Computational Approach to Edge Detection." *IEEE Transactions on Pattern Analysis and Machine Intelligence*, Vol 8, No. 6, Nov 1986.
2. Intel Corp., *Intel® Integrated Performance Primitives for Intel® Architecture (Part 2: Image and Video Processing)*.
3. *MATLAB Image Processing Toolbox User's Guide*, imfilter Function.
4. Pratt, W.K., *Digital Image Processing, 2nd Edition* (John Wiley and Sons, 1991).
5. *MATLAB Creating Graphical User Interfaces (Version 7)*, "Getting Started with GUIDE."
6. *MATLAB Compiler User's Guid.*
7. *MATLAB Function Reference Vol. 2: F-O*, imread Function.
8. Texas Instruments, *DSP/BIOS, RTDX, & Host-Target Communications* (SPRA895.pdf).
9. Keil, Deborah., *Real-Time Data Exchange White Paper* (Texas Instruments, 1998).
10. The Mathworks, Inc., "Link For Code Composer Studio." Retrieved October 2004 from:

http://www.mathworks.com/access/helpdesk/help/toolbox/ccslink

11. *MATLAB Image Processing Toolbox User's Guide*, `imresize` Function.

12. Texas Instruments, *TMS320 DSP/BIOS User's Guide* (SPRU423b.pdf).

13. Spectrum Digital, *TMS3206416 DSK Technical Reference* (505945-001 Rev. A), Chapter 1, section 1.4, *Memory Map*.

14. *MATLAB Image Processing Toolbox User's Guide*, `im2bw` Function.

15. Texas Instruments, *TMS320C64x Technical Overview* (SPRU395b), Appendix B.1, *Threshold Example*.

16. Hoover A., Kouznetsova V., Goldbaum M., "Locating Blood Vessels in Retinal Images by Piecewise Threshold Probing of a Matched Filter Response," *IEEE Transactions on Medical Imaging*, vol. 19 no. 3, March 2000, pp. 203-210.

17. Ridler, T. W., Calvard S., "Picture thresholding using an iterative selection method," *IEEE Transactions on Systems, Man and Cybernetics*, SMC-8:630-632, 1978.

18. Zack, G. W., Rogers, W. E., Latt, S. A., "Automatic measurement of sister chromatid exchange frequency," *Journal of Histochemistry and Cytochemistry*, 25(7):741-753, 1977.

19. Chow, C.K., Kaneko, T., "Automated Boundary Detection of the Left Ventricle from Cineangiograms," *Computers and Biomedical Research*. Vol. 5, No. 4. August 1972, pp. 388-410.

20. *MATLAB Image Processing Toolbox User's Guide*, `graythresh` Function.

21. Otsu, N., "A Threshold Selection Method from Gray-Level Histograms," *IEEE Trans. on Systems, Man, and Cybernetics*, Vol. 9, No. 1, 1979, pp. 62-66.

22. Chassaing, R., *DSP Applications Using C and the TMS320C6x DSK* (Wiley, 2002), Chapter 8, Section 8.4, *Programming Examples Using Code Optimization Techniques*.

23. Texas Instruments, *Nested Loop Optimization on the TMS320C6x* (SPRA519).

24. Texas Instruments, *TMS320C6000 Optimizing Compiler User's Guide* (SPRU187k.pdf).

25. Texas Instruments, *TMS320C6000 Programmer's Guide* (SPRU198G.pdf).

26. Texas Instruments, *TMS320C64x Technical Overview* (SPRU395b), Appendix A.1, *Sum of Products Example*.

27. Texas Instruments, *TMS320C62x/64x FastRTS Library Programmer's Reference* (SPRU653.pdf).

28. Microsoft Developers Network (MSDN), see "What is a COM Object?" Retrieved October 2004 from:
http://msdn.microsoft.com/archive/default.asp?url=/archive/en-us/directx9_c/directx/intro/program/com/whatiscom.asp

29. Alexandrecu, A., "Adapting Automation Vectors to the Standard Vector Interface," *C/C++ User's Journal*, Apr. 1999;
Available online from: http://www.moderncppdesign.com

Chapter 6

WAVELETS
Multiscale Edge Detection and Image Denoising

Wavelets, the "little waves" of signal processing, came to the fore in the early 1990s as an attractive alternative to classical Fourier Transform based signal and image processing. While the underlying concepts behind wavelets have been known for close to a century (Haar described his multiresolution analysis in 1910), the pioneering work of many applied mathematicians have brought new insights into the field and wavelet theory and applications have found use in diverse areas like geophysical signal processing, medical imaging, and information theory. In this chapter, we look at wavelets from the image processing perspective, and develop wavelet-based algorithms for two operations previously studied in this book. The topic of edge detection was covered in 5.1, and in the first part of this chapter we develop a wavelet-based edge detection algorithm, and of course implement it on the DSP. The enhancement of images through various filtering schemes was the focus of Chapter 4, where we utilized linear, non-linear, and adaptive filters to *denoise* images. In the second part of this chapter, the topic of wavelet denoising is introduced and a fixed-point implementation tested on the C6416 is discussed.

Prior to embarking on this abbreviated tour of wavelet-based image processing, we must introduce at least a bare minimum of theory so that the algorithms and code make some sense to the reader. If the following discussion in 6.1 feels too abstract or mathematical in nature, I urge the reader to not get discouraged by its tone – it is not intended to be a treatise on wavelet theory, as this subject has been covered in depth by authors far more qualified than I. A large portion of the mathematical and signal processing background is skipped over in an attempt in an effort to convey the general idea behind the wavelet transform, and introduce enough of the theory to develop wavelet-based models of the two aforementioned image processing algorithms. A full treatment of wavelets easily warrants its own book, and there are many. More complete and mathematically rigorous discussions of wavelet theory can be found in [1-4].

6.1 MATHEMATICAL PRELIMINARIES

Transform-based methods are of fundamental importance in signal and image processing. A transform is a mathematical device that takes a function or signal represented in one domain and converts it to a form based in another domain. For example, in 4.1.3, we briefly glossed over the Fourier transform, the most common transform in signal processing applications. The forward Fourier transform can be said to take a signal in the "time-domain" to the "frequency-domain". Put more concretely, a one-dimensional time-domain signal, if plotted, would have its x-axis labeled as something akin to time, and in the two-dimensional (image) case, such a signal's axes might be labeled as position or pixel location. The Fourier transformed signal allows us to change these labels from time to frequency. Quite often it is the case that a transformed signal highlights certain characteristics we may be interested in analyzing, that are not readily apparent in its original form – such behavior is the essence of transform-based signal/image processing.

A transform is said to be invertible if an inverse transform can be used to recover the original signal from the transformed signal. The Fourier transform is invertible because an inverse transform exists that takes a signal in the frequency-domain back to its original time-domain representation. In general, one-dimensional linear transforms are of the form

$$f(x) = \sum_u a_u \psi_u(x)$$

Here, instead of viewing $y = f(x)$ in the usual Cartesian sense of a sequence of y values plotted against an x-axis, we express $f(x)$ as a linear decomposition of a weighted sum of basis functions $\psi_u(x)$. In the Fourier Transform, these basis functions are complex sinusoids – it has been shown that any signal can be expressed as a (possibly infinite) sum of sinusoids of different frequencies. The transform coefficients a_u are calculated from an inner product of $f(x)$ and $\psi_u(x)$:

$$a_u = \int f(x)\ \psi_u(x)\ dx$$

The key point for this discussion is that the basis functions for the classical linear transform $\psi_u(x)$ have infinite extent, or in other words they are non-zero for all x. Again using the Fourier transform as an example, this means the Fourier coefficients a_u depend on all parts of $f(x)$, even as $|x| \to \infty$. Thus, we have essentially lost all information about the time domain. So if we were to, say, analyze an audio signal consisting of a digitized symphonic work using the Fourier transform, we would obtain excellent frequency resolution and would be able to determine which frequencies (notes) were prevalent in that work. What we would not know, however, is when in time those notes occurred. This lack of *localization* with the Fourier and other related transforms is a major drawback, and is partly what led mathematicians to explore wavelet theory.

They found that in certain respects these classical linear transforms were not the ideal mathematical tool for analyzing signals whose behavior changes with time, or in the case of an image, a two-dimensional signal whose behavior changes with spatial location. Such signals are known as *non-stationary*, and most interesting signals encountered in real life exhibit this characteristic. While there are some means of circumventing the lack of localization in the Fourier transform (e.g., the so-called Short Time Fourier Transform or STFT, which uses overlapping windows and multiple FFTs), it eventually became clear that wavelets offer an extremely attractive alternative for analyzing non-stationary signals, overcoming many of the drawbacks associated with the Fourier transform.

In one dimension, the wavelet transform uses a two-parameter system that gives us the means to decompose a signal into various frequency bands and cut these bands into slices in time. The wavelet transform can be thought of as a specialized case of *subband* filtering, and the discrete wavelet transform (DWT) is given by

$$f(x) = \sum_k \sum_j a_{j,k} \psi_{j,k}(x)$$

where k and j are integer indices, and $\psi_{j,k}(x)$ are the wavelet basis functions. A wavelet expansion of a function $f(x)$ yields location information in both time and scale simultaneously, where scale can be related to frequency. Drawing on the symphony example, if $f(x)$ is the digitized audio signal, the DWT of that signal is akin to a musical score, because it tells us what frequencies were played and when those tones occurred. This, in a nutshell, is the overwhelming advantage the wavelet transform offers.

We proceed by first developing an understanding of how to implement the one-dimensional DWT, as the two-dimensional DWT – which is of most significance to us – follows naturally. The DWT is a *separable* transform, meaning we can implement the two-dimensional transform by performing two one-dimensional transforms in series. The manner in which this is typically done is by first transforming all the rows of an image using the one-dimensional DWT, followed by transforming the columns of the image, using the coefficients obtained from the row transforms.

Calculation of the DWT coefficients is done using a *filter bank*, which is a series of cascading digital filters. In 4.1.2, we described how digital filtering is accomplished via convolution, and in subband filtering the same concept holds. However, in subband filtering we use multiple filters (hence the term "bank") to decompose the input into its constituent parts. Implementing the DWT using a filter bank entails passing the signal through two digital filters: a low-pass or moving average filter (e.g., the box filter used in Algorithm 5-2 to smooth the image histogram) and another complementary high-pass, or moving difference, filter. The dual-channel subband filter that decomposes the input signal into its wavelet coefficients is the "analysis" filter bank and is one way of expressing the forward DWT. In the analysis phase of the DWT, the input signal is passed through both filters and this filtered output is then decimated, or downsampled, by a factor of two. Decimation is the process by which we take every other sample from an input to form the output, i.e. $y[n] = x[2n]$.

The inverse DWT (IDWT) reconstitutes the decomposed signal by passing the wavelet coefficients through a "synthesis" filter bank. The wavelet coefficients are upsampled by a factor of two (zeroes are inserted between every other sample) and then passed through time-reversed low-pass and high-pass filters. These two outputs are added together to reconstruct the original signal. If we let the symbols $*h$ denote convolution of a signal with a filter h, and $\downarrow m$ decimation by a factor of m, then the structure of a generalized one-dimensional two-channel perfect reconstruction orthogonal filter bank is given in Figure 6-1.

In an orthogonal filter bank, the synthesis filters g_{HP} and g_{LP} are time-reversed versions of the analysis filters. That is, if $h[n]$ is a low-pass or high-pass filter, with N the number of filter coefficients in h, then $g[n] = h[N-1-n]$ is the corresponding "flipped" synthesis kernel. Moreover, in an orthogonal filter bank, the analysis high-pass filter h_{HP} is related to the low-pass analysis filter h_{LP} in that h_{HP} is a time-reversed alternating negation of the low-pass kernel. In mathematical terms, we form the high-pass filter from the low-pass filter according to the following relation:

$$h_{HP}[n] = -1^n h_{LP}[N-1-n], \qquad n = 0,\ldots,N\text{-}1$$

where N is the number of taps in the low-pass filter. Thus, for this particular class of filter banks, the entire bank is completely defined by a single filter – the low-pass analysis filter. To summarize, if the length N of an analysis low-pass filter is 4, and

$$h_{LP} = \{h_0, h_1, h_2, h_3\}$$

then the rest of the filters in the perfect reconstruction orthogonal filter bank are defined by the following kernels:

$$h_{HP} = \{h_3, -h_2, h_1, -h_0\}$$
$$g_{LP} = \{h_3, h_2, h_1, h_0\}$$
$$g_{HP} = \{-h_0, h_1, -h_2, h_3\}$$

The obvious next question is where does the low-pass analysis filter h_{LP} come from? Essentially, there is any number of choices and the exact choice comes down to a filter design problem, with some global constraints imposed on the design space. The analysis filters' impulse response is directly tied to the mother wavelet, which as we stated must be of limited duration and furthermore have zero mean. Even with these constraints the design space is vast, with various other factors coming into play, such as smoothness, length of the filter's impulse response, vanishing moments (differentiability, i.e. the more vanishing moments in the wavelet the more differentiable it is), and so on. The derivation of h_{LP} from a mother wavelet can be found in the references, and in the remainder of this chapter we will

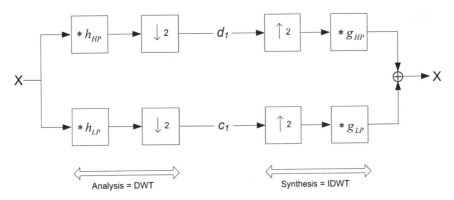

Figure 6-1. Single level, one-dimensional perfect reconstruction orthogonal filter bank. Taken together, c_1 (approximation) and d_1 (detail) are the complete DWT coefficients a_u. Each of these vectors is one-half the length of the input signal X.

use a few of the more common filter banks, in particular the Haar and Daubechies systems.

The left-hand side of Figure 6-1 depicted a single stage of a wavelet decomposition, yielding the wavelet coefficients c_i and d_i at a single scale. We can cascade this filter in a few different fashions to carry the decomposition one level further, with one quite common method known as the pyramid decomposition structure. The pyramid structure is shown in Figure 6-2, where the approximation wavelet coefficients c_1 are fed into the same two-channel filter bank, which in turn emits another set of approximation and detail coefficients. Thus as the cascade proceeds in a *dyadic* (power-of-two) fashion, the DWT effectively tiles the time-frequency space. This tiling is known as a *multi-resolution analysis* (MRA), and really gets to the heart of the primary advantage of the wavelet transform – it solves the localization problem of the Fourier and other classical transforms. The MRA decomposes the signal into nested subspaces, thereby giving us the ability to know not only if a particular event or characteristic occurred, but *when*. Using this formulation of the wavelet decomposition, the output of the analysis cascade is arranged as shown in Figure 6-3, using the example of a three-level wavelet decomposition. The synthesis filter bank reconstitutes the input signal using a similar structure to that of Figure 6-2, but with different filters and the downsampling operation replaced with upsampling of the DWT coefficients (see Figure 6-4).

Another less common decomposition structure, not utilized in the algorithms presented in this book, is the wavelet packet method. In this formulation, both the approximation and detail coefficients are decomposed at each cascade level, with the end result a binary tree (in one dimension). Wavelet packet analysis does offer a richer signal analysis, although at the cost of some complexity.

Now that we have introduced the mechanics behind the wavelet transform, one can now envision how a wavelet-based processing system might work. In a wavelet-based image processing methodology, one can envision first performing the DWT, and then subsequently manipulating the DWT coefficients. After the modifications have been carried out, the altered DWT coefficients are fed into the synthesis filter bank, thereby yielding a processed image. In 6.3, we use this methodology to denoise images using the DWT.

6.1.1   Quadrature Mirror Filters and Implementing the 2D DWT in MATLAB

The oldest and simplest wavelet is the Haar system, defined by the following analysis filters:

$$h_{LP} = \left\{\frac{1}{\sqrt{2}}, \frac{1}{\sqrt{2}}\right\} = \frac{1}{\sqrt{2}}\{1,1\}$$

$$h_{HP} = \left\{\frac{1}{\sqrt{2}}, -\frac{1}{\sqrt{2}}\right\} = \frac{1}{\sqrt{2}}\{1,-1\}$$

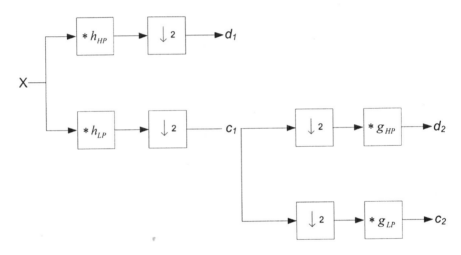

Figure 6-2. Analysis phase of a two-level pyramid wavelet filter bank. This cascading bank of digital filters produces the DWT coefficients as the vectors $\{c_2, d_2, d_1\}$.

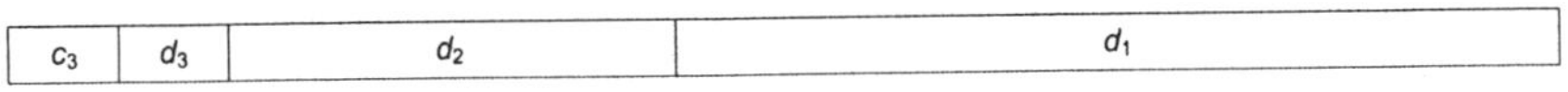

c_3	d_3	d_2	d_1

Figure 6-3. Typical arrangement of the output of a one-dimensional three-level DWT implemented via a pyramid cascade of digital filters.

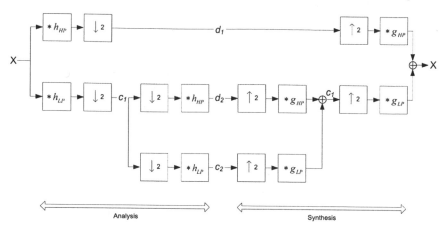

Figure 6-4. Two-level pyramidal filter bank that produces in $\{c_2, d_2, d_1\}$ the two-level DWT.

The filter defined by h_{LP} outputs a normalized moving average of its input, while h_{HP} outputs a normalized moving difference. The Haar basis functions (the mother wavelet) are discontinuous, and for that reason are not used very often in real applications, except for some specialized video processing. However, due to its simplicity, it is instructive to illustrate its usage in the processing of images. The application of more sophisticated wavelets follows from the general idea behind the implementation of the Haar wavelet transform.

The filter bank tree structures shown in Figures 6-1, 6-2, and 6-4 are known as *Quadrature Mirror Filters* (QMF). In [1], it is shown that a QMF tree can be efficiently represented in matrix form. For the forward DWT, the convolution and decimation operations can be combined into a single matrix operation. In general, a *polyphase* matrix H is constructed from the analysis low-pass and analysis high-pass filter coefficients. Remembering that h_{HP} is the time-reversed alternating negation of h_{LP}, if h_i are the analysis low-pass filter coefficients and the length of h is 4 (for the Haar wavelet the length is 2), then the format of the block-diagonal matrix H is:

$$H = \begin{bmatrix} h_0 & h_1 & h_2 & h_3 & & & & & & \\ h_3 & -h_2 & h_1 & -h_0 & & & & & & \\ & & h_0 & h_1 & h_2 & h_3 & & & & \\ & & h_3 & -h_2 & h_1 & -h_0 & & & & \\ & & & & \ddots & & \ddots & & & \\ & & & & & \ddots & & \ddots & & \\ & & & & & & h_0 & h_1 & h_2 & h_3 \\ & & & & & & h_3 & -h_2 & h_1 & -h_0 \\ h_2 & h_3 & & & & & & & h_0 & h_1 \\ h_1 & -h_0 & & & & & & & h_3 & -h_2 \end{bmatrix}$$

The structure of H is such that the length of the output (transformed) signal is exactly the same as the input. Each successive row in H is translated by 2 samples, so that when we multiply the input signal by this forward transform matrix we achieve the downsampling operation. Moreover, multiplication of the input and the transform matrix yields a circular convolution, hence the wrap-around effect at the bottom rows – the input data is treated as a periodic sequence of samples. The forward transform matrix for the Haar system is given below – note that there is no wrap-around in this case because the Haar system consists of only two filter coefficients and we are dealing with dyadic scales:

$$H = \frac{1}{\sqrt{2}} \begin{bmatrix} 1 & 1 & 0 & \cdots & & & 0 \\ 1 & -1 & 0 & \cdots & & & 0 \\ 0 & 0 & 1 & 1 & & & \vdots \\ \vdots & \vdots & 1 & -1 & & & \\ & & & \ddots & \ddots & & \\ & & & & \ddots & \ddots & \\ 0 & \cdots & & & & 1 & 1 \\ 0 & \cdots & & & & 1 & -1 \end{bmatrix}$$

These block-diagonal transform matrices are also known as block-Toeplitz matrices, and in MATLAB the Kronecker tensor product can be used to quickly construct such matrices in an elegant fashion. The built-in

MATLAB function `kron`[5] can be used to form a 32x32 forward Haar transform matrix with the following code:

```
I = eye(16); % 16x16 identity matrix
haar = [ 1 1; 1 -1];
H = (1/sqrt(2)) .* kron(I, haar);
```

If s_i are the samples of a one-dimensional signal s, then we compute one level of the DWT by multiplying H by s:

$$
\begin{bmatrix}
c_{1,0} \\
d_{1,0} \\
c_{1,1} \\
d_{1,1} \\
\vdots \\
\vdots \\
c_{1,N-1} \\
d_{1,N-1}
\end{bmatrix}
=
\begin{bmatrix}
h_0 & h_1 & h_2 & h_3 & & & & & \\
h_3 & -h_2 & h_1 & -h_0 & & & & & \\
& & h_0 & h_1 & h_2 & h_3 & & & \\
& & h_3 & -h_2 & h_1 & -h_0 & & & \\
& & & & \ddots & & \ddots & & \\
& & & & & \ddots & & \ddots & \\
& & & & & h_0 & h_1 & h_2 & h_3 \\
& & & & & h_3 & -h_2 & h_1 & -h_0 \\
h_2 & h_3 & & & & & & h_0 & h_1 \\
h_1 & -h_0 & & & & & & h_3 & -h_2
\end{bmatrix}
\begin{bmatrix}
s_0 \\
s_1 \\
s_2 \\
s_3 \\
\vdots \\
\vdots \\
\\
s_{N-1}
\end{bmatrix}
$$

The vector on the left-hand side of the above equation contains in $c_{1,i}$ and $d_{1,i}$ the approximation and detail wavelet coefficients respectively, but in order to arrange the output in the form shown in Figure 6-3, we need to shuffle the coefficients around. We do this not because it is aesthetically pleasing to do so, but because it is required in order to implement a multi-level DWT. To rearrange the coefficients appropriately, we define the permutation matrix P as a reordered identity matrix that moves all odd-numbered rows to the top-half of the output matrix, and puts all even-numbered rows in the bottom half of the output matrix (reminiscent of the butterfly structure prevalent in the FFT algorithm). Using the definitions of the matrices S, H, and P, a single iteration through one of the cascade stages of the QMF tree is

$$output^{(1)} = PHS$$

Now, to perform one level of the DWT on an image S, we take advantage of the fact that the DWT is separable. We first transform the rows of S using the above equation, which holds for both a one-dimensional Nx1 signal and a two-dimensional MxN signal (image). After applying successive one-dimensional DWTs on the rows of the image, we transform the columns of the image by transposing the matrices, as shown below:

$$output^{(1)} = PHSH^T P^T$$

A variety of MATLAB M-files illustrating how to implement the DWT and IDWT can be found in the `Chap6\wavelet` directory. Listing 6-1 are the contents of three of those, which collectively implement a single-level Haar wavelet decomposition.

Listing 6-1: MATLAB functions for performing the Haar 2D DWT on a square image.

```
function J = haar_2d_dwt(I)

[M N] = size(I);
if M ~= N
    error('Only support square images.');
end

H = haar_basis(N);
P = permutation_matrix(N);
A = P*H;
B = H'*P';
J = A*double(I)*B; % forward transform

%
% haar_basis
%
function H = haar_basis(N)

haar_wavelet = [1 1; 1 -1];
H = (1/sqrt(2)) .* kron(eye(N/2),haar_wavelet);
```

```
%
% permutation_matrix
%
function P = permutation_matrix(N)

I  = eye(N);
P  = zeros(N,N);

odd_indices = 1:2:N;
P(1:N/2,:) = I(odd_indices,:);

even_indices = 2:2:N;
P(N/2+1:N,:) = I(even_indices,:);
```

Implementing the DWT using matrix operations is trivial in MATLAB, as the programming language inherently supports matrix multiplication and inversion (which as we shall soon see is important when performing the reverse transform). In `haar_2d_dwt`, the apostrophe character, which in MATLAB denotes matrix transposition, is used to effectively move the transform over to the columns of the image. This implementation of the Haar DWT is limited to square images where the length of all sides is a power-of-two, and a single level of the Haar wavelet decomposition of an image I may be visualized using `imshow(haar_dwt(I),[])` as shown in Figure 6.5.

The 2D DWT, with its output arranged in the format illustrated graphically in Figure 6-5, effectively splits the input image into four bands: a low-pass filtered approximation "low-low" (LL) subimage, and three high-pass filtered detail subimages, "low-high" (LH), "high-low" (HL), and "high-high" (HH). These subbands are shown in Figure 6-6.

The wavelet approximation coefficients in the $LL^{(1)}$ subband are nothing more than a subsampled version of the input, while the three high-pass filtered subimages $LH^{(1)}$, $HL^{(1)}$, and $HH^{(1)}$ accentuate the horizontal, vertical, and diagonal edges respectively. As one might suspect, we will use this fact to our advantage during implementation of a wavelet-based edge detection scheme in 6.2.2. To continue the pyramidal wavelet decomposition, we recursively perform the same DWT matrix operations on the LL subband, which for a two-level DWT produces an output image like that shown in Figure 6-7. We can continue this process until the $LL^{(n)}$ subimage is nothing more than a 2x2 square that is an average of the entire image. Listing 6-2 is a MATLAB function that performs the 2D Haar pyramid DWT on an image for a given number of decomposition levels. Example output from this function is shown in Figure 6-8.

(a) (b)

Figure 6-5. One level of the Haar wavelet decomposition. (a) Original Lenna image. (b) The result from the `haar_2d_dwt` MATLAB function of Listing 6-1. This matrix, consisting of the 2D DWT coefficients, has been shifted, scaled, and contrast-stretched for optimal display using an 8-bit dynamic range. The upper-left quadrant contains the approximation coefficients, while the other three quadrants are the various detail coefficients.

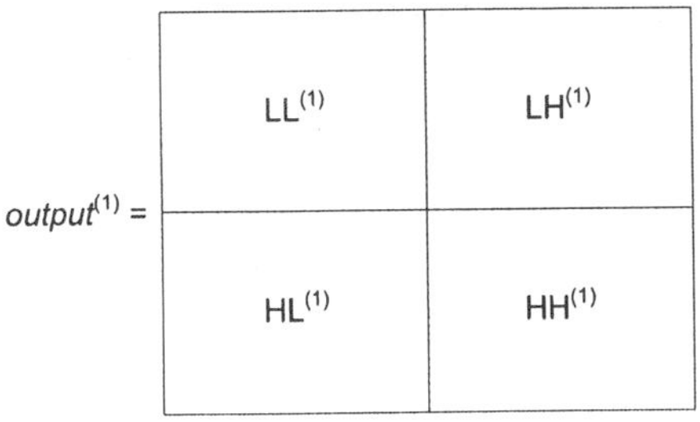

$$output^{(1)} = \begin{array}{|c|c|} \hline LL^{(1)} & LH^{(1)} \\ \hline HL^{(1)} & HH^{(1)} \\ \hline \end{array}$$

Figure 6-6. Decomposing an image into four subbands using the 2D DWT.

LL$^{(2)}$	LH$^{(2)}$	LH$^{(1)}$
HL$^{(2)}$	HH$^{(2)}$	
HL$^{(1)}$		HH$^{(1)}$

$output^{(2)} =$

Figure 6-7. Two-level 2D pyramid DWT.

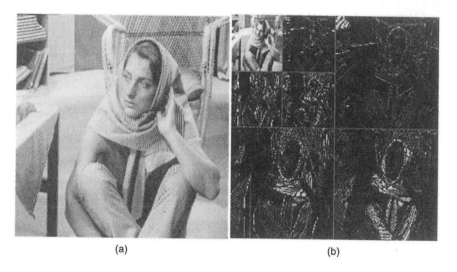

(a)   (b)

Figure 6-8. Two-level Haar wavelet decomposition. (a) Original Barbara image. (b) The result from the `haar_2d_dwt_n` MATLAB function. This 2D DWT coefficient matrix has been contrast-stretched for an 8-bit dynamic range, and artificial gray border lines delineate the seven subbands.

Listing 6-2: MATLAB function for performing a multi-level Haar 2D DWT on a square image.

```
function J = haar_2d_dwt_n(I, N)

I = double(I);
[nrows ncols] = size(I);
if nrows ~= ncols,  error('Only support square images.');, end

num_levels = log2(nrows);
if num_levels ~= floor(num_levels)
```

```
      error('Image must have power-of-2 dimensions.');
   end


   J = I;
   H = haar_basis(2^num_levels);
   for kk = num_levels:-1:num_levels-N+1
      sz = 2^kk;
      P = permutation_matrix(sz);
      J(1:sz,1:sz) = P*H(1:sz,1:sz)*J(1:sz,1:sz)*H(1:sz,1:sz)'*P';
   end
```

Implementing the 2D IDWT proceeds along similar lines to that shown in Figure 6.4. Using the definition of *output*[(1)] in terms of the H and P matrices, and if S is now treated as a matrix of wavelet coefficients, the IDWT is calculated as

$$IDWT = (PH)^{-1} S (H^T P^T)^{-1}$$

To compute the IDWT from an N-level DWT, one must proceed up the filter tree by first reconstituting $LL^{(n)}$, then $LL^{(n-1)}$, and so on all the way up to $LL^{(1)}$. This iterative procedure is illustrated in haar_2d_idwt_n, shown in Listing 6-3.

Listing 6-3: MATLAB function for performing a multi-level Haar 2D IDWT on a square image.

```
   function J = haar_2d_idwt_n(I, N)

   I = double(I);
   [nrows ncols] = size(I);
   if nrows ~= ncols
      error('Only support square images.');
   end


   num_levels = log2(nrows);
   if num_levels ~= floor(num_levels)
      error('Image must have power-of-2 dimensions.');
   end


   if N > num_levels
      error('# of levels exceeds log2(length of image).');
   end
```

```
J = I;
H = haar_basis(2^num_levels);
for kk = N:-1:1
    sz = nrows/(2^(kk-1));
    P = permutation_matrix(sz);
    A = inv( P*H(1:sz,1:sz) );
    B = inv( H(1:sz,1:sz)' * P' );
    J(1:sz,1:sz) = A*J(1:sz,1:sz)*B;
end
```

6.1.2 The Wavelet Toolbox

There are many frequently used wavelet bases, and prototyping wavelet-based algorithms is easily accomplished in MATLAB using the Wavelet Toolbox. This toolbox includes a plethora of functions for dealing with wavelet transforms. Since this book's focus is on image processing, we restrict our discussion to those capabilities pertaining to the 2D DWT and IDWT. However, the Wavelet Toolbox contains numerous other functions for performing one-dimensional wavelet transforms, as well as functions for analyzing signals using the continuous wavelet transform. This section briefly introduces this toolbox, and describes some of the functions one may find useful in image processing. More detailed information can be found in [6].

The `dwt2` function can be used to perform a single-level wavelet decomposition. There are two usage patterns of this function, both of which expect the image matrix to be the first argument. One form of `dwt2` is generic in that the user provides the low-pass and high-pass filter coefficients to the function. The other form allows the user to specify certain canned wavelet decompositions, which are selected by treating the second argument as a string specifying the wavelet basis to be used for the wavelet analysis. Thus

```
[LL, LH, HL, HH] = dwt2(I, 'haar' );
```

where `I` is a variable containing the pixel data, performs the Haar DWT and returns the approximation coefficients in `LL` and the three detail coefficient matrices depicted in Figure 6-6 in `LH`, `HL`, and `HH`. Two other common bases are the "Daubechies-4" (D4) and "Daubechies-6" (D6) wavelets[4]. The D6 wavelet is implemented using six-tap filters, and implementing the 2D D6 DWT using hand-crafted functions similar to Listing 6-2 involves constructing a periodized H matrix that looks like the following:

$$
H_{D6} = \begin{bmatrix}
h_0 & h_1 & h_2 & h_3 & h_4 & h_5 & & & & & & & \\
h_5 & -h_4 & h_3 & -h_2 & h_1 & -h_0 & & & & & & & \\
& & h_0 & h_1 & h_2 & h_3 & h_4 & h_5 & & & & & \\
& & h_5 & -h_4 & h_3 & -h_2 & h_1 & -h_0 & & & & & \\
& & & & & & \ddots & & \ddots & & & & \\
& & & & & & & & h_0 & h_1 & h_2 & h_3 & h_4 & h_5 \\
& & & & & & & & h_5 & -h_4 & h_3 & -h_2 & h_1 & -h_0 \\
h_4 & h_5 & & & & & & & & & h_0 & h_1 & h_2 & h_3 \\
h_1 & -h_0 & & & & & & & & & h_5 & -h_4 & h_3 & -h_2 \\
h_2 & h_3 & h_4 & h_5 & & & & & & & & & h_0 & h_1 \\
h_3 & -h_2 & h_1 & -h_0 & & & & & & & & & h_5 & -h_4
\end{bmatrix}
$$

where $\{h_0, h_1, h_2, h_3, h_4, h_5\}$ are the analysis low-pass filter coefficients for the D6 wavelet[4]. If the Wavelet Toolbox is available, dwt2 can be used instead by providing the appropriate string to indicate the desired wavelet. Upon encountering a string indicating which wavelet family to use, dwt2 actually invokes another Wavelet Toolbox function, wfilters, which returns the actual filter coefficients associated with that particular wavelet family. The following code illustrates both methods of performing the D4 and D6 wavelet decomposition on an image I:

```
[D4LL, D4LH, D4HL, D4HH] = dwt2(I, 'db2'); % Daubechies-4
% or specify the D4 decomposition filters explicitly
[d4_lo_d, d4_hi_d] = wfilters('db2', 'd');
[D4LL, D4LH, D4HL, D4HH] = dwt2(I, d4_lo_d, d4_hi_d);

[D6LL, D6LH, D6HL, D6HH] = dwt2(I, 'db3'); % Daubechies-6
% or specify the D6 decomposition filters explicitly
[d6_lo_d, d6_hi_d] = wfilters('db3', 'd');
[D6LL, D6LH, D6HL, D6HH] = dwt2(I, d6_lo_d, d6_hi_d);
```

The output of dwt2 can be combined into a single matrix via MATLAB's built-in matrix concatenation operations. For example,

```
D6 = [D6LL D6LH; D6HL D6HH];
```

It is oftentimes difficult to visualize the detail coefficient matrices due to the scaling of the coefficients – the wcodemat function is useful for displaying pseudo-color representations of wavelet coefficient matrices. The inverse 2D DWT toolbox function is idwt2 and reconstitutes the original image from the four subbands. For example,

```
[D4LL, D4LH, D4HL, D4HH] = dwt2(I, 'db2'); % Daubechies-4
J = idwt2(D4LL, D4LH, D4HL, D4HH, 'db2'); % synthesis
```

In addition, there is another form of idwt2 that expects the user to pass in the actual synthesis low-pass and high-pass filters for reconstruction of the image. Performing a multi-level wavelet decomposition and corresponding reconstruction is more involved than the single-level case. While it is certainly possible to put dwt2 and idwt2 in a loop, the Wavelet Toolbox provides a few very convenient functions that should be used instead. A multi-level pyramid wavelet analysis can be performed in one fell swoop using wavedec2, an example of which is shown below for a three-level decomposition of an image I:

```
[D6, S] = wavedec2(I, 3, 'db3'); % 3-level Daubechies-6
```

S is a bookkeeping matrix that maintains size information about each level's components. Extraction of approximation and detail coefficients from a multi-level decomposition is accomplished via appcoef2 and detcoef2. Extraction of the approximation coefficients via appcoef2 is straightforward; one passes in the coefficient and bookkeeping matrices returned from wavedec2, the level, and either a wavelet family string or the actual synthesis low-pass and high-pass filter coefficients:

```
D6_LL2 = appcoef(D6, S, 'db3', 2); % LL2 from 3-level D6
```

Extracting the detail coefficients can be done en masse (for a single level), or each of the LH, HH, and HL subbands may be extracted separately. Extraction of each subband is specified in detcoef2 with a string argument, where 'h' refers to the horizontal edge subband (LH), 'v' refers to the vertical edge subband (HL), and 'd' refers to the diagonal edge subband (HH). Some examples follow:

```
D6_HH1 = detcoef2('d', D6, S, 1); % HH1 from 3-level D6
[D6_LH3,D6_HL3,D6_HH3] = detcoef2('all', D6, S, 3);
```

Use wrcoef2 to reconstruct either an entire image or certain subbands from a multi-level decomposition. For example, to reconstruct the original image from a three-level 2D D6 DWT, the equivalent IDWT is achieved using this command:

J = wrcoef2(C, S, 'db3'); % full IDWT from 3-level D6

A form similar to `detcoef2` is used in `wrcoef2` to synthesize a particular subband. The first argument is a string, with the acceptable values being the same as `detcoef2`, except in `wrcoef2` 'a' synthesizes the level-n approximation. For example,

D6_LL1 = wrcoef2('a' , C, S, 'db3' , 1); % level-1 approximation
D6_HL2 = wrcoef2('v' , C, S, 'db3' , 2); % HL2 from 3-level D6

Finally, an extremely powerful and sophisticated graphical user interface for the Wavelet Toolbox is available. To access this GUI, type `wavemenu` at the MATLAB prompt, and for more information refer to [6] or the online help.

6.1.3   Other Wavelet Software Libraries

A variety of free Wavelet software packages and libraries exist for MATLAB. One of the more fully featured of these is the WaveLab package that can be downloaded from http:://www-stat.stanford.edu. The DSP group at Rice University has implemented the Rice Wavelet Toolbox (RWT), portions of which accompany [2] and can be downloaded from http://www-dsp.rice.edu/software/rwt.shtml. Finally, "Amara's Wavelet Page", accessed at http://www.amara.com/current/wavelet.html is a fabulous one-stop shop for wavelet information and software in particular.

The Intel IPP Library includes several functions for implementing the 2D DWT on Pentium-based systems. As is the case with the rest of the library, the functions are heavily optimized and chances are that a faster implementation targeting the Pentium processor does not exist. Of course, the source code is not available, but nonetheless as we have demonstrated in previous applications having an algorithm working on the desktop in a C/C++ environment has its own benefits. For more details, refer to [7].

6.1.4   Implementing the 2D DWT on the C6416 DSK with IMGLIB

The C6x IMGLIB wavelet functions are a perfect example of tailoring the mathematics of an algorithm to a form commensurate with the computing environment. Recall that we implemented the 2D DWT in 6.1.1 by first performing a series of row DWTs, followed by a series of column

DWTs on the transformed data. We expressed this computation in MATLAB rather elegantly by simply transposing the circular convolution and permutation matrices, effectively moving the filtering operation to operate down the columns of the input image. Unfortunately, expressive elegance does not always lend itself to computational efficiency, as was evident in 4.5.4.3 when we saw how inefficient qsort was when used to implement the median filter. Elegance should always take a back seat to swift processing in deeply embedded software.

Performing the 2D DWT on the C6x DSP using IMGLIB is done via the IMG_wave_horz and IMG_wave_vert functions. TI's implementation of the 2D DWT is unique in that the transformation of the image columns is performed in-place, allowing us to avoid an expensive matrix transposition operation, which impacts both storage requirements and computation cycles. Moreover, the functions are constructed in such a manner that making it fairly straightforward to DMA in blocks of pixels during the horizontal (across the rows) and in particular the vertical (down the columns) phases of the DWT. In this fashion we are able to use the 2D DWT in a memory-constrained system, even though the mechanics of the transform diverge somewhat from the standard theoretical formulation. To that end, in this section we develop the forward 2D DWT via IMGLIB on the C6416 DSK in a series of stages, similar to what we did in 4.3 and 4.4 where we optimized linear filtering of images on a step-by-step basis. Our roadmap to the 2D DWT follows:

1. A single-level 2D DWT program that pages blocks of image data in and out of internal memory as needed using memcpy (2d_dwt_single_level.c).
2. Same as (1), except we augment the processing to provide a multi-level 2D DWT (2d_dwt_multi_level.c).
3. A multi-level 2D DWT program that performs some parallel paging of data via DMA operations (2d_dwt_multi_level_dma.c).

The project can be found in the Chap6\IMGLIB_2d_dwt directory, where there are three C source files pertaining to each of the aforementioned programs. To build each project, add the appropriate source file to the project in CCStudio, remembering to remove any of the other C source files from the project to avoid duplication of symbols (failure to do so will result in a linker error). The 256x256 "Lenna" image is included within the image.h file, and due to the in-place nature of the implementation the input and output buffer are one and the same (wcoefs).

The IMGLIB functions limit the length of the low-pass and high-pass wavelet filters to 8 coefficients, which is sufficient for most applications. For those filters defined by less than 8 coefficients, the double-word aligned array containing the coefficients is padded with zeros. These 2D DWT programs all use the D4 wavelet, and the analysis filter coefficients are stored in Q15 format in d4_qmf_Q15 and d4_mqmf_Q15.

Taking advantage of faster on-chip RAM by paging data into buffers strategically placed in on-chip memory is always a priority in data-intensive DSP applications (which is essentially the same as saying it is a priority in all DSP applications), but even more so with the IMGLIB wavelet functions. The IMGLIB wavelet functions are fed 16-bit signed integers, and output the transformed data as 16-bit signed integers. A Q15 format dictates 16-bit output, but the fact that the input must also be 16 bits may seem wasteful at first glance – might it be more efficient for IMG_wave_horz to accept 8 bpp image data? The answer is yes, but only for the case of a single-level wavelet decomposition. To implement a multi-level pyramidal wavelet decomposition, we will continue by repeatedly calling the IMGLIB functions, each time passing them the upper-left quadrant of the previous level's output – the LL subband. Hence, it makes sense to stipulate that all input data be 16 bits. Additionally, the DWT is directly tied to convolution, and we have seen how the convolution operation requires repeated accesses to the data as one slides across the input signal. As a result, it becomes an absolute necessity to construct the implementation so that it is relatively easy to integrate paging of data with the actual processing.

Verification of these programs can be done by spooling the contents of the wcoefs image buffer to disk using the CCStudio File I/O facilities explained in 3.2.2. The MATLAB read_ccs_data_file, which fully supports signed 8 and 16 bit data, can then be used to import the transformed data into MATLAB. The data can then be visualized using either imshow or if the Image Processing Toolbox is not available, imagesc. Note that the buffer length should be set to 32768 when saving the 256x256 16-bit image matrix from within CCStudio, as a 256x256 coefficient matrix contains 64K elements and each line in the CCStudio hex formatted output file consists of a single word, for a total of $(256)(256)/2 = 32$K elements.

6.1.4.1 Single-Level 2D DWT

Listing 6-4 shows the contents of 2d_dwt_single_level.c. A single wavelet decomposition is performed in two steps, transformation of the rows in transform_rows followed by transformation of that output in the orthogonal direction within transform_cols. Passing each row of the input image through the low-pass and high-pass filters constituting the Quadrature Mirror Filter is fairly straightforward. In transform_rows, we page in 8 rows at a time of image data into an on-chip memory buffer wvlt_in_buf with a call to the standard C library function memcpy. We then call IMG_wave_horz on a row-by-row basis, feeding the output to another on-chip memory buffer wvlt_out_buf. These 8 rows of processed horizontal data are then paged back out to a temporary storage buffer horzcoefs, which serves as input data for the next phase of the 2D DWT.

Listing 6-4: 2d_dwt_single_level.c.

```
#define CHIP_6416
#include <dsk6416.h>
#include <string.h> /* memcpy() */
#include "IMG_wave_horz.h"
#include "IMG_wave_vert.h"

#include "image.h" /* image & kernel dimensions,example pixel data */
#pragma DATA_ALIGN (wcoefs, 8);
#pragma DATA_SECTION (wcoefs, "SDRAM");

/* D4 WAVELET ANALYSIS FILTER COEFFICIENTS (Q15) */
#pragma DATA_ALIGN (d4_qmf_Q15, 8);
#pragma DATA_ALIGN (d4_mqmf_Q15, 8);
short d4_qmf_Q15[] = {-4240,7345,27411,15826,0,0,0,0},
      d4_mqmf_Q15[] = {-15826,27411,-7345,-4240,0,0,0,0};

/* BUFFERS USED DURING WAVELET TRANSFORM */
#pragma DATA_ALIGN (wvlt_in_buf, 8);
short wvlt_in_buf[Y_SIZE*8]; /* scratch (input) buffer */

#pragma DATA_ALIGN (wvlt_out_buf, 8);
short wvlt_out_buf[Y_SIZE*8]; /* scratch (output) buffer */

short *pwvbufs[8]; /* IMG_wave_vert() input */

#pragma DATA_ALIGN (horzcoefs, 8);
#pragma DATA_SECTION (horzcoefs, "SDRAM");
short horzcoefs[N_PIXELS]; /* IMG_wave_horz() output */

/* horizontal wavelet transform, output goes into horzcoefs */
void transform_rows()
{
  const int nBlocks = X_SIZE>>3; /* rows/8 */
  int iRow=0, iBlock=0, kk;
```

```
    short *pin, *pout;

    /* pass rows through DWT, in groups of 8 scan-lines*/
    do {
      /* fetch the next group of 8 rows */
      memcpy(wvlt_in_buf,
             &wcoefs[iRow*Y_SIZE],
             8*Y_SIZE*sizeof(short));
      pin = wvlt_in_buf;
      pout = wvlt_out_buf;
      for (kk=0; kk<8; ++kk, pin+=Y_SIZE, pout+=Y_SIZE)
        IMG_wave_horz(pin, d4_qmf_Q15, d4_mqmf_Q15, pout, Y_SIZE);

      /* page out horizontal wavelet coeffs to ext mem storage */
      memcpy(horzcoefs+iRow*Y_SIZE,
             wvlt_out_buf,
             8*Y_SIZE*sizeof(short));
      iRow += 8;
    } while (++iBlock <= nBlocks);
}

/* grab next two lines for IMG_wave_vert, see Figure 6-9 */
inline int fetch_horz_wavelet_scanlines(int r)
{
  short *ptemp1 = pwvbufs[0], *ptemp2 = pwvbufs[1];
  pwvbufs[0] = pwvbufs[2];
  pwvbufs[1] = pwvbufs[3];
  pwvbufs[2] = pwvbufs[4];
  pwvbufs[3] = pwvbufs[5];
  pwvbufs[4] = pwvbufs[6];
  pwvbufs[5] = pwvbufs[7];
  pwvbufs[6] = ptemp1;
  pwvbufs[7] = ptemp2;
  memcpy(pwvbufs[6], horzcoefs+r*Y_SIZE, sizeof(short)*Y_SIZE);
  memcpy(pwvbufs[7], horzcoefs+(r+1)*Y_SIZE, sizeof(short)*Y_SIZE);
  return r+2;
}

/* vertical wavelet transform, output goes into wcoefs */
void transform_cols()
{
  const int nRowsDiv2 = X_SIZE>>1,
```

```
            circular = nRowsDiv2-1;
      int lpRow = nRowsDiv2-3, /* low-pass vert output */
         hpRow = nRowsDiv2,   /* high-pass vert output */
         fetchRow = 2, iRow;
      short *plpvc = wvlt_out_buf, /* ptr to low-pass vert coeffs */
          *phpvc = wvlt_out_buf+Y_SIZE; /* high-pass vert coeffs */

      /* setup scan-lines for DWT down the columns */
      memcpy(wvlt_in_buf, horzcoefs+N_PIXELS-6*Y_SIZE, 6*Y_SIZE);
      memcpy(wvlt_in_buf+6*Y_SIZE, horzcoefs, 2*Y_SIZE);
      pwvbufs[0] = wvlt_in_buf;
      pwvbufs[1] = wvlt_in_buf+Y_SIZE;
      pwvbufs[2] = wvlt_in_buf+2*Y_SIZE;
      pwvbufs[3] = wvlt_in_buf+3*Y_SIZE;
      pwvbufs[4] = wvlt_in_buf+4*Y_SIZE;
      pwvbufs[5] = wvlt_in_buf+5*Y_SIZE;
      pwvbufs[6] = wvlt_in_buf+6*Y_SIZE;
      pwvbufs[7] = wvlt_in_buf+7*Y_SIZE;

      for (iRow=0; iRow<nRowsDiv2; ++iRow) {
        IMG_wave_vert(pwvbufs, d4_qmf_Q15, d4_mqmf_Q15,
                      plpvc, phpvc, Y_SIZE);
        memcpy(wcoefs+lpRow*Y_SIZE, plpvc, sizeof(short)*Y_SIZE);
        memcpy(wcoefs+hpRow*Y_SIZE, phpvc, sizeof(short)*Y_SIZE);
        fetchRow = fetch_horz_wavelet_scanlines(fetchRow);
        lpRow += 1; if (lpRow>circular) lpRow=0;
        hpRow += 1;
      }
    }

    /* single wavelet decomposition */
    void dwt2d()
    {
     transform_rows();
     transform_cols();
    }

    int main(void)
    {
     DSK6416_init(); /* initialize the DSK board support library */
     dwt2d();
    }
```

A simplistic implementation of the 2D DWT could then proceed to follow the example of the MATLAB implementations in Listing 6-2 and 6-3, i.e. just transpose `horzcoefs` and reuse `IMG_wave_horz`. This would work, but a far cheaper method is achieved though the use of `IMG_wave_vert`. This function is repeatedly invoked in `transform_cols`, and it is here where this implementation of the 2D DWT breaks from the traditional model. The usage of `IMG_wave_vert` is thoroughly explained in the IMGLIB documentation[8] – the reason for its somewhat non-intuitive nature is to avoid a matrix transposition. The mechanics of the column-wise transformation is perhaps best garnered through close inspection of the code, but Figure 6-9 illustrates in schematic form how the columns of the `horzcoefs` matrix are processed. A total of 8 rows of `horzcoefs` are passed into `IMG_wave_vert` via an array of pointers (`pwvbufs`), and the vertical filter is then traversed across the width of the entire image, thereby yielding a row of low-pass filtered data and a row of high-pass filtered data. This constitutes the final output for this level of the 2D DWT and both of these rows are then paged back out to the `wcoefs` buffer. The next set of horizontal wavelet coefficients are paged in from external RAM in `fetch_horz_wavelet_scanlines`. With every invocation, six of the eight pointers are shifted up by two slots, and the next two rows are then fetched via `memcpy`. After `transform_cols` completes its march through the `horzcoefs` matrix, a single level of the wavelet decomposition has been completed and resides within `wcoefs`.

Note that we could replace `memcpy` in this and the following program with the `DSP_blk_move` DSPLIB function in order to gain a modest performance increase. The buffers and lengths used in this application meet the requirements (word-alignment and lengths a multiple of 2), but we choose not to do so because in 6.1.4.3 we use DMA as a much more effective optimization technique.

6.1.4.2 Multi-Level 2D DWT

A 2D DWT program that supports multiple wavelet decomposition levels is shown in Listing 6-5. The main complication arising with a multi-level 2D DWT has to do with the bookkeeping associated with the dyadic nature of the DWT. The `transform_rows` and `transform_cols` functions from Listing 6-4 have been modified to accept an argument indicating the current decomposition level and the number of columns in the LL subband. During the first level, the number of columns is simply `Y_SIZE`, and is halved with each decomposition level. In MATLAB, it is a relatively simple matter to employ that language's colon notation to splice out the LL subband at each decomposition level. In C, we of course do not have access to such facilities,

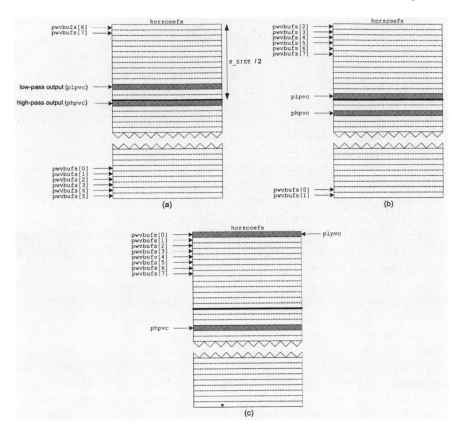

Figure 6-9. IMG_wave_vert in action. Output scan-lines are shaded, and even though they are depicted in this figure in relation to the dimensions of the horzcoefs coefficient matrix, they are actually copied into the 2D DWT output wcoefs. All variables refer to those in Listing 6-4, 6-5, and 6-6. (a) Initial iteration. (b) 3rd iteration. (c) 4th iteration.

and the situation is even more complicated due to the flattened matrix buffers. As a result, in this program we place in separate functions the paging of data into internal memory (fetch_data) and the paging of data out to external memory (page_out_contiguous_block). This modularity serves the purpose of aiding us when we eventually transition to using DMA to move data into and out of on-chip memory.

Listing 6-5: Portions of 2d_dwt_multi_level.c.

```
/* copy a potentially non-contiguous block of data into internal RAM  */
inline void fetch_data(short *pSrc, int nRows, int nCols, short *pDst)
{
  int ii=0;
```

```
    for (; ii<nRows; ++ii, pSrc+=Y_SIZE, pDst+=nCols)
      memcpy(pDst, pSrc, sizeof(short)*nCols);
  }


  /* send data out to wcoefs */
  inline
  void page_out_contiguous_block(short *pSrc, int nElems, short *pDst)
  {
    memcpy(pDst, pSrc, sizeof(short)*nElems);
  }


  /* horizontal wavelet transform, output goes into horzcoefs */
  void transform_rows(int level, int nCols)
  {
    const int nBlockRows = 8<<level;
    const int nBlocks = X_SIZE>>(3+level);
    int iRow=0, iBlock=0, kk;
    short *pin, *pout;

    /* pass rows through DWT, in groups of 8 scan-lines*/
    do {
      /* get next block of rows */
      fetch_data(&wcoefs[iRow*Y_SIZE], nBlockRows,
                 nCols, wvlt_in_buf);

      pin = wvlt_in_buf;
      pout = wvlt_out_buf;
      for (kk=0; kk<nBlockRows; ++kk, pin+=nCols, pout+=nCols)
        IMG_wave_horz(pin, d4_qmf_Q15, d4_mqmf_Q15, pout, nCols);

      /* write horizontal wavelet coeffs out to ext mem storage */
      page_out_contiguous_block(wvlt_out_buf,
                                nBlockRows*nCols,
                                horzcoefs+iRow*nCols);
      iRow += nBlockRows;
    } while (++iBlock <= nBlocks);
  }


  /* grab next two lines for IMG_wave_vert, see Figure 6-9 */
  inline int fetch_horz_wavelet_scanlines(int r, int nCols)
  {
    short *ptemp1 = pwvbufs[0], *ptemp2 = pwvbufs[1];
```

```
    pwvbufs[0] = pwvbufs[2];
    pwvbufs[1] = pwvbufs[3];
    pwvbufs[2] = pwvbufs[4];
    pwvbufs[3] = pwvbufs[5];
    pwvbufs[4] = pwvbufs[6];
    pwvbufs[5] = pwvbufs[7];
    pwvbufs[6] = ptemp1;
    pwvbufs[7] = ptemp2;
    fetch_data(horzcoefs+r*nCols, 1, nCols<<1, pwvbufs[6]);
    return r+2;
}


/* vertical wavelet transform, output goes into wcoefs */
void transform_cols(int level, int nCols)
{
    const int nRows = X_SIZE>>level, nRowsDiv2 = nRows>>1,
              circular = nRowsDiv2-1;
    int lpRow = nRowsDiv2-3, /* low-pass vert output */
        hpRow = nRowsDiv2,   /* high-pass vert output */
        fetchRow = 2, iRow;
    short *plpvc = wvlt_out_buf, /* ptr to low-pass vert coeffs */
          *phpvc = wvlt_out_buf+nCols; /* high-pass vert coeffs */


    /* setup scan-lines for DWT down the columns */
    fetch_data(horzcoefs+(nRows*nCols)-(6*nCols),
                  1,
                  6*nCols,wvlt_in_buf);
    fetch_data(horzcoefs, 1, nCols<<1, wvlt_in_buf+6*nCols);
    pwvbufs[0] = wvlt_in_buf;
    pwvbufs[1] = wvlt_in_buf+nCols;
    pwvbufs[2] = wvlt_in_buf+2*nCols;
    pwvbufs[3] = wvlt_in_buf+3*nCols;
    pwvbufs[4] = wvlt_in_buf+4*nCols;
    pwvbufs[5] = wvlt_in_buf+5*nCols;
    pwvbufs[6] = wvlt_in_buf+6*nCols;
    pwvbufs[7] = wvlt_in_buf+7*nCols;


    /* now march through the image and DWT the columns */
    for (iRow=0; iRow<nRowsDiv2; ++iRow) {
      IMG_wave_vert(pwvbufs,
                      d4_qmf_Q15, d4_mqmf_Q15,
                      plpvc, phpvc, nCols);
```

```
      page_out_contiguous_block(plpvc, nCols, wcoefs+lpRow*Y_SIZE);
      page_out_contiguous_block(phpvc, nCols, wcoefs+hpRow*Y_SIZE);
      fetchRow = fetch_horz_wavelet_scanlines(fetchRow, nCols);
      lpRow += 1; if (lpRow>circular) lpRow=0;
      hpRow += 1;
    }
}


/* multi-level wavelet decomposition */
void dwt2d(int N)
{
  int nCols = Y_SIZE, level;
  for (level=0; level<N; ++level, nCols>>=1) {
    transform_rows(level, nCols);
    transform_cols(level, nCols);
  }
}
```

The `transform_rows` function differs from its predecessor in Listing 6-4 in two respects: with each successive level it expands its operating cache of input data from `wcoefs` (since the total amount of data is a quarter of the previous level), and it uses `fetch_data` to grab these coefficients. Because of the dyadic nature of the decomposition, the LL subband is only a contiguous block in the initial level. Hence in `fetch_data` we must march down each row in the LL subband, extracting `nCols` coefficients, instead of a simple block copy like that used in Listing 6-4. The same concept is also used from within `transform_cols` during initialization of `wvlt_in_buf` and when grabbing the next two scan-lines worth of data in `fetch_horz_wavelet_scanlines`. In contrast, sending the 2D DWT coefficients produced from `IMG_wave_vert` back out to `wcoefs` is a contiguous block copy operation since we page out the data on a row-by-row basis, and for this operation `transform_cols` can invoke `page_out_contiguous_block`.

6.1.4.3 Multi-Level 2D DWT with DMA

Operationally, the program given in Listing 6-6 is identical to that of Listing 6-5. However, with this final IMGLIB-based 2D DWT program we replace the loop in `fetch_data` and the `memcpy` in `page_out_contiguous_block` with a CSL function we have not encountered before, `DAT_copy2d`[9]. This DMA utility function is ideal for copying subregions of matrices over a DMA channel. A second optimization

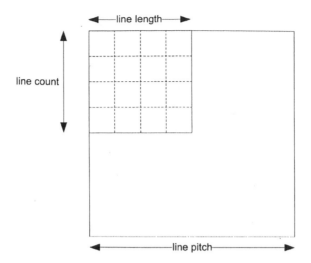

Figure 6-10. Parameters used in DAT_copy2d to perform a DMA copy operation on a matrix of data. The TI documentation refers to the number of rows as lineCnt, the number of bytes per each row as lineLen, and the offset between successive rows as linePitch.

is that we interleave processing and DMA operations in both the transform_rows and transform_cols functions.

DAT_copy2d is a very flexible function allowing one to perform a variety of image copy operations by allowing clients of the function to specify the number of rows and columns in the source buffer as well as the number of bytes between one row and the next. The TI documentation refers to this final argument as the "line pitch", whereas both Intel and Microsoft refer to this quantity as "stride". Figure 6-10 depicts the meaning of the final three arguments of DAT_copy2d. The salient point is that DAT_copy2d removes the need to code a looping structure that kicks off multiple DMA operations, one of which would be needed for *each* row of input data since the data blocks are not necessarily contiguous.

Listing 6-6: Portions of 2d_dwt_multi_level_dma.c. Functions not shown are identical to those given in Listing 6-5.

```
    inline void fetch_data(short *pSrc, int nRows, int nCols, short *pDst) {
    Uint32 id_EDMAin = DAT_copy2d(DAT_2D1D, pSrc, pDst,
                                  nCols*sizeof(short), nRows,
                                  Y_SIZE*sizeof(short));

    DAT_wait(id_EDMAin);
    }
    /* send data out to wcoefs */
```

```
inline
Uint32 page_out_contiguous_block(short *pSrc, int nElems, short *pDst)
{
  return DAT_copy2d(DAT_2D1D, pSrc, pDst,
                    nElems*sizeof(short), 1, nElems*sizeof(short));
}


/* horizontal wavelet transform, output goes into horzcoefs */
void transform_rows(int level, int nCols)
{
  const int nBlockRows = 8<<level;
  const int nBlocks = X_SIZE>>(3+level);
  int iRow=0, iBlock=0, kk;
  short *pin, *pout;
  Uint32  id_EDMAout = DAT_XFRID_WAITNONE;

  /* pass rows through DWT, in groups of 8 scan-lines*/
  do {
    /* get next block of rows */
    fetch_data(&wcoefs[iRow*Y_SIZE], nBlockRows,
               nCols, wvlt_in_buf);

    pin = wvlt_in_buf;
    pout = wvlt_out_buf;
    for (kk=0; kk<nBlockRows; ++kk, pin+=nCols, pout+=nCols)
      IMG_wave_horz(pin, d4_qmf_Q15, d4_mqmf_Q15, pout, nCols);

    /* write horizontal wavelet coeffs out to ext mem storage */
    DAT_wait(id_EDMAout);
    id_EDMAout = page_out_contiguous_block(wvlt_out_buf,
                                           nBlockRows*nCols,
                                           horzcoefs+iRow*nCols);
    iRow += nBlockRows;
  } while (++iBlock <= nBlocks);
}


/* vertical wavelet transform, output goes into wcoefs */
void transform_cols(int level, int nCols)
{
  const int nRows = X_SIZE>>level, nRowsDiv2 = nRows>>1,
            circular = nRowsDiv2-1;
  int lpRow = nRowsDiv2-3, /* low-pass vert output */
```

```
    hpRow = nRowsDiv2,   /* high-pass vert output */
    fetchRow = 2, iRow;
short *plpvc = wvlt_out_buf, /* ptr to low-pass vert coeffs */
      *phpvc = wvlt_out_buf+nCols; /* high-pass vert coeffs */
Uint32 id_EDMAout1 = DAT_XFRID_WAITNONE,
       id_EDMAout2 = DAT_XFRID_WAITNONE;

/* setup scan-lines for DWT down the columns */
fetch_data(horzcoefs+(nRows*nCols)-(6*nCols),
           1, 6*nCols, wvlt_in_buf);
fetch_data(horzcoefs, 1, nCols<<1, wvlt_in_buf+6*nCols);
pwvbufs[0] = wvlt_in_buf;
pwvbufs[1] = wvlt_in_buf+nCols;
pwvbufs[2] = wvlt_in_buf+2*nCols;
pwvbufs[3] = wvlt_in_buf+3*nCols;
pwvbufs[4] = wvlt_in_buf+4*nCols;
pwvbufs[5] = wvlt_in_buf+5*nCols;
pwvbufs[6] = wvlt_in_buf+6*nCols;
pwvbufs[7] = wvlt_in_buf+7*nCols;

/* now march through the image and DWT the columns */
for (iRow=0; iRow<nRowsDiv2; ++iRow) {
  IMG_wave_vert(pwvbufs,
                d4_qmf_Q15, d4_mqmf_Q15,
                plpvc, phpvc, nCols);
  DAT_wait(id_EDMAout1);
  id_EDMAout1 =
    page_out_contiguous_block(plpvc, nCols, wcoefs+lpRow*Y_SIZE);
  DAT_wait(id_EDMAout2);
  id_EDMAout2 =
    page_out_contiguous_block(phpvc, nCols, wcoefs+hpRow*Y_SIZE);
  fetchRow = fetch_horz_wavelet_scanlines(fetchRow, nCols);
  lpRow += 1; if (lpRow>circular) lpRow=0;
  hpRow += 1;
}
}
```

In the DMA-optimized linear filtering program of 4.4.2 (see Listing 4-7), we exploited the asynchronous CSL DMA API to do useful processing work whilst the DMA/EDMA controller was off doing its thing – the same concept is used here. In `transform_rows`, while we do need to block on the DMA read operation performed within `fetch_data`, we do not need to block immediately on the DMA write operation that occurs within `page_out_contiguous_block`. Rather, we simply continue processing and when we are ready to page out to external memory this processed data, only then do we block on this DMA operation (chances are the operation will have long since concluded and thus the call to `DAT_wait` returns immediately). Likewise, within `transform_rows`, the only serialized blocking is due to `fetch_horz_wavelet_scanlines`, because it in turn calls the blocking function `fetch_data`. We do not wait for the output of `IMG_wave_vert` to be DMA'ed out to memory before proceeding on to low-pass and high-pass filter the next chunk of data.

In reality, the overall efficiency gain between this DMA-enabled version of the multi-level 2D DWT implementation and the non-DMA version are rather modest. `IMG_wave_vert` can be used in a more sophisticated fashion than used here, as described in [8]. The documentation describes a scheme where the working buffer is 10 lines, instead of the 8 employed here. Eight lines are preloaded to bootstrap the processing, and then the next 2 lines are fetched in the background during the vertical filter operations. When these filters are done traversing the image, the pointers are moved up by two slots. Meanwhile, the next two lines are fetched via DMA and the actual image processing continues unabated during the fetch. Hence even more of the processing and block memory moves are performed in parallel, which of course will reduce the overall time needed to perform the 2D DWT.

6.2 WAVELET-BASED EDGE DETECTION

In 5.1, we discussed the topic of edge detection from the point-of-view of derivative filters and implemented edge detectors by treating them as modified gradient operators. In this section we take an alternate view of the same problem, and derive a *multiscale* edge detection algorithm that draws upon wavelet theory. Mallat and Zhong published some of the pioneering work in this field in a 1992 paper that presented a numerical approach for the characterization of signals in terms of their multiscale edges[9]. In this section we will show, with some amount of hand-waving, that classical edge detectors can be interpreted as discretized wavelet transforms, where the convolution of the image with an edge detection kernel (e.g., our old friend the Sobel kernel) in fact produces the coefficients of the wavelet transform. Readers interested in a more rigorous treatment of the subject should consult the references. We will improve the performance of classical edge detectors by incorporating scale using a wavelet-based model. MATLAB functions for multiscale edge detection are presented, and a fixed-point edge detection framework suitable for embedded DSP deployment and implemented on the C6416 is also given.

As shown in 5.1, detecting edges boils down to identifying the local irregularities in an image. Perceived edges correspond to those locations where the image undergoes a sharp variation in brightness, which correlates to the maxima of the first derivative of the image. From the multi-level pyramid wavelet decomposition tree, it can be seen that performing the wavelet transform on an image is equivalent to successively smoothing the image and then differentiating it. At each level, the previous level's approximation coefficients are passed through complementary high-pass filters. The output of these high-pass filters are the LL, HL, and HH subbands, and a quick glance at Figures 6-5 and 6-8 confirms that the detail coefficient matrices can be thought of as edge enhanced images pertaining to the edges in the horizontal, vertical, and diagonal directions, respectively. This is evident from the definition of the QMF bank – the high-pass filters produce the detail coefficients and we showed in 4.1.2 and then again in 5.1 that convolution kernels that accentuate edges are high-pass filters. Smoothing and then differentiating the edges results in peaks, corresponding to bright regions in the detail coefficient images in 6-5b and 6-8b, and these peaks can be subsequently segmented by picking out the local maxima of the absolute value of the derivative images.

The Canny edge detector, also introduced in 5.1, uses the first derivative of a Gaussian as the edge-enhancement filter and Mallat generalized Canny's method, linking it to the wavelet transform in [11]. The Canny edge detector uses the modulus of the gradient vector $\nabla f(x,y)$ as a low-pass filtered version of the input image. Points in the image are defined as an edge if the modulus of $\nabla f(x,y)$ are locally maximum in a direction parallel to $\nabla f(x,y)$. Mallat showed that the wavelet transform is proportional to the gradient of a smoothed version of $f(x,y)$, as the detail coefficients are derivatives of the approximation coefficients, which in turn is a low-pass filtered or smoothed version of $f(x,y)$. We can therefore implement edge detection by looking at the *modulus* (absolute value) of the wavelet detail coefficients, and finding the local maxima of the modulus, or the *wavelet modulus maxima*. Multiscale edge detection is based on analysis and segmentation of the wavelet modulus maxima.

Mallat's algorithm makes use of the modulus maxima of the wavelet transform and the angle of the DWT[8]. In addition, only the horizontal and vertical wavelet detail coefficients are used in the computation, and as a result the entire process can be made more efficient. By not computing the diagonal detail coefficients, we lessen the burden on the algorithm at the cost of perfect reconstruction. If we define $W_H(x, y, 2^j)$ to be the horizontal detail coefficients at scale j and $W_V(x, y, 2^j)$ the vertical detail coefficients at scale j, then the wavelet transform modulus $M(x, y, 2^j)$ at scale j is

$$Mf\left(x, y, 2^j\right) = \sqrt{\left|W_H f\left(x, y, 2^j\right)\right|^2 + \left|W_V f\left(x, y, 2^j\right)\right|^2}$$

The angle $Af(x, y, 2^j)$ at scale j is then

$$Af\left(x, y, 2^j\right) = \tan^{-1}\left(\frac{W_V f\left(x, y, 2^j\right)}{W_H f\left(x, y, 2^j\right)}\right)$$

Multiscale edges are points (x', y') where $M(x', y', 2^j)$ is larger than its two neighbors in the direction given by $A f(x', y', 2^j)$. In the digital realm, just like the Canny edge detector, $A f(x', y', 2^j)$ is typically quantized to the set of eight possible directions corresponding to the eight neighbors of the pixel $f(x', y')$.

The algorithm we implement on the TI C6416 DSP is based on the Mallat's algorithm, but uses a simpler segmentation scheme more suitable for an embedded platform. Algorithm 6-1 is a very high-level description of a multiscale edge detection scheme that uses two thresholds to segment a multiscale edge-enhanced image.

Algorithm 6-1: Multiscale Edge Detector
INPUT: Image I, global threshold T, number decomposition levels N
OUTPUT: binary edge image

for each wavelet decomposition level j = 1 … N
    compute DWT coefficients at level j
end
let D_N be the detail coefficients at final level N
compute local modulus maxima of D_N
compute global maxima threshold based on scale N
if (local modulus maxima > T) then mark as edge

Of course, with an algorithm description as high-level as Algorithm 6-1, the devil is always in the details. In subsequent sections we elaborate on the specifics and implement an image processing scheme based on Algorithm 6-1 to edge enhance and subsequently segment images on the C6416 DSP. However, instead of the dyadic 2D DWT derived in 6.1.1, we replace it with a version of the wavelet transform that goes by the moniker "algorithme à trous".

6.2.1 The Undecimated Wavelet Transform

Mallat's fast multiscale edge detector is implemented using an undecimated form of the wavelet transform known as the "algorithme à trous"[12]. This fast dyadic wavelet transform is similar to that depicted in Figure 6-2, except no downsampling is performed. As a consequence, the transform is said to be redundant, or overcomplete, in the sense that superfluous coefficients are retained and successive coefficient matrices are the same size as the input. Recall that with the traditional dyadic DWT, the output from each filter is one-half the length of the input, due to the decimation step. With each decomposition level in the algorithme à trous, the low-pass wavelet filter h_{LP} and high-pass wavelet filter h_{HP} are "stretched" in a sense, and the convolution is performed without any subsampling. This rescaling of the wavelet filters is achieved by inserting 2^j-1 zeros between every filter coefficient (where j is the current decomposition level), hence the name algorithme à trous which means the "holes algorithm" in French. The analysis filter bank for the à trous algorithm is shown in Figure 6-9.

The à trous algorithm outputs N+1 images of the same size, where N is the number of decomposition levels – a single approximation image plus N detail images. This algorithm is very attractive for a number of reasons: it is translation-invariant (a shift in the input simply shifts the coefficients), the discrete transform values are known exactly at every pixel location without the need of any interpolation, and correlation across scales is easily exploited due to its inherent structure. Moreover, an even simpler form of the algorithm exists: through careful choice of the filter h_{LP} we calculate the detail coefficients as differences between consecutive approximation coefficients[13]. This scheme is illustrated graphically in Figure 6-10, where A_j and D_j denote the approximation and detail coefficient matrices, respectively. With this formulation, synthesis is merely the sum of all detail coefficient matrices and the final approximation coefficient matrix. For an N-level decomposition, the reconstruction formula is then

$$\text{reconstructed image} = A_N + \sum_{j=1}^{N} D_j$$

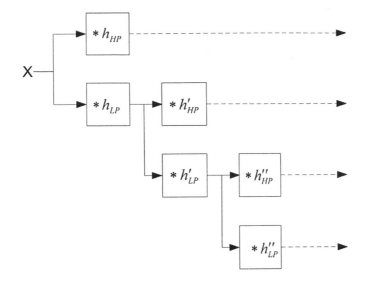

Figure 6-11. Analysis filter bank for the à trous algorithm. With each level, the filters expand by inserting zeros between every coefficient. In contrast to the left-hand side of Figure 6-2, note the lack of decimators.

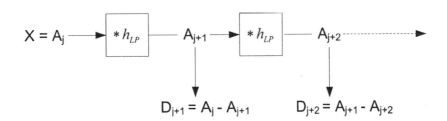

Figure 6-12. Simplified à trous algorithm. Approximation coefficient matrices at level j are denoted by A_j and detail coefficient matrices at level j are denoted by D_j.

6.2.2 Edge Detection with the Undecimated Wavelet Transform

Thus far we have introduced the basics behind multiscale edge detection and have explained the mechanics behind the undecimated wavelet transform. In this section we put the two together and derive a specific edge detection algorithm modeled on the basis of Algorithm 6-1 that can be parameterized by the number of wavelet decompositions.

Recall that there are many choices for wavelet basis functions, and the selection of which basis to use for edge detection is quite important. In particular, the mother wavelet should be symmetric. Symmetry is important in this context because we are essentially differentiating a smoothed image, and therefore a lack of symmetry implies that the location of edges will change as we successively smooth and differentiate the image. With the appropriate choice of wavelet, the locations of edges correspond to the modulus maxima of the wavelet transform at a given scale.

The cubic B-spline (B3) wavelet fits the bill perfectly – it is symmetric and easily derived. A spline of degree 3 is the convolution of four box functions, and in general the spline $\Phi_{N-1}(t)$ of degree N-1 is the convolution of N box functions[1]. Hence to generate the B3 lowpass filter coefficients we convolve (½, ½) with itself four times:

$$B3 = (½, ½) * (½, ½) * (½, ½) * (½, ½) = \frac{1}{16}(1, 4, 6, 4, 1)$$

To generate a low-pass filter kernel h_{LP} suitable for use with the simplified à trous algorithm we extend the above to two dimensions as so:

$$h_{\mathrm{LP}} = (B3)^{\mathrm{T}}(B3) = \begin{bmatrix} \dfrac{1}{256} & \dfrac{1}{64} & \dfrac{3}{128} & \dfrac{1}{64} & \dfrac{1}{256} \\[2mm] \dfrac{1}{64} & \dfrac{1}{16} & \dfrac{3}{32} & \dfrac{1}{16} & \dfrac{1}{64} \\[2mm] \dfrac{3}{128} & \dfrac{3}{32} & \dfrac{9}{64} & \dfrac{3}{32} & \dfrac{3}{128} \\[2mm] \dfrac{1}{64} & \dfrac{1}{16} & \dfrac{3}{32} & \dfrac{1}{16} & \dfrac{1}{64} \\[2mm] \dfrac{1}{256} & \dfrac{1}{64} & \dfrac{3}{128} & \dfrac{1}{64} & \dfrac{1}{256} \end{bmatrix}$$

This kernel is attractive because we can replace the expensive division operations by left bit-shifts since the denominators are all powers of two. We now have enough to prototype the undecimated wavelet transform in MATLAB and see first-hand how it can be used to implement an edge detector. Listing 6-4 is a MATLAB implementation of the simplified à trous algorithm, a_trous_dwt, which does not utilize any functions from the Wavelet Toolbox. The function requires the caller to provide the image and number of decomposition levels, and returns the final approximation image and a three-dimensional array where each "plane" consists of the detail image at that particular scale. We have not encountered higher-dimensional arrays in MATLAB until now, but they work just as one might expect. For example, if one were to perform a three-level decomposition of an image I using a_trous_dwt, accessing the second detail coefficient matrix is accomplished as:

```
[A, D] = a_trous_dwt(I, 3);
D2 = D(:, :, 2); % extract the second plane of the 3D array
```

Listing 6-4: MATLAB function for the simplified à trous algorithm.

```
function [A, D] = a_trous_dwt(I, N)

% generate 2D low-pass filter from cubic B-spline
B3 = [1/16 1/4 3/8 1/4 1/16];
h = B3'*B3;

A = double(I);
for level = 1:N
    approx(:,:,level) = conv2(A, h, 'same');
    D(:,:,level) = A - approx(:,:,level);
    A = approx(:,:,level);
end
```

The simplicity of the reconstruction formula coupled with MATLAB's inherent support of higher dimensional matrices leads to a very simple implementation of the inverse à trous wavelet transform, given in Listing 6-5.

Listing 6-5: a_trous_idwt.m, a MATLAB function for the inverse simplified à trous algorithm.

```
function I = a_trous_idwt(A, D)

I = A + sum(D,3); % sum along the 3rd dimension
```

Now that we have readied an undecimated DWT, we can proceed to graft an edge detector on top of it. As in the case of the classical edge detectors, a thresholding strategy is needed, for even though the wavelet modulus maxima is a decent approximation to the edges of an image, there will exist numerous false positives even in the absence of noise for any reasonably complex image. There are numerous thresholding schemes, ranging from a simple fixed global threshold to more sophisticated multiscale correlation thresholds, as described in [14]. The algorithm we port onto the DSP segments the final detail coefficient image using the two thresholds mentioned in Algorithm 6-1. An estimate of the wavelet modulus maxima that does not factor into account the angle of the DWT is first computed by comparing the wavelet modulus to the local average. This approximation to the wavelet modulus maxima is then compared to a global threshold dynamically calculated from the coefficients of the estimated modulus of the detail coefficients. Listing 6-6 is the MATLAB function that implements this scheme, and forms the basis of the fixed-point C code that we will port to the DSP platform in the next section.

Listing 6-6: edgedet.m, a multiscale edge detector.

```
function J = edgedet(I, N)

% undecimated wavelet transform
[approx, detail] = a_trous_dwt(I, N);

% modulus of the wavelet detail coefficients
D = abs( detail(:,:,N) );

% segment the modulus coefficient matrix
J = (D > filter2(ones(3)/9, D)) .* (D > mean2(D));

% due to the 5x5 convolution kernel and the 3x3
% smoothing operation, zero out the margins
[R C] = size(J);
J(1:3, :) = 0;
J(R-2:R, :) = 0;
J(:, 1:3) = 0;
J(:, C-2:C) = 0;
```

The variable D is the absolute value of the highest level detail coefficient matrix and the subject of the segmentation process that produces a binary image. This makes intuitive sense, as the most negative and most positive coefficients correspond to the stronger edges, and coefficients in this high-pass filtered image close to zero are locations where the approximation image is of constant brightness. The majority of the edge detection work is performed in the third statement of edgedet.m, and warrants a detailed explanation. The call to filter2(ones(3)/9, D) returns an image of the same size as D where pixel values are the local average of the surrounding 3x3 neighborhood. The left-hand side of the statement, (D>filter2(ones(3)/9,D), returns a logical (boolean) coefficient matrix that will have a value of 1 wherever the modulus of the detail coefficient matrix is larger than the average of its neighbors, and 0 otherwise. Hence, we approximate the wavelet modulus maxima without the need for computing an expensive arctangent as Mallat's algorithm does. The right-hand side of the final statement, `(D>mean2(D))`, returns another `logical` matrix with 1s wherever the modulus of the detail coefficient matrix is larger than the mean of the *entire* detail image. This is the global threshold mentioned in Algorithm 6-1 and here we dynamically calculate it. The result of multiplying these two matrices together in an element-by-element fashion (not matrix multiplication) is in effect a logical "and" operation that yields the final binary segmented image.

What makes this algorithm so powerful is that we now have the ability to control the behavior of the edge detection scheme by incorporating scale into the overall framework. With classical edge detectors there is no concept of scale, but with a wavelet model we are free to adjust the scale as desired by controlling the number of wavelet decompositions. Scale controls the significance of the detected edges, as illustrated in Figure 6-13. A careful balance must be struck between the image resolution and the scale of edges: high resolution and small scale (reduced number of wavelet decompositions) results in comparatively noisier and more discontinuous edges, as can be seen from Figure 6-13 when the number of wavelet decompositions is 1. The "peppers" image, in particular, suffers from this problem. However, low resolution coupled with large scale may result in undetected edges.

Sharper edges are more likely to be retained by the wavelet transform across subsequent scales, whereas less significant edges are attenuated as the scale increases. Again referring to Figure 6-13, as the number of wavelet decompositions increases, the boundaries between the objects in the peppers image and the structure present in the butterfly's wings become more pronounced, due to the coarser resolution which has the effect of smoothing out the irregularities in the image. This comes at a price however, because even though the noise and details within the image decrease, the exact location of the edges now becomes more difficult to discern.

State-of-the-art multiscale segmentation schemes are more sophisticated than that employed in `edgedet.m`. They incorporate scale into the criteria for discriminating between real structure and noise. In general, wavelet transform coefficients are likely to increase across scale wherever there are significant edges and likewise, the coefficients are equally likely to decrease across scale at location of noise and details. In fact, we will leverage this behavior during the development of a wavelet-based denoising algorithm in the 6.3.

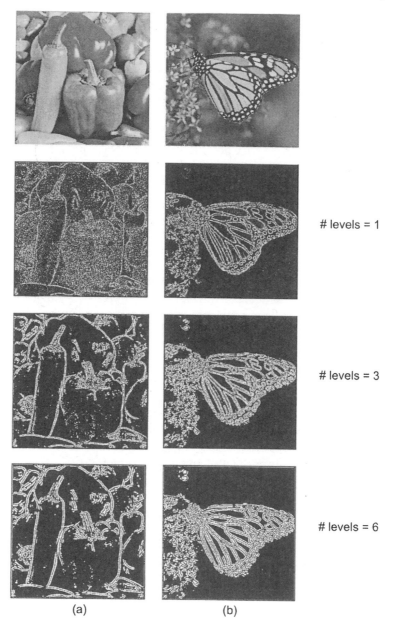

levels = 1

levels = 3

levels = 6

(a)   (b)

Figure 6-13. Wavelet-based edge detection using `edgedet.m`. (a) "Peppers" image, with segmented binary edge images using 1, 3, and 6 decomposition levels. (b) "Monarch" image, again with 1, 3, and 6 decomposition levels.

6.2.3 Multiscale Edge Detection on the C6701 EVM and C6416 DSK

Our initial foray into embedded wavelet-based image processing is a set of programs written and tested on the C6701 EVM and C6416 DSK platforms. These programs are all fixed-point C implementations of the multiscale edge detection scheme shown in MATLAB form in Listing 6-6. The IMGLIB wavelet functions implement the classical 2D DWT as discussed in 6.1, and thus we implement the multiscale edge detector using hand-crafted C functions that implement the algorithme à trous described in the previous section. In the Chap6\wave_edge directory on the CD-ROM are three CCStudio project directories, which are described in more detail in this section: C6701EVM, C6701EVM_HPI, and C6416DSK.

We begin by presenting a wholly unoptimized and standalone fixed-point implementation of the multiscale edge detector built and tested on the C6701 EVM. This project will also run efficiently on a C62xx platform without any code changes because it uses only fixed-point arithmetic. After introducing the fixed-point C implementation of the core algorithm, an interactive system along the lines of the RTDX application of 5.2.5 is given, although this time we employ the services of the Host Port Interface (HPI) to shuttle data to and fro. Finally, the implementation of (1) and (2) is optimized for the C6416 DSK with a combination of both higher-level algorithmic modifications and lower-level optimizations using specialized C compiler intrinsics.

6.2.3.1 Standalone Multiscale Edge Detector (C6701EVM)

The project directory Chap6\wave_edge\C6701EVM is a very simple EVM project that relies on a technique used throughout this book to prototype image processing algorithms – the actual input image is built into the program by inlining the pixel data of the "peppers" image (see Figure 6-11) in the header file image.h. The linker command file is not noteworthy except to say that the .cinit segment has been remapped to external memory due to the fact that the pixel data is compiled into the binary. The contents of the main source file that makes up this program, wave_edge.c, is shown in Listing 6-7.

Listing 6-7: wave_edge.c

```
#include <board.h> /* EVM library */
#include "wave_edge.h"

#pragma DATA_SECTION (out_img, "SBSRAM");
```

```
#pragma DATA_SECTION (in_img, "SBSRAM");

int wave_decomp_a_trous(int nlevels)
{
  int ilevel, irow, jcol, kk;
  for (ilevel=0; ilevel<nlevels; ++ilevel) {
    int iapprox = ilevel+1;

    /*
     * zero out the margins of the current
     * approximation coefficient matrix -
     * 1st and last couple of rows
     */
    for (kk=0; kk<Y_SIZE; ++kk)
      approx[0][kk][iapprox] =
      approx[1][kk][iapprox] =
      approx[X_SIZE-1][kk][iapprox] =
      approx[X_SIZE-2][kk][iapprox] = 0;

    /* first and last couple of columns */
    approx[2][0][iapprox] = approx[2][1][iapprox] = 0;
    approx[2][Y_SIZE-2][iapprox] = approx[2][Y_SIZE-1][iapprox] = 0;
    approx[3][0][iapprox] = approx[3][1][iapprox] = 0;
    approx[3][Y_SIZE-2][iapprox] = approx[3][Y_SIZE-1][iapprox] = 0;

    /* 5x5 convolution for approximation coeffs at level i */
    for (irow=2; irow<(X_SIZE-2); ++irow) {

      approx[irow+2][0][iapprox] = approx[irow+2][1][iapprox] = 0;
      approx[irow+2][Y_SIZE-2][iapprox] = 0;
      approx[irow+2][Y_SIZE-1][iapprox] = 0;

      for (jcol=2; jcol<(Y_SIZE-2); ++jcol) {

                        /* row 1: 1/256 1/64 3/128 1/64 1/256 */
        approx[irow][jcol][iapprox] = (approx[irow-2][jcol-2][ilevel]>>8) +
                        (approx[irow-2][jcol-1][ilevel]>>6) +
                        ((3*approx[irow-2][jcol][ilevel])>>7) +
                        (approx[irow-2][jcol+1][ilevel]>>6) +
                        (approx[irow-2][jcol+2][ilevel]>>8) +
                        /* row 2: 1/64 1/16 3/32 1/16 1/64 */
                        (approx[irow-1][jcol-2][ilevel]>>6) +
```

```
                    (approx[irow-1][jcol-1][ilevel]>>4) +
                    ((3*approx[irow-1][jcol][ilevel])>>5) +
                    (approx[irow-1][jcol+1][ilevel]>>4) +
                    (approx[irow-1][jcol+2][ilevel]>>6) +
                    /* row 3: 3/128 3/32 9/64 3/32 3/128 */
                    ((3*approx[irow][jcol-2][ilevel])>>7) +
                    ((3*approx[irow][jcol-1][ilevel])>>5) +
                    ((9*approx[irow][jcol][ilevel])>>6) +
                    ((3*approx[irow][jcol+1][ilevel])>>5) +
                    ((3*approx[irow][jcol+2][ilevel])>>7) +
                    /* row 4: 1/64 1/16 3/32 1/16 1/64 */
                    (approx[irow+1][jcol-2][ilevel]>>6) +
                    (approx[irow+1][jcol-1][ilevel]>>4) +
                    ((3*approx[irow+1][jcol][ilevel])>>5) +
                    (approx[irow+1][jcol+1][ilevel]>>4) +
                    (approx[irow+1][jcol+2][ilevel]>>6) +
                    /* row 5: 1/256 1/64 3/128 1/64 1/256 */
                    (approx[irow+2][jcol-2][ilevel]>>8) +
                    (approx[irow+2][jcol-1][ilevel]>>6) +
                    ((3*approx[irow+2][jcol][ilevel])>>7) +
                    (approx[irow+2][jcol+1][ilevel]>>6) +
                    (approx[irow+2][jcol+2][ilevel]>>8);

   }  /* end (for each column) */
 }  /* end (for each row) */
 /*
  * end 5x5 convolution, now calc detail coeffs at level i
  * (but things are a bit different the final time through this loop)
  */
 if (ilevel != nlevels-1) { /* not the last time through the outer loop*/

   for (irow=2; irow<(X_SIZE-2); ++irow)
     for (jcol=2; jcol<(Y_SIZE-2); ++jcol)
       detail[irow][jcol][ilevel =
                   approx[irow][jcol][iapprox]-approx[irow][jcol][ilevel];

 } else { /* last time, prep for segmentation of edge image */

   unsigned int sum = 0;
   for (irow=2; irow<(X_SIZE-2); ++irow) {
     for (jcol=2; jcol<(Y_SIZE-2); ++jcol) {
       detail[irow][jcol][ilevel] =
```

```
          (_abs(approx[irow][jcol][iapprox]-approx[irow][jcol][ilevel])) >> 9;
        sum += detail[irow][jcol][ilevel];
      }
    }
    /*
     * calculate average detail coefficient value for the
     * highest decomposition level
     */
    return (sum>>LOG2_N_PIXELS); /* div by (X_SIZE*Y_SIZE) */


    }
  } /* end (for each decomposition level) */
}

void segment_detail_image(int ilevel, int mean)
{
  int p;
  const int M = 3640; /* 1/9 in Q15 format */
  unsigned long avg;
  int irow, jcol;
  ilevel -= 1;
  /*
   * mark those pixels where the detail coefficient
   * is greater than the local mean AND where it is greater
   * than the global mean as an edge.
   */
  for (irow=3; irow<(X_SIZE-3); ++irow) {
    for (jcol=3; jcol<(Y_SIZE-3); ++jcol) {
      p = detail[irow][jcol][ilevel];
      /* find local average using 3x3 averaging filter */
      avg = ( detail[irow-1][jcol-1][ilevel]*M +
              detail[irow-1][jcol][ilevel]*M +
              detail[irow-1][jcol+1][ilevel]*M +
              detail[irow][jcol-1][ilevel]*M +
              p*M +
              detail[irow][jcol+1][ilevel]*M +
              detail[irow+1][jcol-1][ilevel]*M +
              detail[irow+1][jcol][ilevel]*M +
              detail[irow+1][jcol+1][ilevel]*M ) >> 15;
      out_img[irow*Y_SIZE+jcol] = (p>avg && p>mean) ? 255 : 0;
    }
  }
}
```

```
     }

  int main(void)
  {
    const int nlevels = 3;
    int irow=0, jcol=0;
    int mean;

    evm_init(); /* initialize the board */

    for (irow=0; irow<X_SIZE; ++irow)
      for (jcol=0; jcol<Y_SIZE; ++jcol)
        /* scale input for fixed-point arithmetic */
        approx[irow][jcol][0] = (in_img[irow*Y_SIZE+jcol] << 9);

    mean = wave_decomp_a_trous(nlevels);
    segment_detail_image(nlevels, mean);
  }
```

One thing to keep in mind is that the implementation of Listing 6-7 is sub-optimal for a variety of reasons – it is meant to illustrate some of the details behind a fixed-point wavelet-based image processing system and for the sake of clarity and time many optimizations are omitted. Some of these optimizations are applied in the C6416 DSK project, and in particular the DMA paging optimization of 4.3 would go a long way towards improving performance. All of the image buffers are placed in external memory, and consequently we incur a large cost attributed to the EMIF as we access external memory during the processing loops.

One aspect of this program that differs from all of the other previous C image processing algorithms is that the wavelet arrays are not "flattened" and are explicitly defined as multi-dimensional arrays in wave_edge.h, as can be seen in Listing 6-8. This simplification is merely for the purpose of clarity – the optimized implementation described in 6.2.3.3 uses flattened one-dimensional image buffers which then enables us to use certain compiler intrinsics to our advantage.

Listing 6-8: wave_edge.h, which contains definitions for the approximation and detail coefficient matrices used in Listing 6-7.

```
  #define MAX_LEVELS 6

  #pragma DATA_SECTION (approx, "SBSRAM");
  #pragma DATA_ALIGN (approx, 8)
```

```
#pragma DATA_SECTION (detail, "SBSRAM");
#pragma DATA_ALIGN (detail, 8)
int approx[X_SIZE][Y_SIZE][MAX_LEVELS+1],
    detail[X_SIZE][Y_SIZE][MAX_LEVELS];
```

The approximation and detail coefficient are defined as three-dimensional signed integer matrices, and the length of the third dimension of `approx` is one more than `MAX_LEVELS` because the first approximation image matrix is the input image pixels shifted to the left by 9 bits, for fixed-point considerations. As we have seen previously, when developing fixed-point algorithms the programmer must always remain cognizant of the balancing act between dynamic range and resolution. With wavelet algorithms, this problem becomes more acute because the range of possible values increases in proportion to wavelet scale, due to the presence of the high-pass filter. At the same time, we strive to retain adequate resolution and this is the reason for the use of 32-bit integers for storage of the approximation and detail coefficient matrices – a minimum of 22 bits is required to avoid overflow, because we multiply by 1/9 in Q15 format (3640) within `segment_detail_image`. With one bit reserved for sign, that leaves 9 "binal" (as opposed to decimal) points, which accords us sufficient resolution for this particular algorithm. The copy and scaling operation is performed in `main`, where the input data is copied from `in_img` to `approx` and then bit-shifted.

The first step is to perform the wavelet analysis, which is done in `wave_decomp_a_trous`. This function contains a loop which first low-pass filters the current approximation image, and then forms the detail image by subtracting two successive approximation images. Recall that the key reason for choosing the B3 spline filter as the wavelet is its suitability for fixed-point arithmetic – this fact is clearly illustrated in the convolution inner loop where all the divisions are replaced by corresponding right bit shifts. Upon reaching the end of the final iteration through the outer-most loop within `wave_decomp_a_trous`, the mean absolute value of the final detail image is computed and returned back to the caller. Moreover, we segment the absolute value of this final detail image, and thus we right-shift those coefficients back by 9 bits. The `_abs` compiler intrinsic is used in lieu of the C ternary operator to quickly calculate the absolute value of the deltas.

Upon completion of `wave_decomp_a_trous`, the wavelet decomposition has been carried out to the specified number of levels, and what is left is to binarize the final detail coefficient image via the dynamic threshold segmentation algorithm developed in Listing 6-6. This operation is carried out in `segment_detail_image`, and as in the MATLAB version of the algorithm, the wavelet modulus maxima is estimated by comparing each detail coefficient at location (irow, icol) with the average of its surrounding 3x3 neighborhood. This local average is in essence a 2D box filter and a fixed-point implementation of this process was discussed in full detail in 4.3.2. The segmented edge image, where 0 signifies a non-edge pixel and 255 an edge, is placed into out_img and the result can be visualized using CCStudio's image graphing feature, as discussed in 4.3.1.

6.2.3.2 HPI Interactive Multiscale Edge Detector Application with Visual Studio and the TI C6701 EVM

In 5.2 we developed an interactive combined host/target application, with an MFC GUI running on the host PC communicating with a DSP target running an edge detector and image segmentation back-end. Here, we build upon the multiscale edge detector from the previous section and develop an interactive application that communicates to the EVM over the PCI bus using TI's host port interface (HPI)[15,16]. HPI is similar to RTDX in that it enables communication between the host and a target, but it differs in that it removes the CCStudio IDE as a middle-man, thereby enabling *direct* connectivity between the host and the DSP's memory space. HPI has numerous advantages over RTDX, including:

- It is faster than RTDX, although there is an enhanced "high-speed" RTDX (HS-RTDX[17,18]) that provides better performance than the standard version of RTDX we used in Chapter 5. Nevertheless, TI reports benchmarks of 20-30 Mbytes/sec (depending on reading or writing) to SDRAM on a C6701 DSP[19]. Practically speaking, those benchmarks refer to a direct host to HPI connection, whereas with the EVM a PCI bridge is in the loop. As a result, one can expect throughput within the range of 10-12 MB/sec over a 33 MHz PCI bus, which is still superior to RTDX. HPI is the fastest means of transferring large blocks of data between the DSP target and a host PC.
- RTDX only works if the Code Composer Studio IDE is running on the host machine. Obviously in production-level systems the IDE is not going to be running and thus HPI can be used instead.
- The programming model for HPI versus RTDX is decidedly simpler, especially for C/C++ programmers. The host side of HPI is accessed through a relatively simple C API hosted within a few Windows DLLs and consequently the programmer need not concern themselves with all of the nuances of COM, which as we witnessed in 5.2.5.2 can get quite verbose and frankly abstruse in the C and C++ languages. This architecture has performance ramifications on the host side as well; since HPI allows direct access to the DSP's memory, block memory copies are

comparatively much simpler than with RTDX hosts, where one must remain cognizant of the COM underpinnings and marshal the SAFEARRAY data into C/C++ arrays.

- HPI does appear to work more reliably than RTDX, at least on the development platform used in this book. As we shall see, we do not need to tailor the host/target communication protocol due to issues relating to the reliability of communication channel. Recall that with RTDX, while we were able to send 256x256 8-bit images in one block to the target from the host over RTDX, the reverse operation failed. There is one significant HPI problem that impacts Window 2000/XP host, emanating from the EVM Windows device driver.

Since the API is not provided as a COM object, certain alternative programming languages such as Visual Basic 6 are precluded when using HPI on the host. More importantly however, HPI is not available for all development platforms, in particular certain DSKs. For example, on the standard C6416 DSK, the HPI port is unpopulated and the USB emulation does not support HPI – thus the HPI applications in this chapter will not run on those development platforms.

Figure 6-14 is a screen-shot of the host application. To use this application:

1. Load in the COFF binary executable (you can point the host application to the .out file by clicking on the "Browse" button). Recall that with the RTDX applications in Chapter 5, we either relied on Link for Code Composer Studio MATLAB functions to start the CCStudio IDE or we assumed that the CCStudio IDE was up and the user had already manually started the target via **Debug|Run.**
2. Initialize the target, and perform the host/target handshake by clicking on the "Init" button.
3. Load an input image using the **File|Load** menu selection.
4. Select the number of wavelet decompositions and click on the "Process" button. The image is sent to the DSP, processed, and then sent back to the host and displayed.

The EVM board can be temperamental at times, thus you will sometimes need to reset the EVM prior to starting the host application, particularly after having run the CCStudio IDE (this can be done by running the evm6xrst.bat batch file in the TI bin directory). This demo program follows a tried-and-true template used by many HPI-DSP applications – a similar example can be found in [20]. Broadly speaking, what typically occurs is that the host sends a message to the DSP by writing to a memory address on the DSP (remember, with HPI the host has direct access to the DSP's memory), thereby signifying that data is ready for processing. The host sends the data to the target, and the DSP processes this data. To send a completion message from the DSP to the host, the DSP interrupts the host, which then reads the data from the DSP. The exact protocol used for this application is shown in Figure 6-15; in contrast to the equivalent diagram for an RTDX application (see Figure 5-9), note that we send large blocks of data in a full-duplex fashion using single HPI messages.

6.2.3.2.1   C6701 EVM TARGET

The C6701 EVM target program uses PCI functions from the EVM DSP Support Software, defined in the header `pci.h`, to implement its end of the

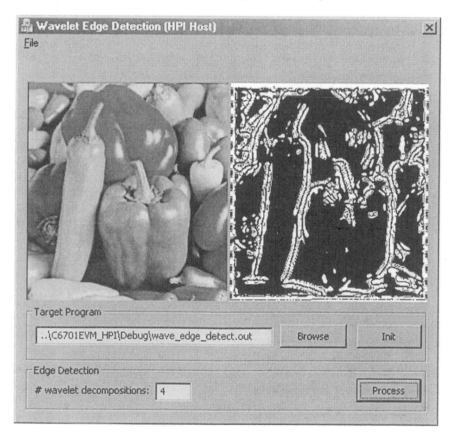

Figure 6-14. Interactive HPI multiscale edge detector demo application.

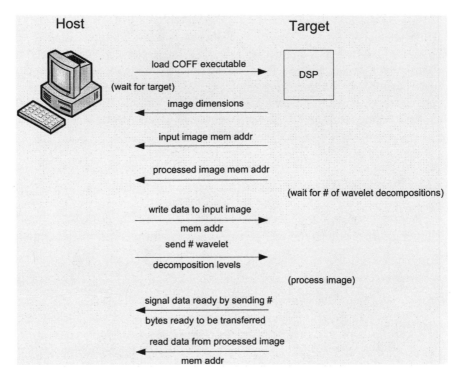

Figure 6-15. Data flow protocol between host and target communicating over HPI for the wavelet-based edge detection application.

HPI protocol depicted in Figure 6-13. The target code may be found in the Chap6\wave_edge\C6701EVM_HPI directory, and Listing 6-9 is the contents of wave_edge.c that handles HPI communication with the host. Only main has changed, and the core image processing from Listing 6-7 remains intact.

Listing 6-9: HPI communication infrastructure in wave_edge.c.

```
int main(void)
{
  unsigned int nlevels = 0;
  int mean;

  evm_init(); /* initialize the board */
  pci_driver_init(); /* call before using any PCI code */

  /*
   * handshake with host, send image size and
```

```
 * memory addresses to host.
 */
{
  int sts;
  unsigned int image_size = (X_SIZE<<16) - Y_SIZE;
  /*
   * target will initialize HPI, therefore we
   * should keep attempting to send the image
   * dimensions until the HPI message is successfully
   * sent.
   */
  do {
    sts = amcc_mailbox_write(2, image_size);
  } while (ERROR == sts);

  /*
   * comm is up now, now send input buf mem addr
   * and processed image buf mem addr
   */
  pci_message_sync_send((unsigned int)in_img, FALSE);
  pci_message_sync_send((unsigned int)out_img, FALSE);
}
/*
 * wait for the host to signal that they
 * have sent us an image.
 */
{
int irow, jcol;
pci_message_sync_retrieve(&nlevels);
if (nlevels > MAX_LEVELS)
  nlevels = MAX_LEVELS;

/*
 * now that we have the image data, we need to
 * scale the pixel values in preparation for
 * fixed-point arithmetic
 */
for (irow=0; irow<X_SIZE; ++irow)
   for (jcol=0; jcol<Y_SIZE; ++jcol)
   approx[irow][jcol][0] = (in_img[irow*Y_SIZE+jcol] << 9);
}
```

```
/* process the image */
mean = wave_decomp_a_trous(nlevels);
segment_detail_image(nlevels, mean);

/* signal that segmented edge image ready */
{
  int sts;
  unsigned int bytes = N_PIXELS;

  /* tell host that N_PIXELS worth of data ready */
  do {
    sts = amcc_mailbox_write(2, bytes);
  } while (ERROR == sts);
}
```

This implementation is somewhat unique in that C scoping is used so that variables can be defined after the beginning of main, simulating the situation in C++ where variables can be defined anywhere within a function definition. The program uses 32-bit mailbox registers, located on the PCI controller, for communication between the target and the host PC. The mailbox is a useful abstraction for exchanging small amounts of data such as integers over the PCI bus and is covered quite thoroughly in [18] and sections 1.6.1-1.6.3 and 1.8.5-1.8.8 of [21]. Fortunately, much of the requisite "bit-banging" associated with dealing with these registers is encapsulated by the EVM DSP Support Software API functions within pci.h, and equivalent functionality is available through the HPI module in the Chip Support Library.

Much of the operation of the DSP, with regards to the HPI communication, should be evident from an understanding of Figure 6-13 and Listing 6-9. For further information, the reader is referred to the previously mentioned references as well as the useful comments in the pci.h header file. Finally, this particular implementation is good for only one go-around – the majority of main would need to be put in a loop for the program to operate in a fashion similar to the RTDX targets discussed in 5.1.2.2 and 5.2.5.1. This is demonstrated in Appendix A (see A.2).

6.2.3.2.2   VISUAL STUDIO .NET 2003 HOST APPLICATION

The host application is found on the CD-ROM in the directory Chap6\wave_edge\Host. In stark contrast to an RTDX host, the host portion of HPI is accessed through a comparatively much simpler API defined in the evm6xdll.h header file (the host project includes this file in the precompiled header stdafx.h), and the primary functionality resides within three DLLs: evm6x.dll, evm6xdm.dll, and evm6xmsg.dll. In the RTDX host application of 5.2.5.2, we used the Visual C++ #import directive to bring the RTDX COM library into the Visual Studio project – here we explicitly link to the evm6x.lib static library. The majority of the code associated with this project is very similar to the previous Visual Studio applications presented in this book. An MFC/GDI+ framework is used to read and display images, and the Intel IPP library is used for resizing the image to the dimensions the target is expecting. Just as in the RTDX host, the host-side HPI communication is encapsulated within a class named Qureshi::HPI (again a C++ namespace is used to prevent future name clashes) – its interface and implementation is shown in Listing 6-10.

Listing 6-10: Interface and implementation for C++ class that encapsulates the HPI host-side communication.

```
//
// Interface for HPI class (HPI.h)
//
namespace Qureshi {

class HPI {

  public:

    HPI(std::string filename);
    ~HPI();

    // connects to EVM, performs handshake, returns image dimensions
    std::pair<int, int> init();

    // also the # of wavelet decompositions
    void sendImage(Image8bpp *pImage, int nLevels);

    // client is responsible for deallocating returned object!
    Image8bpp *readImage();

  private:

    // pair.first = # rows, pair.second = # cols
    std::pair<int, int> handshake();
```

```cpp
    std::string m_programFile;

    HANDLE m_hEvm, m_hEvent;

    LPVOID m_hHpi;

    std::pair<int,int> m_imageRowsCols;

    void *m_pTargetInputBuf,
         *m_pTargetOutputBuf;

    // scratch storage used for sending/receiving pixel data
    std::vector<Ipp8u> m_buffer;

  };

} // namespace Qureshi

//
// HPI.cpp
//
HPI::HPI(string filename) : m_programFile(filename), m_hEvm(NULL),
                            m_hHpi(NULL), m_hEvent(NULL)
{
}

HPI::~HPI()
{
  ::CloseHandle(m_hEvent);
  evm6x_hpi_close(m_hHpi);
  evm6x_close(m_hEvm);
}

pair<int,int> HPI::init()
{
  m_hEvm = evm6x_open(0, FALSE);
  if (INVALID_HANDLE_VALUE==m_hEvm)
    throw runtime_error("Failed to open connection to EVM.");

  evm6x_set_timeout(m_hEvm, 1000);
  evm6x_abort_write(m_hEvm);
  evm6x_clear_message_event(m_hEvm);
```

```
        ostringstream ostr;
        ostr<<EVM6X_GLOBAL_MESSAGE_EVENT_BASE_NAME<<"0";
        m_hEvent = ::CreateEvent(NULL, FALSE, FALSE, ostr.str().c_str());
        if (NULL == m_hEvent)
          throw runtime_error("Could not open event!");

        evm6x_reset_dsp(m_hEvm, HPI_BOOT);
        evm6x_init_emif(m_hEvm, NULL);
        char *sCOFF = (char *)m_programFile.c_str();
        if (!evm6x_coff_load(m_hEvm, NULL, sCOFF,
                             FALSE, FALSE, FALSE))
          throw runtime_error(string("Failed to load .out file"));
        evm6x_unreset_dsp(m_hEvm);

        if (NULL == (m_hHpi = evm6x_hpi_open(m_hEvm)))
          throw runtime_error("Failed to initialize HPI.");

        return this->handshake();
      }

    void HPI::sendImage(Image8bpp *pImage, int nLevels)
    {
      if (pImage->getHeight()!=m_imageRowsCols.first ||
          pImage->getWidth()!=m_imageRowsCols.second)
          pImage->resize(m_imageRowsCols.first, m_imageRowsCols.second);

      Ipp8u *pPixels = pImage->getPixels();
      unsigned long ulBytes = pImage->getHeight()*pImage->getWidth(),
                    expectedBytes = ulBytes;

      if (pImage->getStride() != pImage->getWidth()) {
        // remove padding
        pPixels = &m_buffer[0];
        for (int iRow=0; iRow<pImage->getHeight(); ++iRow) {
          Ipp8u *pScanLine=pImage->getPixels()+iRow*pImage->getStride(),
                *pDest = pPixels + iRow*pImage->getHeight();
          copy(pScanLine, pScanLine+pImage->getWidth(), pDest);
        }
      }

      // write directly to EVM memory
```

```
    if (!evm6x_hpi_write(m_hHpi,
                         (PULONG)pPixels,
                         (PULONG)&ulBytes,
                         (ULONG)m_pTargetInputBuf))
      throw runtime_error("Failed to write image data.");

    if (ulBytes != expectedBytes) {
      ostringstream ostr;
      ostr << "HPI only wrote " << ulBytes
           << " bytes, expecting " << expectedBytes;
      throw runtime_error(ostr.str().c_str());
    }


    // inform evm that we're done transferring data by telling
    // it the # of wavelet decompositions to use for the
    // multi-scale edge detection.
    if (!evm6x_send_message(m_hEvm, (PULONG)&nLevels))
      throw runtime_error("Write completion signal to EVM failed!");
  }

Image8bpp *HPI::readImage()
{
    int wait = 10000; // give it some time to process the image
    while (!::WaitForSingleObject(m_hEvent, 1))
      if (0 == --wait)
        throw runtime_error("Error waiting for handshake from EVM");

    // already know the # of bytes to read, however we're using this data
    // mainly as a signal from the target that the processed data can be
    // now read now over the HPI interface
    unsigned long nBytes;
    for (int ii=0; ii<200; ++ii) {
      if (!evm6x_mailbox_read(m_hEvm, 2, (unsigned long *)&nBytes)) {
      TRACE("Waiting for target to complete (%d) ...\n", ii);
      ::Sleep(100);
      }
      else
        break;
    }
    if (200 == ii)
      throw runtime_error("Failed to retrieve # bytes to read");
```

```
  // read the image data
  if (!evm6x_hpi_read(m_hHpi,
                      (PULONG)&m_buffer[0],
                      &nBytes,
                      (ULONG)m_pTargetOutputBuf))
    throw runtime_error("Failed to read image data.");

  // marshal into Image8bpp object
  if (nBytes != m_buffer.size()) {
    ostringstream ostr;
    ostr << "HPI only read " << nBytes
         << " bytes, expecting "<< m_buffer.size();
    throw runtime_error(ostr.str().c_str());
  }

  return new Image8bpp(m_imageRowsCols.first,
                       m_imageRowsCols.second,
                       m_imageRowsCols.second,
                       &m_buffer[0]);
}

pair<int, int> HPI::handshake()
{
  int wait = 1000;
  while (!::WaitForSingleObject(m_hEvent, 1))
    if (0 == --wait)
      throw runtime_error("Error waiting for handshake from EVM");

  unsigned long dims;
  if (!evm6x_mailbox_read(m_hEvm, 2, (unsigned long *)&dims))
    throw runtime_error("Failed to retrieve image dimensions");
  unsigned short xSize = dims>>16, ySize = dims;

  if (!evm6x_retrieve_message(m_hEvm,
                              (unsigned long *)&m_pTargetInputBuf))
    throw runtime_error("Error : target input image buffer mem addr.");
  ::Sleep(100);
  if (!evm6x_retrieve_message(m_hEvm,
                              (unsigned long *)&m_pTargetOutputBuf))
    throw runtime_error("Error : target output image buffer mem addr.");

  m_imageRowsCols = pair<int,int>(xSize, ySize);
```

```
    m_buffer.resize(xSize*ySize);
    return m_imageRowsCols;
}
```

The HPI functions used in `Qureshi::HPI` are documented in section 2.3 of [19] and in various examples that accompany CCStudio. One point of difference between this code and the CCStudio HPI examples is that the examples were written and tested on Windows 95/98 and will not work without a modification in the way the Win32 `HANDLE` is created. HPI host applications use the Win32 API function `WaitForSingleObject`[22] to pend on incoming HPI messages emanating from the target. `WaitForSingleObject` references a Win32 `HANDLE`, and the TI example code uses `OpenEvent` to create the `HANDLE` object, whereas with Windows 2000 or Windows XP `CreateEvent` should be used – see `Qureshi::HPI::init`. Under Windows 2000/XP, calling `OpenEvent` and passing it the name of the EVM event results in an invalid `HANDLE` returned from the API call. Thus subsequent Win32 calls referencing this `HANDLE` return immediately with a failure code. The HPI host can be made to function, by replacing `OpenEvent` with `CreateEvent`, however not in the way TI designed it to work. The problem occurs within the `readImage` method, where the hope is to use `WaitForSingleObject` in such a way that the host blocks until the target has completed the processing and then signals this completion by sending the host the amount of bytes to read (see Figure 6-13). If this truly worked, the host could theoretically be performing some other useful work while it waited for the target; it would rely on the operating system to raise a Win32 event when the target completed its processing, at which point the host application would then read the data. Thus, in the ideal case we might use the following function call, perhaps in another thread:

```
    DWORD dwStatus = ::WaitForSingleObject(m_hEvent, INFINITE);
```

Instead what we have in `Qureshi::HPI::readImage` is the following:

```
    int wait = 10000; // give it some time to process the image
    while (!::WaitForSingleObject(m_hEvent, 1))
      if (0 == --wait)
        throw runtime_error("Error waiting for handshake from EVM");
```

In effect, what we've done is turn this into a blocking call that effectively "sleeps" for 10,000 milliseconds. The underlying problem is due to the EVM driver and Windows 2000/XP; unfortunately the driver was never tested under Windows 2000/XP. One would surmise that the EVM driver (`evm6x.sys`) would call `CreateEvent` under the covers when the EVM is opened – and subsequently the user application would then call `OpenEvent` and everything should proceed smoothly from there. This is a serious problem, and unfortunately it appears that TI is in no hurry to fix the driver, which dates from 1998. Now that the C6701 EVM has been discontinued, it is something that developers will need to live with. There are similar third-party EVM-like development platforms – some of these include those products sold by Mango DSP (www.mangodsp.com), Signalogic (www.signalogic.com), and many others. Chances are their drivers are not afflicted with such issues, although in TI's defense the EVM boards were released in 1998 prior to the advent of Windows 2000/XP. Theoretically, the Win32 `HANDLE` issue does not constitute a deal-breaker, because since the DSP is running in a deterministic real-time environment, it should be possible to measure exactly how much time it takes to perform the processing and code these delays into the host. While not an ideal solution, it is a feasible workaround.

Thankfully, aside from these shenanigans, the remainder of `HPI.cpp` is reasonably straightforward. The DSP sends the image dimensions and memory addresses for the input image buffer and the processed (edge) image buffer during the host/target handshake, and these are then used when calling `evm6x_hpi_write` in the `sendImage` method and when calling `evm6x_hpi_read` in the `readImage` method. Prior to shipping the input image down to the target, the image is resampled to the image dimensions specified by the DSP using `Image8bpp::resize` (see Listing 5-14). The processed image buffer is read into an array, from which an `Image8bpp` object is instantiated and then displayed on the screen. Note that since the target takes care of making sure that all pixels in the segmented image are 0 or 255 (see `segment_detail_image` in Listing 6-7), no further post-processing on the host is required prior to displaying the segmented image – it is simply rendered as a bitmap via GDI+.

6.2.3.3 Standalone Multiscale Edge Detector (`C6416DSK`)

At this point, C6416 DSK users might be feeling a bit forlorn, as they have been left out in the cold in our pursuit of an embedded multiscale edge detector. In this section, we rectify the situation by presenting an optimized implementation of the wavelet-based edge detector that takes advantage of some of the unique features of the C6416. Our optimizations are two-fold: high-level modifications involving a change in the overall algorithm and general structure of the implementation, and lower-level optimizations that employ C intrinsics. The project files are located on the CD-ROM in the `Chap6\wave_edge\C6416DSK` directory.

One facet of the implementations in the previous two sections that begs for attention is that they are needlessly memory consumptive and actually perform superfluous image subtractions. For starters, there is no reason why in Listing 6-7, the detail image needs to be calculated for each wavelet decomposition. This particular segmentation algorithm calls for only segmenting the final detail image, and we since we never perform the inverse simplified à trous wavelet transform (see Figure 6-10 and the reconstruction formula), we need only calculate the last detail image. Thus, we can do away with the three-dimensional detail matrix as defined in Listing 6-8. Of course, if we happened to be using a more sophisticated segmentation scheme that perhaps looked at wavelet coefficients across different scales, we could no longer take this approach. Yet with this change in hand, it is clear that storing each and every approximation coefficient matrix is also hugely wasteful as well, for even if we did in fact need to compute the inverse transform, we only require storage for the current and previous approximation images (where the initial approximation image is the actual input image). The updated `wave_edge.h` header file is given in Listing 6-11 and reflects these changes. Note that in addition to the drastically reduced memory footprint, we also go back to storing the image buffers as flattened arrays. This aspect has a few advantages: it becomes easier to page in blocks of image data from external RAM to fast on-chip RAM, and the data layout is such that it is more suitable for packed data processing using specialized C64x instructions via C intrinsics.

Listing 6-11: The updated C6416 DSK version of `wave_edge.h`, featuring flattened arrays and substantially less storage than that defined in Listing 6-7.

```
#include "image.h"

#pragma DATA_ALIGN (approx1, 8)
#pragma DATA_SECTION (approx1, "SDRAM")
#pragma DATA_ALIGN (approx2, 8)
#pragma DATA_SECTION (approx2, "SDRAM")
#pragma DATA_ALIGN (detail, 8)
#pragma DATA_SECTION (detail, "SDRAM")
int approx1[N_PIXELS],
    approx2[N_PIXELS],
    detail[N_PIXELS];
```

In the new `wave_decomp_a_trous` function, we alternate between using `approx1` and `approx2` as input into the 2D B3 low-pass filter, with the corresponding other buffer used as output for the filter. Then at the end of the outer-most loop (per each wavelet decomposition level) we simply feed those two arrays into `calc_detail`. That function then creates the final detail coefficient image by calculating the absolute value of the delta between the final two approximation images, returning the mean of that image which is then used during segmentation. The three functions that constitute the meat of this program are given in Listing 6-12.

Listing 6-12: Optimized versions of the `wave_decomp_a_trous`, `calc_detail`, and `segment_detail_image` functions from the C6416 DSK version of `wave_edge.c`.

```
unsigned int calc_detail(int *p_approx1, int *p_approx2, int *p_detail)
{
  int *pa1 = p_approx1 + (Y_SIZE<<1) + 2,
      *pa2 = p_approx2 + (Y_SIZE<<1) + 2,
      *pd  = p_detail  + (Y_SIZE<<1) + 2;
  int irow, jcol;
  unsigned int sumlo = 0, sumhi = 0;

  for (irow=2; irow<(X_SIZE-2); ++irow) {
    for (jcol=2; jcol<(Y_SIZE-2); jcol+=2) {
      pd[0] = /* |pa2[0]-pa1[0]| >> 9 */
        _sshvr(_abs(_lo(_amemd8(pa2))-_lo(_amemd8(pa1))),9);
      sumlo += pd[0];
      pd[1] = /* |pa2[1]-pa1[1]| >> 9 */
        _sshvr(_abs(_hi(_amemd8(pa2))-_hi(_amemd8(pa1))),9);
      sumhi += pd[1];
      pa1 += 2;
      pa2 += 2;
      pd += 2;
    }
    pa1 += 4;
    pa2 += 4;
    pd += 4;
  }
  return (sumlo+sumhi)>>LOG2_N_PIXELS; /* sum/# pixels */
}
```

```
int wave_decomp_a_trous(int nlevels)
{
  int ilevel, irow, jcol, kk,
      cpi; /* "cp" = center pixel index */
  int *p_approx1 = approx1,
      *p_approx2 = approx2,
      *p;

  for (ilevel=0; ilevel<nlevels; ++ilevel) {
    /*
     * zero out the margins of the current approximation coefficient
     * matrix - 1st and last couple of rows
     */
    p = p_approx2;
    for (kk = 0; kk < Y_SIZE; kk+=2)
      _amemd8(&p[kk]) =
      _amemd8(&p[kk+Y_SIZE]) =
      _amemd8(&p[kk+N_PIXELS-(Y_SIZE<<1)]) =
      _amemd8(&p[kk+N_PIXELS-Y_SIZE]) = 0;

    /*
     * first and last couple of columns ·(just  the next 2 rows, the
     * remainder are done in the main convolution loop)
     */
    p += (Y_SIZE<<1);
    _amemd8(p) = 0;
    _amemd8(p+Y_SIZE-2) = 0;
    p += Y_SIZE;
    _amemd8(p) = 0;
    _amemd8(p+Y_SIZE-2) = 0;
    p += Y_SIZE;

    /* 5x5 convolution for approximation coeffs at level i */
    for (irow=2; irow<(X_SIZE-2); ++irow) {

      /* clearing out 1st & last two cols */
      _amemd8(p) = 0;
      _amemd8(p+Y_SIZE-2) = 0;
      p += Y_SIZE;

      for (jcol=2; jcol<(Y_SIZE-2); ++jcol) {
```

```
        cpi = (irow<<LOG2_Y_SIZE) + jcol;
                    /* row 1: 1/256 1/64 3/128 1/64 1/256 */
        p_approx2[cpi] = (p_approx1[cpi-(Y_SIZE<<1)-2]>>8) +
                    (p_approx1[cpi-(Y_SIZE<<1)-1]>>6) +
                    ((3*p_approx1[cpi-(Y_SIZE<<1)])>>7) +
                    (p_approx1[cpi-(Y_SIZE<<1)+1]>>6) +
                    (p_approx1[cpi-(Y_SIZE<<1)+2]>>8) +
                    /* row 2: 1/64 1/16 3/32 1/16 1/64 */
                    (p_approx1[cpi-Y_SIZE-2]>>6) +
                    (p_approx1[cpi-Y_SIZE-1]>>4) +
                    ((3*p_approx1[cpi-Y_SIZE])>>5) +
                    (p_approx1[cpi-Y_SIZE+1]>>4) +
                    (p_approx1[cpi-Y_SIZE+2]>>6) +
                    /* row 3: 3/128 3/32 9/64 3/32 3/128 */
                    ((3*p_approx1[cpi-2])>>7) +
                    ((3*p_approx1[cpi-1])>>5) +
                    ((9*p_approx1[cpi])>>6) +
                    ((3*p_approx1[cpi+1])>>5) +
                    ((3*p_approx1[cpi+2])>>7) +
                    /* row 4: 1/64 1/16 3/32 1/16 1/64 */
                    (p_approx1[cpi+Y_SIZE-2]>>6) +
                    (p_approx1[cpi+Y_SIZE-1]>>4) +
                    ((3*p_approx1[cpi+Y_SIZE])>>5) +
                    (p_approx1[cpi+Y_SIZE+1]>>4) +
                    (p_approx1[cpi+Y_SIZE+2]>>6) +
                    /* row 5: 1/256 1/64 3/128 1/64 1/256 */
                    (p_approx1[cpi+(Y_SIZE<<1)-2]>>8) +
                    (p_approx1[cpi+(Y_SIZE<<1)-1]>>6) +
                    ((3*p_approx1[cpi+(Y_SIZE<<1)])>>7) +
                    (p_approx1[cpi+(Y_SIZE<<1)+1]>>6) +
                    (p_approx1[cpi+(Y_SIZE<<1)+2]>>8);


    }
}

/* prep for next iteration, 1st swap pointers */
p = p_approx2;
p_approx2 = p_approx1;
p_approx1 = p;

} /* end (for each decomposition level) */
```

```
    return calc_detail(p_approx1, p_approx2, detail);
}

void segment_detail_image(int mean)
{
    int cp; /* cp = center pixel */
    const int M = 3640; /* 1/9 in Q.15 format */
    unsigned long avg;
    int irow, jcol;
    /* pointers to 3 scan-lines for 3x3 filtering */
    int *pd1 = detail+(2*Y_SIZE+2), /* detail row 1 */
        *pd2 = detail+(3*Y_SIZE+2), /* detail row 2 */
        *pd3 = detail+(4*Y_SIZE+2); /* detail row 3 */
    unsigned char *pout = out_img+(3*Y_SIZE+3); /* output img */

    /*
     * mark those pixels where the detail coefficient
     * is greater than the local mean AND where it is greater
     * than the global mean as an edge.
     */
    for (irow=3; irow<(X_SIZE-3); ++irow, pout+=6) {
      for (jcol=3; jcol<(Y_SIZE-3); ++jcol, ++pd1, ++pd2, ++pd3) {
        cp = pd2[1];
        /* find local average using 3x3 averaging filter */
        avg = ( pd1[0]*M + pd1[1]*M + pd1[2]*M +
                pd2[0]*M + cp*M     + pd2[2]*M +
                pd3[0]*M + pd3[1]*M + pd3[2]*M ) >> 15;
        *pout++ = (cp>avg && cp>mean) ? 255 : 0;
      }
      /* move pointers over to start of the next row */
      pd1 += 6; pd2 += 6; pd3 += 6;
    }
}
```

The first two of the modified functions, calc_detail and wave_decomp_a_trous, make extensive use of various compiler intrinsics. These are covered in more detail in Appendix B (see B.2). In segment_detail, we optimize the performance of the 3x3 averaging filter by replacing array subscripting with pointer arithmetic. We made this tradeoff between readability and performance extensively in Chapter 4, but Kernighan and Ritchie said it best in [23] that "any operation that can be done by array subscripting can also be done with pointers. The pointer version will be faster but, at least to the uninitiated, somewhat harder to understand." Hopefully, if the reader has gotten this far into the book, he or she no longer falls under the ranks of the uninitiated! This optimization could (and should) also be applied in the 5x5 convolution loop in `wave_decomp_a_trous`. The number of operations in the filtering double-loop in `segment_detail` could also be further reduced by employing the technique used in Listing 5-10 (`calc_hist`) and particularly Listing 4-19 (`collect_local_pixel_stats`), by "shifting out" the left-most 1x3 column via subtraction and only adding the next 1x3 column for every iteration across the column dimension of the image (the innermost loop). This optimization flows out of the fact that we are using a box filter with constant coefficients to compute the localized average – with the 2D B3 spline filter this condition does not hold and hence we cannot apply the same optimization in `wave_decomp_a_trous`.

6.3 WAVELET DENOISING

The use of non-linear adaptive filters in noise removal was discussed in 4.6. In this section, we will show how the discrete wavelet transform can be used to remove noise from a corrupted image using a process known as *wavelet denoising*. This technique, initially proposed by Donoho,[24,25,26] relies on the idea of thresholding in the wavelet domain and remains very much in vogue and the subject of active research. The accolades attributed to wavelet denoising are certainly impressive – advocates claim that it "offers all that we might desire of a technique, from optimality to generality."[27] While there may be some amount of hyperbole as to it's efficacy, Donoho did show in his seminal work that the technique attains near-optimal noise reduction properties, in the minimax sense, for a large class of signals – although it should be stressed that much of this work assumed the corrupted signal suffered from additive Gaussian noise. One of the driving factors behind use of the technique is its inherent simplicity. For these reasons, after image compression, wavelet denoising represents perhaps the second most common class of wavelet-based algorithms with regards to image processing. In this section, the theory behind wavelet denoising is introduced, a MATLAB prototype is discussed, and the use of the Wavelet Toolbox for image denoising is also briefly touched upon. Following these preliminaries, we take a quick detour into explaining how to implement an efficient 2D IDWT algorithm in C. Wavelet denoising requires the inverse transform and as we shall soon see, both the C62x/C67x and C64x versions of IMGLIB do not provide inverse wavelet transform functions. Once we have C functions for the 2D IDWT, we can proceed to implement wavelet denoising on the C6416.

Consider an image *f(x,y)* corrupted with additive noise *n(x,y)*. Then the observed image *g(x,y)* is

$$g(x,y) = f(x,y) + n(x,y)$$

We wish to recover from the noisy image *g* an approximation $\hat{f}$ to the original *f*, which minimizes the mean squared error between the approximation and *f*. Let W[·] and W^{-1}[·] denote the 2D DWT and 2D IDWT, respectively. Now let T[·,λ] be a wavelet thresholding operator with threshold λ. Given these definitions, a wavelet based denoising procedure is summarized in Algorithm 6-2.

Algorithm 6-2: Wavelet denoising
INPUT: Noisy image g(x,y), threshold λ
OUTPUT: Denoised image f(x,y)

w(x,y) = W[g(x,y)]	*(perform 2D discrete wavelet transform)*
z(x,y) = T[w(x,y), λ]	*(threshold DWT coefficients)*
f(x,y) = W^{-1}[z(x,y)]	*(inverse 2D discrete wavelet transform)*

Although some researchers have reported slightly better results using the overcomplete wavelet transform we used in the preceding section for multi-scale edge detection,[28,29] the typical means of implementing wavelet denoising of images is to use the classical 2D DWT, as described in 6.1. The algorithme à trous has comparatively larger storage requirements due to the use of the undecimated transform, and concomitantly more comparisons needed for the thresholding operator. These two facets combine for a substantial increase in computations that limit the practicality of utilizing the undecimated wavelet transform in image denoising algorithms.

Thresholding in the wavelet domain works to remove noise due to the "energy compaction" property of the wavelet transform. This property is an integral part of what makes the DWT such a useful tool in signal processing, and while we apply it here in the context of image denoising it is equally vital in image compression. What this property means is that as we proceed with the wavelet transform, signal energy concentrates in the LL band, while the noise energy dissipates throughout the other detail subbands. Hence we use the DWT to separate the actual signal content from the noise, where the low-frequency "base" of the image is contained in the LL wavelet coefficients. If the coefficients in the detail subbands are significant (large in

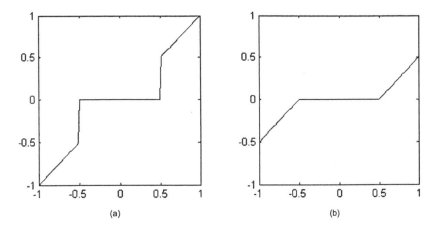

Figure 6-16. Hard threshold versus soft threshold using input data normalized to the interval [-1,1]. (a) Hard threshold transfer function, with $\lambda = 0.5$. (b) Soft threshold transfer function, with $\lambda = 0.5$.

magnitude), they are attributed to the original signal. We exploited this fact in the multiscale edge detection algorithms of the previous section, as we saw that in a multi-level wavelet decomposition, the number of wavelet coefficients with significant magnitude is small, and the significant coefficients correspond to high-frequency detail and edges. Insignificant detail coefficients are assumed to be noise and by discarding these coefficients corresponding to $n(x,y)$ we are hopefully able to reconstruct $f(x,y)$ without loss of detail.

This procedure works so long as the noise power is much smaller than the signal power, the signal is assumed to have low frequency components, and $n(x,y)$ is white[1]. Although the above description may make it appear that it is a form of smoothing, wavelet denoising is non-linear and *not* a smoothing operator. A smoothing operator destroys high-frequency content at the expense of low-frequency content, while wavelet denoising "attempts to remove whatever noise is present and retain whatever signal is present regardless of the frequency content of the signal."[27]

Algorithm 6-2 leaves out two important facets to the denoising algorithm: the details of the thresholding operator T and selection of the threshold λ. Generally speaking, $T[\cdot,\lambda]$ comes in two flavors, a so-called *hard* threshold or *soft* threshold, examples of which are plotted in Figure 6-16. Hard thresholding of wavelet coefficients is a form of thresholding similar in form to that described in 5.2.1 and implemented in various functions within IMGLIB and the Intel IPP library. This operator retains all coefficients greater and than λ and sets the remaining coefficients to zero:

$$T_{hard}(x) = \frac{x, |x| \geq \lambda}{0, |x| < \lambda}$$

Inspection of Figure 6-16a shows an abrupt discontinuity in the hard threshold transfer function at $|x| = \lambda$. This discontinuity tends to manifest itself as annoying artifacts in the denoised image, and motivated researchers to propose the soft threshold scheme, otherwise known as "shrinkage". This transfer function is more continuous and aims to shrinks those coefficients above λ in absolute value. The soft thresholding operator sets to zero all wavelet coefficients x if $|x| < \lambda$, and otherwise:

$$T_{soft}(x) = \frac{x - \lambda, x \geq \lambda}{x + \lambda, x \leq -\lambda}$$

In general, there are three different methodologies for performing the threshold operation on the DWT coefficients. Most of these do not apply any thresholding on the LL (approximation) subimage:

1. **Universal Threshold**: a single threshold is applied globally to all detail coefficients, at all scales. Donoho and Johnstone used this methodology in [25] as part of their *VisuShrink* wavelet shrinkage denoising algorithm. This threshold has been shown to be a good choice for use with soft-thresholding on images corrupted with additive Gaussian noise, although it does tend to produce over-smoothed images.
2. **Level-Dependent Threshold**: The *SureShrink*[26] and *BayesShrink*[30] methods attempt to improve upon a single threshold λ by making it subband-adaptive. A different λ is computed for each detail subband in each wavelet scale.
3. **Spatially Adaptive Threshold**: Akin to the adaptive filtering schemes of 4.6, these operators build upon (2) by tailoring λ based on the local characteristics of the subband. By their very nature, they usually scale-dependent.

Determination of the actual value(s) used for λ is critical to the performance of wavelet denoising, and is not an easy task. Here we find ourselves in the same boat as in Chapter 5, where it is highly desirable to have a data-driven means of calculating λ. Many schemes abound, most of which compute λ based on an estimate of the signal energy and noise variance σ^2 (see 4.6.1 for an explanation of this parameter). Donoho and Johnstone proposed in [25] as part of *VisuShrink* the universal threshold

$$\lambda = \sigma\sqrt{2\log M}$$

where M is the sample size, or total number of image pixels in our case. *VisuShrink* uses a robust estimation of σ based on the median absolute value of the DWT coefficients:

$$\sigma = \text{median}(\ |W_c|\ )\ /\ .6745$$

where W_c denotes the wavelet coefficients. While we use this estimate of the noise power in the MATLAB prototypes that follow, it is difficult to make a case for it in an embedded DSP wavelet denoising implementation. DSPs may excel at certain numerical computations; however they begin to lose some of their luster when utilized in more logic-based operations. Sorting is one such operation, and in order to use the median estimator we must sort a potentially rather large set of numbers so as to calculate the median wavelet coefficient value. The partial sorting optimization described in 4.5.4.3 does not apply here because we are not sorting a fixed, small neighborhood of values. There are parallel sorting algorithms that are designed to sort large arrays of data in chunks. Some variant of this procedure would be needed to *efficiently* sort the DWT coefficients. The best means would be to split the data into blocks small enough to fit in internal RAM, sort each of these blocks, and then merge the results into a single sorted array.

A popular level-dependent variation on the universal threshold theme is:

$$\lambda_j = \sigma\sqrt{2\log M}\, 2^{j-J/2}, j = 0,\ ...,\ J$$

again where M is the total number of signal samples and j is the wavelet scale level[31]. The algorithm we implement on the DSP is a subband-adaptive, hard thresholding denoising algorithm that uses the standard deviation of one of the detail subbands (HH) as an estimate of the noise power. This procedure is summarized in Algorithm 6-3.

Algorithm 6-3: Level dependent hard threshold wavelet denoising
    INPUT: Noisy image I, # decomposition levels N, multiplier s
    OUTPUT: Denoised image K

    W = DWT2D(I, N)   *(N-level 2D discrete wavelet transform)*
    W' = W   *(processed DWT coefficients go in W')*

Algorithm 6-3 (continued)
for each wavelet decomposition level j = 1 ... N

(*determine threshold per each level*)
HH_j = extract_HH(W, j)   (*extract level j HH subband from W*)
λ_j = s[std(HHj)]   (*standard deviation of HH_j multiplied by s*)

(*hard threshold the HH subband*)
for each coefficient c in HH_j
    if |c| < λ_j then c = 0
end
replace_HH(W', j, HH_j)   (*update 2D DWT with processed HH_j*)

(*hard threshold the LH subband*)
LH_j = extract_LH(W, j)   (*extract level j LH subband from W*)
for each coefficient c in LH_j
    if |c| < λ_j then c = 0
end
replace_LH(W', j, LH_j)   (*update 2D DWT with processed LH_j*)

(*hard threshold the HL subband*)
HL_j = extract_HL(W, j)   (*extract level j HL subband from W*)
for each coefficient c in HL_j
    if |c| < λ_j then c = 0
end
replace_HL(W', j, HL_j)   (*update 2D DWT with processed HL_j*)

end (*for each wavelet decomposition level*)

K = IDWT2D(W', N)   (*denoised image is the inverse DWT of W'*)

6.3.1 Wavelet Denoising in MATLAB

The Image Processing Toolbox function `imnoise` has been introduced in previous chapters as a convenient means of generating test data for image processing simulations. If this toolbox is available, the following commands can be used to generate pixel data suitable for image enhancement via wavelet denoising:

```
sigma = 0.01; % use this to specify the magnitude of the noise
J = imnoise(I, 'gaussian', 0, sigma);
```

This MATLAB code snippet contaminates the image I with zero-mean Gaussian white noise of a specific noise variance, determined by the variable sigma. If the Image Processing Toolbox is not available, the M-file add_zero_mean_gwn.m, located in the Chap6 directory on the CD-ROM, can be used as a replacement. This M-file uses the built-in MATLAB function randn, which returns zero mean Gaussian white noise with unit variance. The input pixel data is scaled appropriately, the output from randn multiplied by the square root of σ, and finally the two are added together before the returned data is casted back to the original MATLAB class type. This function, shown in Listing 6-13, was used to generate all of the test data for the wavelet denoising simulations presented in this section.

Listing 6-13: MATLAB function for generating zero-mean Gaussian white noise of a specific magnitude.

```
function J = add_zero_mean_gwn(I, sigma)

if isa(I, 'uint8')
    max_pixel = 255;
elseif isa(I, 'uint16')
    max_pixel = 65535;
else
    error('Only support uint8 or uint16 images.');
end

% normalize pixel values to range [-1,1]
J = double(I)./max_pixel;

% randn returns gaussian noise with zero mean and unit variance
noise = sqrt(sigma).*randn(size(I));


% add noise with specified variance to the image
J = max_pixel .* (J + noise);

% returned image should have same class as input image
eval(sprintf('J = %s(J);', class(I)));
```

Two MATLAB wavelet denoising prototypes are located in the Chap6 directory. Both functions use the multi-level D4 system (see 6.2) for the DWT and draw on some of the functions introduced and developed in 6.1.1. The MATLAB 2D DWT code in 6.1.1 utilized a function haar_basis (see Listing 6-1) that returned the polyphase matrix *H*, constructed in such a manner that convolution followed by downsampling by 2 was achieved via matrix multiplication of H and the input signal. The D4 wavelet transform is defined by the following four tap low-pass filter[4]:

$$
h_{LP}^{D4} = \{h_0, h_1, h_2, h_3\} = \left\{ \frac{1+\sqrt{3}}{4\sqrt{2}}, \frac{3+\sqrt{3}}{4\sqrt{2}}, \frac{3-\sqrt{3}}{4\sqrt{2}}, \frac{1-\sqrt{3}}{4\sqrt{2}} \right\}
$$

Hence a similar periodized polyphase matrix *H* is needed that contains the circular shifted D4 filter coefficients, in order to utilize a linear algebra based implementation of the DWT. This matrix will be of the form explained in 6.1.1, and the MATLAB function daub4_basis (Listing 6-14) creates this matrix, which is then used to implement the forward and inverse D4 transform.

Listing 6-14: MATLAB function for the creation of the D4 forward transform matrix.

```
function D4 = daub4_basis(N)

h0=(1+sqrt(3))/(4*sqrt(2)); h1=(3+sqrt(3))/(4*sqrt(2));
h2=(3-sqrt(3))/(4*sqrt(2)); h3=(1-sqrt(3))/(4*sqrt(2));
daub4 = [ h0 h1 h2 h3; h3 -h2 h1 -h0 ];
D4 = zeros(N);

for ii=1:2:N-2

    row1 = zeros(1,N);
    row1(ii:ii+3) = daub4(1,:);
    row2 = zeros(1,N);
    row2(ii:ii+3) = daub4(2,:);

    D4(ii,:) = row1;
    D4(ii+1,:) = row2;

end

% periodization (wrap-around)
row1 = zeros(1,N); row1(1:2) = [h2 h3]; row1(N-1:N)  = [h0 h1];
row2 = zeros(1,N); row2(1:2) = [h1 -h0]; row2(N-1:N) = [h3 -h2];
D4(N-1:N,:) = [ row1; row2 ];
```

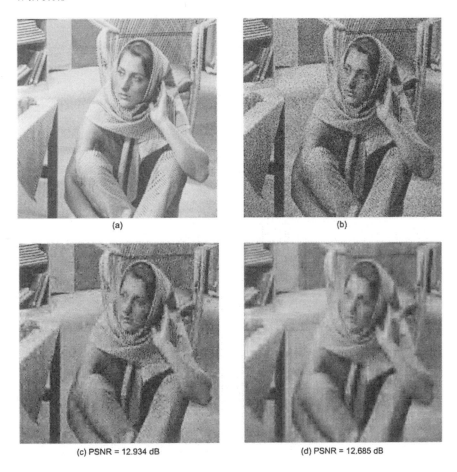

(a)

(b)

(c) PSNR = 12.934 dB

(d) PSNR = 12.685 dB

Figure 6-17. Wavelet denoising with the D4 wavelet on the "Barbara" image corrupted with additive Gaussian white noise. Images (c) and (d) were generated via denoise_daub4_universal, using a three-level wavelet transform and threshold multiplier of 0.5. (a) Original image. (b) Corrupted image with a noise variance of .005, created using the add_zero_mean_gwn function from Listing 6-13. (c) Hard thresholding result, with a threshold derived using the universal threshold. (d) Soft thresholding result, using the same threshold value as (c).

Both MATLAB prototypes implement hard and soft thresholding of DWT coefficients. In denoise_daub4_universal, a global threshold similar in concept to *VisuShrink* is used to perform wavelet denoising. The value of λ is calculated from the finest HH subband, using the median estimator of the noise variance. Clients of this function may pass in a scalar multiplier (which defaults to 1) in order to adjust this estimate, if so desired. Furthermore, the function accepts the number of wavelet decomposition levels and a threshold type string (either 'H' for hard thresholding or 'S' for soft thresholding). This function is given in Listing 6-15 and some results using the "Barbara" image are shown in Figure 6-17.

Listing 6-15: MATLAB M-file for multi-level wavelet denoising using a global threshold and the D4 wavelet.

```
function K = denoise_daub4_universal(I, N, type, mult)

I = double(I);

if nargin<4, mult = 1;, end % default to mult = 1
if nargin<3, type='H';, end % default to hard thresholding

% sanity checks omitted (check for double, square matrix, etc.)
[nrows ncols] = size(I);

% N-level D4 forward wavelet transform on the image I
num_levels = N;
J = I;
for level = num_levels:-1:num_levels-N+1
    sz = 2^(level-1); % assumes # rows = # cols
    H = daub4_basis(sz);
    P = permutation_matrix(sz);
    J(1:sz,1:sz) = P*H*J(1:sz,1:sz)*H'*P';
end
% denoising step (i.e. thresholding in the wavelet domain)
lambda = mult*universal_threshold(J, N);

m = nrows/(2^N);
n = ncols/(2^N);
if type == 'H' % hard threshold

    J(:,n+1:ncols) = hard_thresh(J(:,n+1:ncols), lambda);
    J(m+1:nrows,1:n) = hard_thresh(J(m+1:nrows,1:n), lambda);

elseif type == 'S' % soft threshold

    J(:,n+1:ncols) = soft_thresh(J(:,n+1:ncols), lambda);
    J(m+1:nrows,1:n) = soft_thresh(J(m+1:nrows,1:n), lambda);

end
```

```
% N-level D4 inverse wavelet transform
K = J;
sz = nrows/2^num_levels;
for level = N:-1:1
    sz = sz*2;
    H = daub4_basis(sz);
    P = permutation_matrix(sz);
    K(1:sz,1:sz) = inv(P*H)*K(1:sz,1:sz)*inv(H'*P');
end


function lambda = universal_threshold(J, N)

% 1st order of business is to estimate the variance

% extract the finest HH subband
[nr nc] = size(J);
m = nr/2;
n = nc/2;
HH = J(m+1:nr,n+1:nc);

% sigma estimate
sigma = median( abs( HH(:) ) ) / .6745;

lambda = sigma*sqrt(2*log(nr*nc));

function thresholded = hard_thresh(detail_coefs, T)

% simple, "keep or kill": anything above T keep and
% anything below T set to zero.
thresholded = (abs(detail_coefs) > T) .* detail_coefs;
function thresholded = soft_thresh(detail_coefs, T)

% In the text we define the soft threshold operator as so:
%
%              x-T,  if x >= T
% Tsoft(x) = 0,    if x < T
%              x+T, if x <= -T
%
% An equivalent means of stating the above is to set to zero
% any coefficient below T (in absolute value), and
%
```

% Tsoft(x) = sgn(x)(|x|-T), if |x|>T

```
absd = abs(detail_coefs);
thresholded = sign(detail_coefs).*(absd >= T).*(absd - T);
```

The `denoise_daub4_subband_adaptive.m` M-file is a subband-adaptive wavelet denoising function that implements the scheme mapped out in Algorithm 6-3. This function uses the standard deviation of each HH subband to derive λ for the current level. The built-in MATLAB function `std2`, which returns the standard deviation of a matrix of values, is used as a convenient means of calculating the threshold value. Results for the "Elaine" image are shown in Figure 6-18, and relevant portions of the M-file are given in Listing 6-16. It should be noted that both Listings 6-16 and 6-17 leave plenty of room for optimization. As we shall soon see when we take a closer look at an efficient C implementation of the IDWT, sub-optimal wavelet implementations perform superfluous multiplications by zero, and our MATLAB linear algebra based approach suffers from the same problem. The DWT and IDWT are actually examples of *multi-rate signal processing* and there are functions available in the Signal Processing Toolbox that can be used to efficiently pass signals through QMF banks. The `upfirdn`[32] function is just what the doctor ordered when it comes to implementing the IDWT in an optimized fashion. This function upsamples, FIR filters, and then downsamples a signal, precisely the operations we need to perform the IDWT. However the goal of these M-files is not to illustrate an optimized MATLAB-centric means of performing the DWT and IDWT, rather they are meant to provide a fairly easy-to-read reference implementation of Algorithm 6-3. As we have previously seen, a straightforward reference implementation is worth its weight in gold during development and deployment to an embedded platform.

Listing 6-16: Portions of a MATLAB subband-adaptive wavelet denoising implementation. See Listing 6-15 for the contents of the `hard_threshold` and `soft_threshold` functions.

```
function K = denoise_daub4_subband_adaptive(I, N, type, mult)

% default argument setup & initialization omitted (see Listing 6-15)

% N-level D4 forward wavelet transform on the image I (see Listing 6-15)

% subband-adaptive denoising step
for level = num_levels:-1:num_levels-N+1
    sz = 2^level; % assumes # rows = # cols
```

```
HH = J(sz/2+1:sz, sz/2+1:sz);
% calculate threshold based on current HH subband,
% using standard deviation estimate of noise variance
lambda = mult*std2(HH);
% threshold HH, LH, and HL subbands
LH = J(1:sz/2, sz/2+1:sz);
HL = J(sz/2+1:sz, 1:sz/2);
if type == 'H' % hard threshold
    HH = hard_thresh(HH, lambda);
    LH = hard_thresh(LH, lambda);
    HL = hard_thresh(HL, lambda);
elseif type == 'S' % soft threshold
    HH = soft_thresh(HH, lambda);
    LH = soft_thresh(LH, lambda);
    HL = soft_thresh(HL, lambda);
end
% finally, replace existing detail coefficients
J(sz/2+1:sz, sz/2+1:sz) = HH;
J(1:sz/2, sz/2+1:sz) = LH;
J(sz/2+1:sz, 1:sz/2) = HL;
end

% N-level D4 inverse wavelet transform (see Listing 6-15)
```

The Wavelet Toolbox includes a gamut of useful functions for wavelet denoising of signals and images. These range from catch-all functions like wdencmp that perform everything pertinent to denoising (controlled via a list of string input arguments) to lower-level functions such as wnoisest (noise estimation) and wthresh (soft or hard thresholding). By piecing together various Wavelet Toolbox functions it is possible to very easily prototype any number of denoising schemes. In addition, the Wavelet Toolbox includes a GUI program that puts a user-friendly face onto the myriad number of wavelet denoising options available through the toolbox. Interested readers should consult [6] for further in-depth information regarding these functions and the GUI application.

6.3.2 Wavelet Denoising on the C6x

In order to provide a practical wavelet denoising system running on the C62x/C67x/C64x DSP platform, we require efficient implementations of the three operations enumerated in Algorithm 6-2: the forward wavelet transform, threshold operator, and inverse wavelet transform. In 6.1.4, the use of the IMGLIB forward wavelet transform functions was discussed, and a program that utilized two-dimensional DMA paging was presented. Unfortunately, as of version 1.02 of the C62x/C67x IMGLIB and version 1.4 of the C64x IMGLIB, there is no inverse wavelet transform functionality, which is somewhat odd since the forward DWT in and of itself is of rather limited utility. Interestingly enough, the C55x version of IMGLIB includes a plethora of both forward and inverse optimized wavelet functions, as described in [33].

(a)

(b)

(c) PSNR = 11.577 dB

(d) PSNR = 13.051 dB

Figure 6-18. Wavelet denoising with the D4 wavelet on the "Elaine" image corrupted with additive Gaussian white noise, using the method of Algorithm 6-3. Images (c) and (d) were generated via `denoise_daub4_subband_adaptive`, using a three-level wavelet transform and threshold multiplier of 0.5. (a) Original image. (b) Corrupted image with a noise variance of .005. (c) Hard thresholding result, with λ derived on a per-level basis. (d) Soft thresholding result, using the same λ as (c).

Thus in this section we will take a slight detour and show how to implement an efficient IDWT in C on the C6416. The wavelet denoising algorithm introduced in the preceding section and implemented here utilizes the D4 wavelet. This choice of wavelet accords us the opportunity to craft a forward wavelet transform that will be faster than `IMG_wave_horz` and `IMG_wave_vert`, due to the reduced number of filter coefficients. Recall that the IMGLIB functions in question presuppose filters consisting of eight coefficients, and with the functions presented in this section we are able to reduce the number of MAC operations in the row and column convolutions by a factor of two. An additional rationale behind the development of these functions is that they provide a template for further wavelet implementations, and the hope is that with this code in hand the wavelet functions are not treated as a black box with little understanding of what is going on underneath the covers. As the old saying goes: "Give a man a fish, and you have fed him for today. Teach a man to fish, and you have fed him for a lifetime." Substitute fishing for programming (a fair trade if there ever was one), and hopefully the techniques described in the forthcoming section can be readily adapted in fairly short order to implement efficient DWT/IDWT functions for wavelets of any lengths (e.g. the D6 wavelet or any number of wavelets characterized by filters with more than eight coefficients, which are not supported by the C62x/C67x/C64x IMGLIB functions). There are numerous sources that provide C implementations of the forward and inverse wavelet transform,[34,35] however most of these are very general, which is probably a good thing for implementations targeting desktop or workstation PCs. On resource constrained embedded systems however, this type of generality is typically eschewed if in fact one can glean substantial performance improvements by tailoring the algorithm exactly to the task at hand.

In particular, any decent IDWT implementation avoids needless multiplications by zero during the upsampling and convolution processes, as we shall soon see. In addition, in order to code generic DWT and IDWT implementations (i.e. the number of filter coefficients is variable and passed into the functions from the client), various looping structures are needed during the convolution and resampling phases of both the DWT and IDWT. Texas Instruments compilers are good, but not *that* good, and it is shown in Appendix B that through some clever loop unrolling and other techniques we can make better use of the C64x instruction set when implementing the DWT/IDWT. We can only apply these "no-loop" optimizations if we know in advance the number of filter coefficients. Finally, both the forward and inverse 2D wavelet transforms presented here are inspired by the IMGLIB forward transforms discussed in 6.1.4, in that there is no matrix transposition involved.

6.3.2.1   D4 DWT and IDWT functions on the C6416

The `Chap6\daub4` directory contains a C6416 DSK project that demonstrates the use of efficient multi-level D4 2D DWT and 2D IDWT functions. The usage of the horizontal and vertical DWT functions, `wave_horz` and `wave_vert`, is modeled closely after the IMGLIB functions `IMG_wave_horz` and `IMG_wave_vert` and is illustrated in Figure 6-19 for the case of a single 2D wavelet decomposition. As shown in this figure, two buffers are used to first transform the image and then reconstruct it using the 2D IDWT functions `invwave_vert` `invwave_horz`. For a multi-level pyramidal 2D DWT the process cascades down the LL subband, with the converse occurring in the IDWT to reconstruct the image.

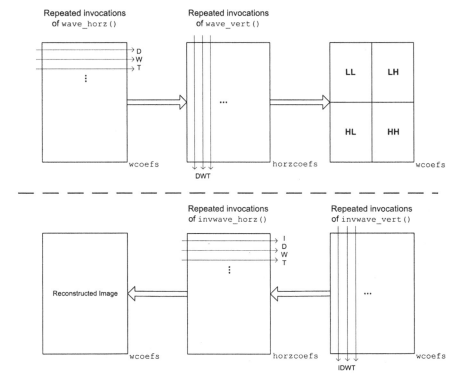

Figure 6-19. High-level data flow for the `daub4` C6416 program given in Listing 6-17. This figure depicts the operations performed for a single-level wavelet analysis and subsequent wavelet synthesis (i.e. `levels=1` in `main`). In the figure, labels in `courier` font are actual symbols from the code. The upper portion of the figure corresponds to the operations occurring within the `dwt2d` function, and the lower corresponds to `idwt2d`.

The code is given in Listing 6-17 (see Appendix B, section B.1.4, for further information on how this code can be optimized). For the sake of clarity, this code does not implement the DMA optimization as described in 6.1.4, and directly accesses memory residing in the SDRAM external memory segment. However, the functions are constructed in the same manner as the IMGLIB functions, and so it should be reasonably straightforward to apply those same memory optimizations here.

Just as in the case of Listings 6-4 and 6-5, the initial phase of the 2D DWT entails the 1D wavelet transform on each row of the input image. This means that each input row is convolved with both low-pass and high-pass D4 analysis filters, and the filtered output decimated by a factor of two. Except for periodization, which is handled as a special case, the two convolution and decimation processes are performed using two loops in wave_horz. Both of these are unrolled to the extent that the kernels explicitly lay out the multiplication and addition operations needed to convolve the filter coefficients with four image row pixels, instead of performing the sum of vector products via an additional nested loop. We can do this because the code assumes four coefficients – a completely generalized function would require a loop of some sort.

Listing 6-17: daub4.c (see B.1.4 for further optimizations of this code).

```
#define Qr 16384          /* Q15 round (0.5) */
#define MAX_LEVELS 6 /*min(X_SIZE_LOG_2,Y_SIZE_LOG_2)-2*/
/* forward DWT along the horizontal (row) direction */
void wave_horz
(
  short *in_data, /* row of input pixels */
  const short *hLP, /* low-pass 4-tap D4 decomposition filter */
  const short *hHP, /* high-pass 4-tap decomposition filter */
  short *out_data, /* row of output data */
  int   cols     /* length of input */
)
{
  int      ii, iters = (cols>>1) - 2;
  const short *xptr;
  int      sum;

  /* ---------------------------------------------------------------------
   * Convolve input with 4 tap low-pass filter, followed by a
   * downsampling by 2x. Periodize the signal by moving the
   * last three pixels to the beginning of the sequence.
   * --------------------------------------------------------------------- */
```

```
/* periodization: 1st two times through the loop */
*out_data++ = (hLP[3]*in_data[cols-3] + hLP[2]*in_data[cols-2] +
               hLP[1]*in_data[cols-1]  + hLP[0]*in_data[0]) >> 15;
*out_data++ = (hLP[3]*in_data[cols-1] + hLP[2]*in_data[0] +
               hLP[1]*in_data[1]      + hLP[0]*in_data[2]) >> 15;

/* rest of the low-pass filter + decimation loop */
xptr = in_data + 1;
for (ii=0; ii<iters; ++ii, xptr+=2) {
  sum = Qr +
     xptr[0]*hLP[3] +
     xptr[1]*hLP[2] +
     xptr[2]*hLP[1] +
     xptr[3]*hLP[0];

     *out_data++ = sum>>15;
}

/* Same as above but this time convolve with the high-pass filter. */

*out_data++ = (hHP[3]*in_data[cols-3] + hHP[2]*in_data[cols-2] +
               hHP[1]*in_data[cols-1] + hHP[0]*in_data[0]) >> 15;
*out_data++ = (hHP[3]*in_data[cols-1] + hHP[2]*in_data[0] +
               hHP[1]*in_data[1]      + hHP[0]*in_data[2]) >> 15;

xptr = in_data + 1;
for (ii=0; ii<iters; ++ii, xptr+=2) {
  sum  = Qr +
       xptr[0]*hHP[3] +
       xptr[1]*hHP[2] +
       xptr[2]*hHP[1] +
       xptr[3]*hHP[0];

  *out_data++ =  sum>>15;
 }
}

/*  forward DWT along the vertical (col) direction */
void wave_vert
(
 const short *
```

```
    const          *in_data,       /* array of row pointers */
    const short *gLP,              /* low-pass 4-tap D4 decomposition filter */
    const short *gHP,              /* high-pass 4-tap D4 decomposition filter */
    short          *out_approx, /* approximation coefficients */
    short          *out_detail,  /* detail coefficients */
    int            cols            /* length of rows to process */
    )
{
    int ii;
    int approx, detail;


    /*-------------------------------------------------------------------------
     * In comparison to wave_horz, much simpler (most of the
     * complications are dealt outside of the function, by the callee).
     * Simply run across the current row, passing 4 pixels in the
     * vertical (column) direction through the low-pass and high-pass
     * decomposition filters.
     *-------------------------------------------------------------------------*/

    for (ii=0; ii<cols; ++ii) {
      approx = gLP[3]*in_data[0][ii] + gLP[2]*in_data[1][ii] +
                 gLP[1]*in_data[2][ii] + gLP[0]*in_data[3][ii];
      *out_approx++ = approx>>15;
      detail = gHP[3]*in_data[0][ii] + gHP[2]*in_data[1][ii] +
                 gHP[1]*in_data[2][ii] + gHP[0]*in_data[3][ii];
      *out_detail++ = detail>>15;
    }
}


/* inverse DWT along the horizontal (row) direction */
void invwave_horz(
    short *in_data,      /* row of input pixels */
    const short *gLP, /* low-pass 4-tap reconstruction filter  */
    const short *gHP, /* high-pass 4-tap reconstruction filter */
    short *out_data,  /* row of output data */
    int    cols          /* length of input */
    )
{
    int   out1, out2, ii;
    short *pa = in_data, /* start of approx coefs */
           *pd = in_data + (cols>>1), /* start of detail coefs */
           *in_detail = pd,
```

```
        *pout = out_data;

    /* ----------------------------------------------------------------
     * This loop is equivalent to splicing out the approximation and detail
     * coefficients out of   in_data, upsampling both by a factor of 2 (by
     * inserting zeros in between every sample), passing both upsampled
     * signals through their respective reconstruction filter, and finally
     * adding the results.
     * ---------------------------------------------------------------- */

    /* this loop is unrolled by a factor of 2 */
    for (ii=0; ii<cols-4; ii+=2, ++pa, ++pd) {
      out1 = gLP[3]*pa[0]+gLP[1]*pa[1] + /* filtered approximation */
             gHP[3]*pd[0]+gHP[1]*pd[1];  /* filtered detail */
      *pout++ = out1>>15;
      out2 = gLP[2]*pa[1]+gLP[0]*pa[2] + /* filtered approximation */
             gHP[2]*pd[1]+gHP[0]*pd[2];  /* filtered detail */
      *pout++ = out2>>15;
    }


    /* periodization (wrap-around with final 4 samples) */
    out1 = (gLP[3]*pa[0]+gLP[1]*pa[1]) +
           (gHP[3]*pd[0]+gHP[1]*pd[1]);
    *pout++ = out1>>15;
    out2 = (gLP[2]*pa[1]+gLP[0]*in_data[0]) +
           (gHP[2]*pd[1]+gHP[0]*in_detail[0]);
    *pout++ = out2>>15;
    out1 = (gLP[3]*pa[1]+gLP[1]*in_data[0]) +
           (gHP[3]*pd[1]+gHP[1]*in_detail[0]);
    *pout++ = out1>>15;
    out2 = (gLP[2]*in_data[0]+gLP[0]*in_data[1]) +
           (gHP[2]*in_detail[0]+gHP[0]*in_detail[1]);
    *pout++ = out2>>15;
}


/* inverse DWT along the vertical (col) direction */
void invwave_vert
(
  const short *
  const      *in_data,  /* array of row pointers */
  short      *gLP,      /* low-pass 4-tap reconstruction filter  */
  short      *gHP,      /* high-pass 4-tap reconstruction filter */
```

```
short      *out_data1, /* 1st output row */
short      *out_data2, /* 2nd output row */
int         cols        /* length of rows to process */
)
{
  int out1, out2, ii;

  /*-----------------------------------------------------------------------
   * This loop works in a similar fashion to that of wave_vert. For each
   * column in the two rows pointed to by out_data1 and out_data2,
   * perform the inverse DWT in the vertical direction by convolving
   * upsampled approximation and detail portions of the  the input with
   * the respective reconstruction filters. Except for a single periodization
   * case, out_data1 and out_data2 will point consecutive rows in the
   * output image.
   *-----------------------------------------------------------------------*/

  for (ii=0; ii<cols; ++ii) {
    out1 = gLP[2]*in_data[0][ii]+gLP[0]*in_data[1][ii] +
           gHP[2]*in_data[2][ii]+gHP[0]*in_data[3][ii];
    *out_data1++ = out1>>15;
    out2 = gLP[3]*in_data[0][ii]+gLP[1]*in_data[1][ii] +
           gHP[3]*in_data[2][ii]+gHP[1]*in_data[3][ii];
    *out_data2++ = out2>>15;
  }
}

/* multi-level wavelet decomposition (1st horz, then vert) */
void dwt2d(int nLevels)
{
  /*
   * D4 scaling filter multiplied by 2^15 (Q15 format)
   * High-pass filter is the flipped low-pass filter
   */
  const short d4_decomp_lp_Q15[] = {15826,27411,7345,-4240},
              d4_decomp_hp_Q15[] = {-4240,-7345,27411,-15826};

  int nlr = X_SIZE, /* # rows in current level */
      nlc = Y_SIZE, /* # cols in current level */
      ilevel = 0, irow, ii;

  /* for wave_vert */
```

```
short *pwvbufs[4]; /* ptrs to input scan-lines */
short *plopass, *phipass; /* output ptrs */
int nrow, scol;

for (; ilevel<nLevels; ++ilevel, nlr>>=1, nlc>>=1) {

  /* transform the rows */
  for (irow=0; irow<nlr; ++irow)
    wave_horz(wcoefs+irow*Y_SIZE, /* input (row n) */
            d4_decomp_lp_Q15, d4_decomp_hp_Q15,
            horzcoefs+irow*Y_SIZE, /* output */
            nlc);

  /* transform the cols */
  nrow = nlr-3;

  /* periodization: 1st time through input rows N-2, N-1, N, and 1st */
  pwvbufs[0]=horzcoefs+nrow*Y_SIZE;
  pwvbufs[1]=pwvbufs[0]+Y_SIZE;
  pwvbufs[2]=pwvbufs[1]+Y_SIZE;
  pwvbufs[3]=horzcoefs;
  plopass=wcoefs; phipass=wcoefs+(nlr>>1)*Y_SIZE;
  wave_vert(pwvbufs,
            d4_decomp_lp_Q15, d4_decomp_hp_Q15,
            plopass, phipass,
            nlc);
  plopass += Y_SIZE; phipass += Y_SIZE;

  /* 2nd time throw last row and 1st three rows */
  pwvbufs[0]=horzcoefs+(nrow+2)*Y_SIZE;
  pwvbufs[1] = horzcoefs;
  pwvbufs[2]=pwvbufs[1]+Y_SIZE;
  pwvbufs[3]=pwvbufs[2]+Y_SIZE;
  wave_vert(pwvbufs,
            d4_decomp_lp_Q15, d4_decomp_hp_Q15,
            plopass, phipass,
            nlc);
  plopass += Y_SIZE; phipass += Y_SIZE;

  /* and the rest of 'em through this loop */
  scol = 1; /* start column */
  for (ii=0; ii<(nlr>>1)-2; ++ii,
```

```
        plopass+=Y_SIZE, phipass+=Y_SIZE) {
      pwvbufs[0] = horzcoefs+(scol*Y_SIZE);
      pwvbufs[1] = pwvbufs[0]+Y_SIZE;
      pwvbufs[2] = pwvbufs[1]+Y_SIZE;
      pwvbufs[3] = pwvbufs[2]+Y_SIZE;
      wave_vert(pwvbufs,
                 d4_decomp_lp_Q15, d4_decomp_hp_Q15,
                 plopass, phipass,
                 nlc);
      scol += 2;
    }

  } /* end (for each wavelet decomposition level) */
}

void idwt2d(int nLevels)
{
  /* inverse D4 filter coefficients in Q15 format */
  short d4_synth_lp_Q15[] = {-4240,7345,27411,15826},
        d4_synth_hp_Q15[] = {-15826,27411,-7345,-4240};

  int nlr = X_SIZE>>(nLevels-1), /* # rows in current level */
      nlc = Y_SIZE>>(nLevels-1), /* # cols in current level */
      ilevel, irow, ii;

  /* for invwave_vert */
  short *pwvbufs[4];
  short *pidwt1, *pidwt2, *pnext;

  for (ilevel=0; ilevel<nLevels; ++ilevel, nlr<<=1, nlc<<=1) {

    /*
     * first perform the vertical transform, 2D DWT coefficients
     * are in the wcoefs buffer, send inverse into horzcoefs
     */

    /*
     * periodization: 1st time through input
     * rows 1, 2, half-way, half-way+1
     */
    pwvbufs[0]=wcoefs;              pwvbufs[2]=wcoefs+(nlr>>1)*Y_SIZE;
    pwvbufs[1]=pwvbufs[0]+Y_SIZE;   pwvbufs[3]=pwvbufs[2]+Y_SIZE;
```

```
pidwt1 = horzcoefs+(nlr-1)*Y_SIZE;          pidwt2 = horzcoefs;
invwave_vert(pwvbufs,
             d4_synth_lp_Q15, d4_synth_hp_Q15,
             pidwt1, pidwt2,
             nlc);

/*
 * interior portion of the vertical convolutations with
 * no periodization effects
 */
pidwt1=horzcoefs+Y_SIZE; pidwt2=pidwt1+Y_SIZE;
for (ii=0; ii<nlr-2;
     ii+=2, pidwt1+=(Y_SIZE<<1), pidwt2+=(Y_SIZE<<1)) {

  /* shift 1st two up by 1 and fetch next row */
  pnext = pwvbufs[1]+Y_SIZE;
  pwvbufs[0]=pwvbufs[1]; pwvbufs[1]=pnext;
  /* shift 2nd two up by 1 and fetch next row */
  pnext = pwvbufs[3]+Y_SIZE;
  pwvbufs[2]=pwvbufs[3]; pwvbufs[3]=pnext;

  invwave_vert(pwvbufs,
               d4_synth_lp_Q15, d4_synth_hp_Q15,
               pidwt1, pidwt2,
               nlc);

}

/*
 * periodization: last time input rows are half-way point and 1st,
 * for each of the two sections of wcoefs pointed to by pwvbufs.
 */
pwvbufs[0]=pwvbufs[1]-Y_SIZE;    pwvbufs[2]=pwvbufs[3]-Y_SIZE;
pwvbufs[1]=wcoefs;              pwvbufs[3]=wcoefs+(nlr>>1)*Y_SIZE;
invwave_vert(pwvbufs,
             d4_synth_lp_Q15, d4_synth_hp_Q15,
             pidwt1-(Y_SIZE<<1), pidwt2-(Y_SIZE<<1),
             nlc);

/*
 * done with vertical inverse transform,
 * horizontal direction much simpler.
```

```
    */
    for (irow=0; irow<nlr; ++irow)
      invwave_horz(horzcoefs+irow*Y_SIZE,
                   d4_synth_lp_Q15, d4_synth_hp_Q15,
                   wcoefs+irow*Y_SIZE,
                   nlc);

  } /* end (for each wavelet scale) */
}

int main(void)
{
  int levels = 3;
  DSK6416_init(); /* initialize the DSK board support library */
  dwt2d(levels);
  idwt2d(levels);
}
```

Figure 6-20 is a diagram of `wave_horz`, and essentially a pictorial explanation of what occurs for each one of the horizontal lines in the upper portion of Figure 6-19. As we saw in the MATLAB code responsible for constructing the D4 polyphase filter matrix (see the end of Listing 6-14), periodization can be somewhat of a hassle. For code optimization purposes, it is best to handle the periodization (a.k.a "wrap-around" or "circular buffer management") as special cases outside of the main convolution loop. This strategy avoids a highly detrimental branch operation in the middle of the loop kernel. We avoid any extra loops for the periodization by explicitly mapping out the circular shifts of the input, which we can do because we know in advance exactly how many samples from the end of each row to shift over to the beginning of each sequence. In general, for a convolution of a signal with an N-tap filter, a circular shift that moves the last N-1 samples over to the beginning of the input signal is required. As shown in Figure 6-20, this means that for the D4 analysis stage, we need to shift the last three pixels in every row over to the beginning of each sequence, in order to properly compute the horizontal wavelet coefficients a_0, a_1, d_0, and d_1.

The next step in the 2D DWT is to pass the columns of the transformed rows through the D4 analysis filters. This is done in `dwt2d` by repeatedly calling `wave_vert`, in a fashion similar to how `IMG_wave_vert` was utilized (see Figure 6-9). The big change here is that the total number of row pointers into `horzcoefs` is exactly one-half of what is needed for `IMG_wave_vert`. And once again, periodization is the cause of a programming inconvenience. In contrast to `wave_horz`, where these

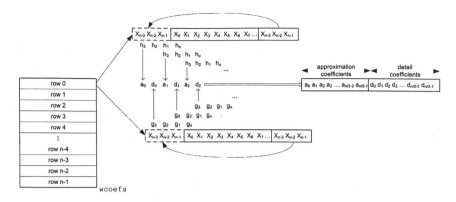

Figure 6-20. Single iteration of `wave_horz`, referenced in Listings 6-17 to pass each row of an image through the D4 low-pass (h) and high-pass (g) analysis filters. Dotted portions of the input sequences correspond to the circular shifts required for convolution.

$a_0 = h_3 x_{n-3} + h_2 x_{n-2} + h_1 x_{n-1} + h_0 x_0$, $d_0 = g_3 x_{n-3} + g_2 x_{n-2} + g_1 x_{n-1} + g_0 x_0$, and so on.

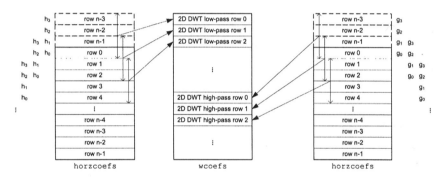

Figure 6-21. Mechanics of `wave_vert`. The last three rows are shifted to the top of the image to account for periodization of the input, and then `wave_vert` is called n/2 times, with each invocation producing a low-pass and high-pass filtered output row. Once we have cycled through all of `horzcoefs`, the 2D D4 DWT is complete.

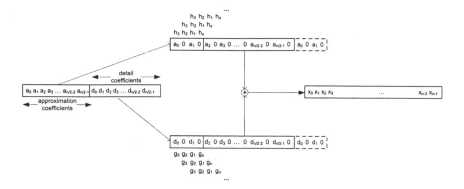

Figure 6-22. Inverse D4 DWT. The approximation and detail coefficients are upsampled and then convolved with the D4 synthesis low-pass filter (h) and high-pass filter (g). The filtered outputs are added together to yield the IDWT. An efficient polyphase implementation avoids multiplications by zero. Dotted portions of the upsampled approximation and coefficient vectors again represent circular shifts of the data.

circular shifts are handled within the innards of the actual horizontal DWT function, here we must handle them outside of `wave_vert`, in `dwt2d`.

In the case of a multi-level pyramidal wavelet decomposition (i.e. `nLevels ≠ 1` in `dwt2d`), the code then proceeds to halve the number of rows and columns with each level so that `wave_horz` and `wave_vert` operate on the previous scale's LL subband. Multi-level wavelet synthesis is performed within `idwt2d`, and starting here we do not have the luxury of existing IMGLIB functions to fall back on. Figure 6-22 shows the steps to perform a single-level inverse DWT on a one-dimensional signal, and this is one area where the matrix multiplication formulation of the IDWT, employed heavily in the MATLAB code, is not as efficient as a properly coded polyphase C implementation. We need to upsample the approximation and detail coefficients by a factor of two, and then pass these two upsampled signals through the wavelet synthesis low-pass and high-pass filters. Referring to Figure 6-22, the convolution computations needed to filter the upsampled approximation and detail coefficients in order to synthesize vector x are

$$x_0 = a_0h_3+0h_2+a_1h_1+0h_0 + d_0g_3+0g_2+d_1g_1+0g_0$$
$$= a_0h_3+a_1h_1 + d_0g_3+d_1g_1$$
$$x_1 = 0h_3+ a_1h_2+0h_1+a_2h_0 + 0g_3+ d_1g_2+0g_1+ d_2g_0$$
$$= a_1h_2+a_2h_0 + d_1g_2+d_2g_0$$
$$x_2 = a_1h_3+0h_2+a_2h_1+0h_0 + d_1g_3+0g_2+d_2g_1+0g_0$$
$$= a_1h_3+a_2h_1 + d_1g_3+d_2g_1$$

...

$$x_{n-3} = 0h_3 + a_{n/2-1}h_2 + 0h_1 + a_0h_0 + 0g_3 + d_{n/2-1}g_2 + 0g_1 + d_0g_0$$
$$= a_{n/2-1}h_2 + a_0h_0 + d_{n/2-1}g_2 + d_0g_0$$
$$x_{n-2} = a_{n/2-1}h_3 + 0h_2 + a_0h_1 + 0h_0 + d_{n/2-1}g_3 + 0g_2 + d_0g_1 + 0g_0$$
$$= a_{n/2-1}h_3 + a_0h_1 + d_{n/2-1}g_3 d_0g_1$$
$$x_{n-1} = 0h_3 + a_0h_2 + 0h_1 + a_1h_0 + 0g_3 + d_0g_2 + 0g_1 + d_1g_0$$
$$= a_0h_2 + a_1h_0 + d_0g_2 + d_1g_0$$

There is a distinct pattern to this upsampling followed by convolution process, and in both `invwave_horz` and `invwave_vert` this pattern is exploited to avoid any unnecessary multiplications by zero. Aside from this optimization, the innards of both functions are quite similar in nature to that of the corresponding forward wavelet transform functions, and their usage from the context of `idwt2d` is roughly the same. And now with both the 2D DWT and 2D IDWT in place, we can move on to actual wavelet denoising.

6.3.2.2 A C6416 Wavelet Denoising Implementation

Once IDWT functions are available, implementing wavelet denoising is not particularly difficult. In fact, most of the complications in this particular implementation arise from the fixed-point standard deviation computation. In the `Chap6\denoise` directory, a C6416 DSK project that implements Algorithm 6-3 can be found. This project is mostly the same as the one described in 6.3.2.1, except for the addition of two new functions, `denoise` and `calc_threshold`, which are shown in Listing 6-18. The `denoise` function is called right after `dwt2d`, and it uses `calc_threshold` to compute λ in a level-dependent fashion using the standard deviation of the HH subband and then thresholds each of the three current detail subbands. All of the computations are performed using fixed-point arithmetic, except for the final calculation of the square root of the variance (the standard deviation). This program uses the FastRTS function `ulongsp` to convert the variable containing the variance (`sumsq`) into a single-precision quantity, which is subsequently fed into the standard C library function `sqrt`. Alternatively, an integer arithmetic square root algorithm (see [36]) could be used instead of FastRTS, but these algorithms are typically plagued with numerical stability issues. After `denoise` finishes thresholding the detail subbands, `idwt2d` is then called to reconstitute the denoised image. Testing of this project can be done in a similar fashion to the program described in 6.1.4, that is, use the CCStudio file export facility and then import the image into MATLAB via `read_ccs_data_file`.

Listing 6-18: Two functions from `denoise.c`, `calc_threshold` and `denoise`, together which implement the thresholding scheme detailed in Algorithm 6-3.

```c
/* calculate level-dependent threshold for wavelet denoising */
int calc_threshold(int level)
{
  const float threshold_multiplier = 1.f;
  float std_dev = 0.f;
  const int Q = 6, R = 1 << (Q-1);
  const int subband_rows = X_SIZE >> (level+1),
        subband_cols = Y_SIZE >> (level+1),
        n_subband_log2 = N_PIXELS_LOG_2 - ((level+1)<<1);

  /* index into the HH subband for this level */
  short *p_subband = wcoefs + subband_rows*Y_SIZE + subband_cols,
        *pHH = p_subband;
  int sum = R, irow, jcol, mean, delta;
  unsigned long sumsq = R;

  /* compute mean wavelet coefficient value in this band */
  for (irow=0; irow<subband_rows; ++irow, pHH+=Y_SIZE)
    for (jcol=0; jcol<subband_cols; ++jcol)
      sum += pHH[jcol];
  mean = sum >> (n_subband_log2-Q);

  /* compute variance */
  pHH = p_subband;
  for (irow=0; irow<subband_rows; ++irow, pHH+=Y_SIZE) {
    for (jcol=0; jcol<subband_cols; ++jcol) {
      delta = (pHH[jcol]<<Q) - mean;
      sumsq += (delta*delta);
    }
  }

  /* bit shift by 2Q due to multiplication, shift by rest for avg */
  sumsq >>= (Q<<1) + n_subband_log2;

  /* standard deviation */
  std_dev = sqrt(ulongsp(sumsq));

  /* wavelet denoise threshold */
  return (threshold_multiplier*std_dev)+0.5f;
```

```
}

void denoise(int nLevels)
{
  int subband_rows = X_SIZE,
      subband_cols  = Y_SIZE,
      colsx2        = Y_SIZE<<1,
      n_subband     = subband_rows*subband_cols,
      T, irow, jcol, level;
  short *p_subband;

  for (level=0; level<nLevels; ++level)
  {
    subband_rows >>= 1; subband_cols >>= 1;
    colsx2 >>= 1;
    n_subband >>= 2;

    T = calc_threshold(level); /* level-dependent threshold */

    /* only keep those coefficients above the threshold */

    /* threshold LH subband */
    p_subband = wcoefs + subband_cols;
    for (irow=0; irow<subband_rows; ++irow, p_subband+=Y_SIZE) {
      for (jcol=0; jcol<subband_cols; ++jcol) {
        if (_abs(p_subband[jcol])<=T)
          p_subband[jcol] = 0;
      }
    }

    /* threshold HL & HH subbands */
    p_subband = wcoefs + subband_rows*Y_SIZE;
    for (irow=0; irow<subband_rows; ++irow, p_subband+=Y_SIZE) {
      for (jcol=0; jcol_colsx2; ++jcol) {
        if (_abs(p_subband[jcol])<=T)
          p_subband[jcol] = 0;
      }
    }
  } /* end (for each level) */
}
```

REFERENCES

1. Gilbert, S., Nguyen, T., *Wavelets and Filter Banks* (Wellesley-Cambridge Press, 1997).
2. Burrus,C.S., Gopinath, R.A., Guo, H., *Introduction to Wavelet and Wavelet Transforms: A Primer* (Prentice Hall, 1998).
3. Meyer, Y., *Wavelets: Algorithms & Applications* (SIAM, 1994).
4. Daubechies, I., *Ten Lectures on Wavelets* (SIAM, 1992).
5. The Mathworks, Inc., "imread." Retrieved October 2004 from:
   http://www.mathworks.com/access/helpdesk/help/techdoc/ref/imread.html
6. Misiti, M., Misiti, Y., Oppenheim, G., Poggi, J.M., *Wavelet Toolbox User's Guide* (The MathWorks, Inc., 2000).
7. Intel Corp., *Intel® Integrated Performance Primitives for Intel® Architecture (Part 2: Image and Video Processing)*, Chapter 13, *Wavelet Transforms*.
8. Texas Instruments, *TMS320C64x Image/Video Processing Library Programmer's Reference* (SPRU023a.pdf), Chapter 5, section 5.3, *IMG_wave_vert*.
9. Texas Instruments, *TMS320C6000 Chip Support Library API User's Guide* (SPRU401f.pdf), Chapter 6, Section 6.2, *DAT_copy2d*.
10. Mallat, S., Zhong, S., "Characterization of Signals from Multiscale Edges," *IEEE Transactions on Pattern Analysis and Machine Intelligence*, 1992. 14(7): p. 710-732.
11. Mallat, S., Hwang, W.L., "Singularity Detection and Processing with Wavelets," *IEEE Transactions of Information Theory*, Vol. 38, No.2, March 1992.
12. Mallat, S., *A Wavelet Tour of Signal Processing* (Academic Press, 1999).
13. Bijaoui, A., Starck J.L., Murtagh, F., "Restauration des images multiéchelles par l'algorithme à trous," *Traitement du Signal*, 11:229-243, 1994.
14. Gunatilake, P., Siegel, M., Jordan, A., Podnar, G., "Image Understanding Algorithms for Remote Visual Inspection of Aircraft Surfaces," *Machine Vision Applications in Industrial Inspection V*, Vol. 3029, February 1997, pp. 2 - 13.
15. Texas Instruments, *TMS320C6000 Peripherals Reference Guide* (SPRU190d.pdf), Chapter 7, *Host-Port Interface*.
16. Texas Instruments, *TMS320C6000 Chip Support Library API User's Guide* (SPRU401f.pdf), Chapter 12, *HPI Module*.
17. Texas Instruments, *Real-Time Digital Video Transfer via High-Speed RTDX* (SPRA398.pdf).
18. Texas Instruments, *How To Use High-Speed RTDX Effectively*, (SPRA821.pdf).
19. Texas Instruments, *TMS320C6201/6701 DSP Host Port Interface (HPI) Performance*, (SPRA449a.pdf).
20. Dahnoun, N., *Digital Signal Processing Implementation using the TMS320C6000 DSP Platform* (Prentice-Hall, 2000), Chapter 9, Section 9.5, *TMS320C6201 EVM-PC host communication*.
21. Texas Instruments, *TMS320C6201/6701 Evaluation Module Technical Reference* (SPRU305.pdf).
22. Microsoft Developers Network (MSDN), see "WaitForSingleObject" Retrieved October 2004 from:
    http://msdn.microsoft.com/library/default.asp?url=/library/enus/dllproc/base/waitforsingle object.asp
23. Kernighan, B., Ritchie, R., *The C Programming Language: 2ⁿᵈ Edition*, (Prentice Hall, 1988).
24. Donoho, D., "Denoising by Soft Thresholding," *IEEE Transactions on Information Theory*, 41(3):613-627, May 1995.

25. Donoho, D., Johnstone, I., "Ideal Spatial Adaptation by Wavelet Shrinkage," *Biometrika*, 81(3):425-455, August 1994.
26. Donoho, D., Johnstone, I., "Adapting to Unknown Smoothness via Wavelet Shrinkage," *Journal of the American Statistical Association*, 90(432):1200-1224, Dec. 1995.
27. Taswell, C., "The What, How, and Why of Wavelet Shrinkage Denoising," *Computing in Science and Engineering*, pp. 12-19, May/June 2000.
28. Gyarouva, A., Kamath, C., Fodor, I., "Undecimated wavelet transforms for image denoising", *LLNL Technical Report*, UCRL-ID-150931, 2002.
29. Atkinson, I., Kamalabadi, F., Jones, D., Do, M., "Adaptive Wavelet Thresholding for Multichannel Signal Estimation," *Proc. of SPIE conference on Wavelet Applications in Signal and Image Processing X*, San Diego, USA, August 2003.
30. Chang, S,. Yu, B., Vetterli, M., "Adaptive wavelet thresholding for image denoising and compression," *IEEE Transactions on Image Processing*, 9(9):1532–1546, Sep. 2000.
31. Donoho, D., "Wavelet Thresholding and W.V.D.: A 10-minute Tour", *Int. Conf. on Wavelets and Applications*, Toulouse, France, June 1992.
32. The Mathworks, Inc., "upfirdn." Retrieved November 2004 from: http://www.mathworks.com/access/helpdesk/help/toolbox/signal/upfirdn.html
33. Texas Instruments, *Wavelet Transforms in the TMS320C55x* (SPRA800.pdf).
34. Press, W., Teukolsky, S., Vetterling, W., Flannery, B., *Numerical Recipes in C: The Art of Scientific Computing (2nd Ed.)*, Chapter 13, Section 13.10, *Wavelet Transforms* (Cambridge University Press, 1992).
35. Morgan, D., "The Fast Wavelet Transform," *Embedded Systems Programming*, pp. 105-109, Mar. 1998.
36. Crenshaw, J., *Math Toolkit for Real-Time Programming* (CMP Books, 2000).

Appendix A

PUTTING IT TOGETHER: A STREAMING
VIDEO APPLICATION
C6701 EVM, HPI, MATLAB, and MEX-files

Throughout this book, a variety of image processing algorithms and host/target data transfer techniques have been discussed and used to prototype "test-bench" applications. In this appendix, a few of these disparate pieces are put together as a demonstration of how a streaming video application can be implemented with the C6701 EVM and MATLAB. In some respects, video processing can be thought of as adding another dimension – time – to image processing. If we think of a video stream as a time sequence of image *frames*, then certain video processing techniques might be conceptualized as applying some image processing algorithm to each successive frame in a video sequence. There are some real-world applications that make use of this concept, for example Motion JPEG. Motion JPEG is a light-weight video *codec* (compressor-decompressor) used in Video CD (VCD) and security camera systems. In Motion JPEG, a video stream is compressed by JPEG-compressing each individual frame, which in code is simply treated as an image. Playback of the video is then accomplished by decompressing each frame in the bit stream and rendering the image to the display device. The demo application in this appendix does not involve codecs, but rather builds on some of the techniques developed in Chapters 4, 5 and 6. Almost all of the previous image processing programs have used somewhat contrived image dimensions, for example square images of length 256. In this project we use more realistic dimensions (240 rows by 320 columns) to process video captured from a digital camera.

In 4.5.2, the median filter was shown to be highly effective in the reduction of impulsive noise, and a few different implementations were given. In 6.2.3.2, an image processing front-end that used the Host Port Interface (HPI) to transmit and receive pixel data to the EVM was explained.

Finally, in 5.1.1, the use of MEX-files as a bridge between MATLAB and C/C++ was illustrated in the context of a replacement for imread. What is presented here is how to tie all of these disparate pieces together to form a fairly complex application, certainly more complex from a programming perspective than anything else previously encountered in this book. While this application has only been tested and verified on a C6701 EVM, the techniques and code used to implement the processing are readily transferable to any other TI DSP development board with HPI support. With a change to the linker command file and use of the Chip Support Library (see 4.3.4), the code presented in this appendix can serve as a boilerplate for future development.

The front-end UI for this application is shown in Figure A-1, and all code can be found on the accompanying CD-ROM in the AppendixA directory. To use the application, first close any instances of Code Composer Studio and reset the EVM board using the evm6xrst.bat batch file. From within MATLAB, execute median_filter_video, which is located in AppendixA\median_filter_hpi_video\MATLAB. Nothing can be run until the COFF binary is loaded onto the EVM, which is done by pointing to median_filter_hpi.out via the "Browse" button and then clicking on "Init". C6701 debug and release binaries can be found under AppendixA\median_filter_hpi_video\C6701EVM. The "Run" button becomes enabled after initializing the EVM and host/target communication, and after loading in a movie the video streaming is ready to begin. Use the **File|Load (.avi)** menu selection to read in an AVI (Audio-Video Interlaced) movie file, which is subsequently converted to monochrome and contaminated with shot noise. A sample AVI movie of cars driving by is included on the book CD-ROM in the misc directory. This directory also contains instructions on how to convert an MPEG-1 bit stream to the AVI format, along with an M-file to aid in this task.

The interactions between the various elements that comprise the application are shown in Figure A-2. The MATLAB GUI uses built-in MATLAB functions to import a Windows AVI file, and then streams noisy video frames down to the C6701 EVM. Individual frames are transmitted using HPI via a MEX-file that implements a communication protocol similar in nature to that shown in Figure 6-15. The C6701 EVM processes the individual frames using the IMGLIB function IMG_median_3x3 (see 4.5.4.3) and then sends the denoised frame back to MATLAB, again using HPI. The same MEX-file marshals the processed pixel buffer into a MATLAB object and then yields control back to the MATLAB GUI, which was built using the GUIDE GUI designer tool (see 5.1.1).

Figure A-1. MATLAB/HPI/C6701 EVM streaming video application. The left-most display renders the current pristine video frame. The middle display shows the corrupted frame, using the noise density from the edit box directly underneath it. The processed video frame, upon arrival from the EVM, is rendered in the right-hand display.

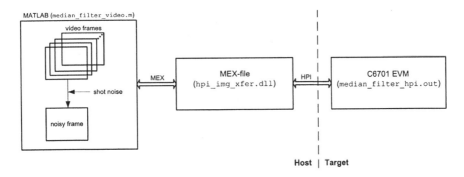

Figure A-2. Data-flow relationship between the various entities that comprise the streaming video application.

We start off the exploration of this software with a brief foray into the creation of MEX-files using Visual Studio .NET 2003. Following this, the implementation of the MEX-file that interfaces MATLAB with HPI and enables the transfer video frames is outlined. Finally, the implementations of the MATLAB front-end and C6701 EVM programs are examined.

A.1 CREATION AND DEBUGGING OF MEX-FILES IN VISUAL STUDIO .NET 2003

MEX-files are a flexible facility for adding new custom features to MATLAB. MEX-files are hosted within a Windows DLL, and they must follow a few rules so that they can be properly hooked into the MATLAB language. While there are some references describing how to create such DLLs[1,2], the process can be tedious and is not very well documented, especially as it pertains to starting from scratch in Visual Studio .NET 2003. This section outlines a procedure for how to create a Visual Studio .NET 2003 Win32 DLL project that can be used to create MEX-files. There will be slight adjustments needed for other versions of Visual Studio or alternate compilers, but for most Windows C/C++ compilers, the general steps are very similar. Then, as an introduction to programming MEX-files, we will briefly discuss `import_grayscale_image`, the MEX-file used in 5.1.1 as a replacement for `imread`.

The steps required to create a skeleton MEX-file project in Visual Studio .NET 2003 are:

1. Create a new project via **File|New|Project**.
2. Expand the "Visual C++ Projects" tree, select "Win32" and choose "Win32 Console". This selection is a bit misleading as we shall *not* be building a console application.
3. In the dialog that follows, select "Application Settings", and choose DLL for the application type. Also check the "Empty project" option. Click on the "Finish" button.
4. You need to add a `.def` file which the build system uses to define the public C functions exported by this DLL. The format of this file is shown below, where you should replace *DLLname* with the name of your DLL, minus the `.dll` extension. The *DLLname* specification is set and can be viewed from the project properties dialog (**Project|Properties**), by expanding the "Linker" tree and then selecting "General". The name of the output DLL is then shown in the "Output File" field. After creating this file, add it to the project.

```
LIBRARY   DLLname
EXPORTS
  mexFunction
```

5. Create a new C or C++ source file that will contain the DLL entry-point function `DllMain` and MEX-file gateway function `mexFunction`. Add this source file into the project.

6. Update the include file settings so that Visual Studio knows where to look for the MATLAB external interfaces header files. For MATLAB R14, this directory is [MATLABROOT]\extern\include, where [MATLABROOT] is typically C:\MATLAB7sp1, C:\MATLABR14, or wherever you installed MATLAB on your local machine. You set the include file path from the project properties dialog by expanding "C\C++", selecting "General", and then adding the directory to the field marked "Additional Include Directories". Note that the "C\C++" field will not appear within the project properties until the project contains at least one C/C++ source file.

7. Now you are ready to start coding. Listing A-1 contains a nearly empty shell of the bare minimum, DllMain and mexFunction, needed for a MEX-file.

8. The project will actually compile as of now, but if your C/C++ source file is like that of Listing A-1, it will fail to link. We can fix this by adding linker dependencies to the libmex.lib, libmx.lib, and libut.lib libraries. All three of these are not required, but they are a decent starting point for any non-trivial MEX-file (additional static libraries may be required, depending on the implementation of the MEX-file). You add these to the project by first bringing up the project properties dialog, expanding the "Linker" tree, selecting "Input", and then adding the aforementioned three libraries in the "Additional Dependencies" field. You also need to tell the Visual Studio linker where these static libraries are located, which is also done within the "Linker" project settings sub-tree, but under the "General" branch. In this portion of the project properties, there is a field "Additional Library Directories" where you will need to add the MATLAB external library directory [MATLABROOT]\extern\lib\win32\microsoft\msvc71.

Listing A-1: Shell for MEX-file implementation. Remove the extern "C" statements if your MEX-file is implemented in C.

```
#include <windows.h>
#include "mex.h" // MATLAB external interfaces header file

extern "C" BOOL APIENTRY DllMain(HANDLE hModule,
                                 DWORD ul_reason_for_call,
                                 LPVOID lpReserve)

{
  return TRUE; // boiler-plate
}
```

```
extern "C" void mexFunction(int nlhs, mxArray *plhs[],
                            int nrhs, const mxArray *prhs[])
{
    mexPrintf("Hello World");
}
```

This new MATLAB function takes the name of whatever is defined in the `.def` file. So if the DLL is `my_mex.dll`, the `.def` file should then specify `my_mex` as the library name, and thus from within MATLAB, issuing a call to `my_mex` results in the string "Hello World" being printed. The DLL has to be somewhere in the MATLAB path so that MATLAB can find it. There are two ways to deal with this, either add the directory to the MATLAB path or from within MATLAB, change directories to where the MEX-file happens to reside.

A quick note on the debugging of MEX-files is useful prior to studying the implementation of some non-trivial MEX-files. One means of initiating a Visual Studio debug session is to make "active" whatever project you are interested in running in the debugger, and then selecting **Debug|Start** from within Visual Studio. If the project happens to be a DLL, then Visual Studio will complain as there is no executable for it to run, and we need to tell it which executable will be dynamically loading in our DLL, at runtime. Thus, to debug MEX-files set the executable as MATLAB itself. To do this, bring up the project properties dialog, and select "Debugging". In this dialog, under the "Action" category, there is a "Command" field where you can point the IDE to the MATLAB executable. For Release 14 of MATLAB, this executable is found at `[MATLABROOT]\bin\win32\MATLAB.exe`. You can now set a break-point in your MEX-file and "run" it in the debugger, even though any break-points you set will initially change to the "?" symbol upon starting `MATLAB.exe` (because MATLAB does not load the DLL into memory until needed, the first time it gets called). Ignore any warnings about the executable not containing debug symbols, and the MATLAB IDE will then start normally. Then when you call your MEX-file from within MATLAB, Visual Studio breaks into the debugger at the break-points set in the MEX-file project, provided of course the build configuration (either "Debug" or "Release") specifies the inclusion of debug symbols.

Another hint concerns what to be on the lookout for in the event of a "??? Invalid MEX-file" error when attempting to invoke a MEX-file function. What this typically means is that there is some dependency on other DLLs that MATLAB is unable to locate. For example, the HPI MEX-file discussed in A.2 relies on three TI EVM DLLs. MATLAB will complain with this error message if it fails to locate any of the dependent DLLs, during the dynamic loading of the HPI MEX-file.

A.1.1 The `import_grayscale_image` MEX-file

In each of the Visual Studio image processing applications discussed in this book, a combination of GDI+ and the Intel IPP library (encapsulated within the `Image8bpp` class) was used to handle the duties of importing image files. The RTDX MATLAB host application from 5.1.1 used the `import_grayscale_image` as a replacement for `imread`, and this MEX-file is a perfect example of a moderately sized, but simple to digest, MEX-file that leverages work developed and debugged by someone else – in this case Microsoft, the makers of the GDI+ library. While `imread` is one of the most useful MATLAB functions, there may be cases in the future where it would be useful to read an image file format not supported by `imread`. One notable case is JPEG-2000, a state-of-the-art image codec that is heavily based on wavelet techniques. Of course, MATLAB code could be written to parse the JPEG-2000 bit stream, but if there were an available C/C++ library, then the MEX-file concept and programming strategy makes it straightforward to integrate such code in MATLAB.

Listing A-2 show the majority of the C++ source file for the `import_grayscale_image` MEX-file. The project directory is located in `AppendixA\import_grayscale_image_mex`. An example of its use is

> I = import_grayscale_image('some_file.bmp');

As its name suggests, this MEX-file returns an 8-bit monochrome image, and the code currently supports only a very limited number of color conversions. The code also does not make use of the `Image8bpp` class, for this class is meant to facilitate interoperability between Intel IPP and GDI+, whereas here we are interested in integrating MATLAB and GDI+. The first order of business is to understand the signature of `mexFunction`, which is where each MEX-file begins execution:

> void mexFunction(int nlhs, mxArray *plhs[],
>   int nrhs, const mxArray *prhs[])

The first two arguments are the number of left-hand side arguments and an array of pointers to MATLAB objects that comprise the return arguments to this MATLAB function. Remember that in the MATLAB language we have the ability to code functions that can return any number of arguments, as opposed to C/C+ where functions must return either no arguments or a single argument.

For example in MATLAB we can design a MEX-file that has the following signature:

[out_arg1, out_arg2] = foo(in_arg);

In this case the implementation for `foo` should check that `nlhs` is two and if that is indeed the case, then `plhs[0]` points to the first (non-initialized) output argument and `plhs[1]` points to the second. Likewise, the same meaning is attached to `nrhs` and `prhs`, except that now we are referring to the right-hand side arguments, or what goes between the parentheses of the MATLAB function call. So in the case of `import_grayscale_image`, the code checks that both `nlhs` and `nrhs` are equal to one, since it accepts exactly one input and returns a single output.

Listing A-2: `import_grayscale_image.cpp`.

```
#include <windows.h>
#include <vector>
#include <gdiplus.h>
#include "mex.h"

using namespace std;
using namespace Gdiplus;

BOOL APIENTRY DllMain(HANDLE hModule,
                      DWORD  ul_reason_for_call,
                      LPVOID lpReserved)
{
   return TRUE;
}

static void print_gdiplus_err_and_exit(Status s)
{
  switch (s) {
    case Ok: mexErrMsgTxt("no error");
    case GenericError: mexErrMsgTxt("generic error");
    // rest omitted ...
  }
}
```

```
void read_image(const char *filename, int buflen, mxArray *plhs[])
{
  // ASCII -> UNICODE conversion
  vector<wchar_t> wName(buflen+1);
  mbstowcs (&wName[0], filename, buflen);
  wName[buflen] = L'\0';

  // use this temporary object purely to deal with
  // the machinations of performing file I/O to parse
  // the binary image data
  Status s = Ok;
  Bitmap bmp(&wName[0]);
  if (Ok != (s = bmp.GetLastStatus()))
    print_gdiplus_err_and_exit(s);

  // allocate MATLAB matrix for gray-scale image
  plhs[0] = mxCreateNumericMatrix(bmp.GetHeight(), bmp.GetWidth(),
                            mxUINT8_CLASS, mxREAL);
  unsigned char *pOutput = (unsigned char *)mxGetPr(plhs[0]);

  BitmapData bmpData;
  Rect rect(0, 0, bmp.GetWidth(), bmp.GetHeight());
  PixelFormat fmt = bmp.GetPixelFormat();
  if (PixelFormat24bppRGB == fmt) { // RGB to grayscale conversion

    s = bmp.LockBits(&rect, ImageLockModeRead,
                    PixelFormat24bppRGB, &bmpData);
    if (Ok != s)
      print_gdiplus_err_and_exit(s);

    // color conversion here (remember MATLAB is
    // column-major which is opposite from C)
    unsigned char *pInput  = (unsigned char*)bmpData.Scan0;
    for (UINT iCol=0; iCol<bmp.GetWidth(); ++iCol) {
      for (UINT iRow=0; iRow<bmp.GetHeight(); ++iRow) {
        unsigned char *pRGB = &pInput[iRow*bmpData.Stride + iCol];
        *pOutput++ = (unsigned char)(pRGB[0]*0.114 +
                              pRGB[1]*0.587 +
                              pRGB[2]*0.299);

      }
    }
```

```
  } else if (PixelFormat8bppIndexed == fmt) { // no color conversion

    // get to raw bits comprising the temporary GDI+ bitmap object
    s = bmp.LockBits(&rect, ImageLockModeRead,
                     PixelFormat8bppIndexed, &bmpData);
    if (Ok != s)
      print_gdiplus_err_and_exit(s);

    // copy from temporary GDI+ object into MATLAB matrix
    unsigned char *pInput = (unsigned char*)bmpData.Scan0;
    for (UINT iCol=0; iCol<bmp.GetWidth(); ++iCol)
      for (UINT iRow=0; iRow<bmp.GetHeight(); ++iRow)
        *pOutput++ = pInput[iRow*bmpData.Stride + iCol];

  }
  else
    mexErrMsgTxt("Only support 8bpp indexed or 24bpp RGB images.");
}

void mexFunction(int nlhs, mxArray *plhs[],
                 int nrhs, const mxArray *prhs[])
{
  int buflen = 0;
  char *filename = NULL;

  /* Check for proper number of arguments */
  if (nrhs != 1)
    mexErrMsgTxt("One input argument required.");
  else if (nlhs != 1)
    mexErrMsgTxt("One output argument required.");

  if (!mxIsChar(prhs[0]))
    mexErrMsgTxt("IMPORT_GRAYSCALE_IMAGE requires "
                 "a string input argument.");

  /* extract the filename string from the Matlab object */
  buflen = (mxGetM(prhs[0]) * mxGetN(prhs[0])) + 1;
  filename = (char *)mxCalloc(buflen, sizeof(char));
  if(0 != mxGetString(prhs[0], filename, buflen))
    mexWarnMsgTxt("Not enough space. String is truncated.");
```

```
/* initialize GDI+ */
  GdiplusStartupInput gdiplusStartupInput;
  ULONG_PTR gdiplusToken;
  if (Ok != GdiplusStartup(&gdiplusToken,
                           &gdiplusStartupInput,
                           NULL))
    mexErrMsgTxt("Failed to initialize GDI+");

  read_image(filename, buflen, plhs);

  /* shutdown GDI+ */
  GdiplusShutdown(gdiplusToken);
}
```

The key is making use of the `mxArray` pointers. These pointers reference a C *opaque type*, which is a term given to a data structure appearing only in a forward declaration (so that the internal layout of the data structure is obscured from external prying eyes) and can *only* be manipulated by functions accepting pointers to these opaque types. The MATLAB external interfaces libraries provide a slew of C-callable functions for creating, initializing, accessing, deleting, and manipulating `mxArray` pointers. For a complete discussion, see [1] and [3].

The logical organization of Listing A-2's `mexFunction` can be gleaned from the comments and a perusal of [3] or searching for the functions beginning with either `mex` or `mx` in the MATLAB online help. The real meat of the MEX-file occurs within `read_image`, which is more or less an adaptation of `Image8bpp`'s constructor (see Listing 3-8). The main difference between `read_image` and `Image8bpp`'s constructor is that in `read_image`, the data contained in the GDI+ Bitmap object is marshaled into an `mxArray` object, whereas with `Image8bpp`, the pixel data goes into an aligned array allocated with an Intel IPP function. `mxArray` is *polymorphic* in the sense that it reflects a MATLAB variable, which can hold a scalar, multi-dimensional matrix, string, structure, cell array, or other MATLAB data types. There are separate MATLAB external interface APIs for dealing with an `mxArray` in each of these contexts. Moreover, while the default format for a MATLAB number is a double-precision complex number, this is overkill for monochrome images and we prefer to deal with MATLAB integer matrices (at least until we need to apply certain numerical functions to the pixel data). Since `import_grayscale_image` returns a UINT8 image matrix, we specify `mxUINT8_class` in the call to `mxCreateNumericMatrix` to allocate the MATLAB matrix pointed to by `plhs[0]`. After the `mxArray` is initialized, a pointer to the underlying storage can be obtained via a call to mxGetPr ("mx get a pointer to the real data"). Since the default numeric format is double precision, mxGetPr returns a pointer to a double array. We then need to cast this pointer to whatever is appropriate. For example, if we want the return argument to be a 16-bit image, we would use mxUINT16_CLASS in the call to mxCreateNumericMatrix and cast the returned pointer from mxGetPr to a pointer to an unsigned short array for access to the data. Finally, MATLAB belies its FORTRAN heritage by storing data in column-major format, as well as using one-based, instead of zero-based, indices. In what turns out to be a recurring theme in MATLAB-to-C/C++ interoperability, the column-major versus row-major discrepancy has to be taken into account. The difference in storage formats is handled in read_image by exchanging the order of the double loops that walk through the GDI+ Bitmap object and copy the pixel data into the mxArray object.

A.1.2 A MEX-file for HPI communication between MATLAB and the C6x EVM

The project for the MEX-file that enables the high-speed transfer of images between MATLAB and the C6701 EVM, a la the communication protocol explained in 6.2.3.2, can be found on the CD-ROM under the directory AppendixA\hpi_mex. The DLL hpi_img_xfer.dll hosts the MEX-file and has dependencies on the three EVM host library DLLs (evm6xmsg.dll, evm6xdm.dll, and evm6x.dll). As a result, all four DLLs need to be placed in a location where MATLAB can find them, either in a directory in the MATLAB path or in the current working directory. There are two forms of usage for hpi_img_xfer. The first usage pattern initializes the board and HPI communication infrastructure, and requires the name of a COFF file to load onto the DSP. This form of the function takes a single string input argument and does not return anything:

    hpi_img_xfer('Release\median_filter_hpi.out');

After the board is initialized and the HPI communication infrastructure is up and running, sending an image down to the DSP and waiting for the processed output is accomplished as so:

    msg2target = uint32(1); % can be any UINT32 value
    processed_img = hpi_img_xfer(input_img, msg2target);

The second input argument to hpi_img_xfer can be used to propagate application-specific parameters to the DSP, for example the number of wavelet decompositions to use in a denoising or edge detection algorithm. The MEX-file is generic in the sense that the image dimensions, while fixed, come from the program running on the EVM. As described in the next section, as part of the handshake procedure the target sends these expected image dimensions back to the host, and `hpi_img_xfer` then caches this information to enforce the correct size of `input_img`. This makes for a flexible data transfer protocol, as all that is needed to change the dimensions of the video frame is to alter the `X_SIZE` and `Y_SIZE` definitions in the EVM source code, and everything else flows from there. `hpi_img_xfer` is *blocking*, in that MATLAB remains suspended while the image processing takes place. Implementing a non-blocking MEX-file is possible and potentially quite useful, although it may be somewhat difficult. There is a mechanism for MEX-files to call back into MATLAB, leading to a potential scheme whereby the MEX-file spins off a child thread which does the work of waiting on the DSP for the video frame processing. When this processing completes, the MEX-file would invoke a MATLAB callback function that would then display the output.

The gateway function for `hpi_img_xfer`, `mexFunction`, is shown in Listing A-3. Since the data transfer protocol is quite similar to that shown in Figure 6-15, `hpi_img_xfer` makes extensive use of a slightly altered version of the `Qureshi::HPI` class presented in 6.2.3.2.2. As in `read_image` from the preceding section, one of the changes entails the replacement of `Image8bpp` with the equivalent functionality provided by MATLAB's `mxArray`. Moreover, the discrepancy between storage of two-dimensional buffers between MATLAB (column-major) and C/C++ (row-major) is again the source of some trouble. This version of the `Qureshi::HPI` class must be cognizant of this difference and take appropriate actions, by performing a column-major to row-major transformation just prior to transmitting the image and then the reverse transformation right after receiving the processed pixel data from the target. See 6.2.3.2.2 for an in-depth discussion of the `Qureshi::HPI` class.

Listing A-3: `mexFunction` from `hpi_img_xfer.cpp`.

```
void mexFunction(int nlhs, mxArray *plhs[],
                 int nrhs, const mxArray *prhs[])
{
  static pair<int,int> frameSize; // from HPI::init()

  try {

    // sanity check function arguments
```

```cpp
if (1==nrhs && mxCHAR_CLASS==mxGetClassID(prhs[0])) {
  // initialization
  if (pHPI) delete pHPI;
  char *pathname = mxArrayToString(prhs[0]);
  if (!pathname)
    throw runtime_error("Failed to extract pathname string.");
  pHPI = new Qureshi::HPI(pathname);
  frameSize = pHPI->init();
  return;
}
else if (2==nrhs) { // arg 1 = image, arg 2 = app-specific data
  mxClassID dataTypeArg1 = mxGetClassID(prhs[0]),
            dataTypeArg2 = mxGetClassID(prhs[1]);
  if (mxUINT8_CLASS != dataTypeArg1 ||
      mxUINT32_CLASS != dataTypeArg2)
    throw runtime_error("Expect image to be UINT8 and "
                        "2nd arg to be UINT32 ");

  if (nlhs != 1)
    throw runtime_error("One output argument required.");

  // image processing occurs below ...
}
else
  throw runtime_error("Invalid combination of input arguments");

// verify size of video frame
int nr = mxGetM(prhs[0]), nc = mxGetN(prhs[0]);
if (frameSize.first!=nr || frameSize.second!=nc) {
  ostringstream ostr;
  ostr << "Expecting " << frameSize.first << "x" << frameSize.second
       << " frame, got " << nr << "x" << nc;
  throw runtime_error(ostr.str().c_str());
}

// if we got this far, send msg to DSP
unsigned int msg2DSP = *mxGetPr(prhs[1]);
pHPI->sendImage(prhs[0], msg2DSP);

// allocate output (denoised) frame
plhs[0] =
        mxCreateNumericMatrix(nr, nc,
```

mxUINT8_CLASS, mxREAL);

```
// and now we wait while the EVM does its thing
pHPI->readImage(plhs[0]);

}
catch (exception &e) {
  mexErrMsgTxt(e.what());
}
}
```

A.2 THE C6701 EVM PROGRAM

If you have read previous sections of this book, in particular 4.5.2 and 6.2.3.2, then the code shown here should be fairly self-explanatory. Essentially this program, whose source code may be found in the AppendixA\median_filter_hpi_video directory, sits in an infinite loop waiting for an image. Upon receiving an image, it then uses an IMGLIB function to process the frame with a 3x3 median filter (see 4.5.2). The code presented in 6.2.3.2.1 was an example of how a C6701 EVM program used HPI to perform the entire wait/receive/process/transmit sequence a single time. The code in Listing A-4 performs this procedure forever, and is basically Listing 6-9 in an infinite loop with median filtering replacing wavelet edge detection.

Listing A-4: C6701 EVM program that implements the back-end processing for the streaming video application (median_filter_hpi.c).

```
#include <common.h>
#include <board.h>  /* EVM library */
#include <pci.h>
#include <img_median_3x3.h>

// each video frame is 240 rows by 320 cols
#define X_SIZE 240
#define Y_SIZE 320
#define N_PIXELS (X_SIZE*Y_SIZE)

#pragma DATA_ALIGN (in_img, 4);
#pragma DATA_SECTION (in_img, "SBSRAM");
unsigned char in_img[N_PIXELS];
```

```c
#pragma DATA_ALIGN (out_img, 4);
#pragma DATA_SECTION (out_img, "SBSRAM");
unsigned char out_img[N_PIXELS]; /* filtered image */

/*
 * Faster than memset(), count must be a multiple of
 * 8 and greater than or equal to 32
 */
void memclear( void * ptr, int count )
{
  long * lptr = ptr;
  _nassert((int)lptr%8==0);
  #pragma MUST_ITERATE (32);
  for (count>>=3; count>0; count--)
    *lptr++ = 0;
}

/* 3x3 median filter */
void medfilt()
{
  int irow = 0;
  unsigned char *pin = in_img+Y_SIZE,
              *pout = out_img+Y_SIZE;
  for (; irow<X_SIZE-2; ++irow, pin+=Y_SIZE, pout+=Y_SIZE)
    IMG_median_3x3(pin, Y_SIZE, pout);
}

/* host | target handshake, transmit image size and mem addrs */
void handshake()
{
  unsigned int image_size = (X_SIZE<<16) | Y_SIZE;
  int sts;
  /*
   * target will initialize HPI, therefore we keep attempting to send the
   * image dimensions until the HPI message is successfully sent.
   */
  do {
    sts = amcc_mailbox_write(2, image_size);
  } while (ERROR == sts);

  /* comm is up now, now send input buf mem addr
   * and processed image buf mem addr */
```

```
    pci_message_sync_send((unsigned int)in_img, FALSE);
    pci_message_sync_send((unsigned int)out_img, FALSE);
}

/* wait for anything from the host */
void wait_4_frame()
{
    unsigned int msg;
    pci_message_sync_retrieve(&msg);
}

/* tell host that N_PIXELS worth of data ready */
void processing_complete()
{
    int sts;
    unsigned int bytes = N_PIXELS;
    do {
        sts = amcc_mailbox_write(2, bytes);
    } while (ERROR == sts);
}

int main(void)
{
    evm_init(); /* initialize the board */
    pci_driver_init(); /* call before using any PCI code */
    handshake(); /* comm init */

    memclear(out_img, Y_SIZE);
    memclear(out_img+N_PIXELS-Y_SIZE, Y_SIZE);

    for (;;) {
        /* wait for host to signal that they have sent a frame. */
        wait_4_frame();

        /* process the video frame */
        medfilt();

        /* signal to host that denoised frame ready */
        processing_complete();
    }
}
```

A.3 MATLAB GUI

The final piece to the puzzle is the MATLAB front end, which can be found in AppendixA\median_filter_hpi_video. As described in 5.1.1, creation of a MATLAB GUI using the GUIDE designer results in two MATLAB files being created, a figure file with the .fig extension and an M-file with the actual code for each GUI dialog in the application. In this case, the M-file for the GUI is median_filter_video.m, and the interesting callbacks are the ones that load in an AVI file from the **File|Load (.avi)** menu selection, and the main processing loop (the function called when the "Run" button is clicked). Both of these are shown in Listing A-5. The callback for the "Init" button is not shown – all it does is open a file browser dialog and then call hpi_img_xfer with the selected DSP program, as described in A.1.2.

Listing A-5: MATLAB GUI code for the streaming video example (median_filter_video.m). The first function is what gets invoked when **File|Load (.avi)** is selected, and the second is the main processing loop, which is linked to the "Run" button.

```
function MenuFileLoad_Callback(hObject, eventdata, handles)
% hObject    handle to MenuFileLoad (see GCBO)
% eventdata  reserved - to be defined in a future version of MATLAB
% handles    structure with handles and user data (see GUIDATA)
% load in AVI File

[filename, pathname] = uigetfile('*.avi', 'Pick AVI file');
if ~isequal(filename, 0)
    try
        handles.mov = aviread(fullfile(pathname,filename));
        guidata(hObject, handles); % Update handles structure
        % display 1st frame
        subplot(handles.input_frame);
        image(rgb2gray(handles.mov(1).cdata));
        axis off;
        set(handles.original_text, 'String', ...
            sprintf('Original Frame: 1 of %d', length(handles.mov)));
    catch
        errordlg(lasterr, 'AVI file I/O error');
    end
end
```

```
% --- Executes on button press in run_pushbutton.
function run_pushbutton_Callback(hObject, eventdata, handles)
% hObject    handle to run_pushbutton (see GCBO)
% eventdata  reserved - to be defined in a future version of MATLAB
% handles    structure with handles and user data (see GUIDATA)

if isempty(handles.mov)
    errordlg('Please load in a movie from the File menu');
    return;
end

msg2target = uint32(256*256); % could be anything

how_much_noise = str2num(get(handles.noise_density_edit, 'String'));

% loop through every frame, and send down to the DSP
nframes = length(handles.mov);
try
    for ii=1:nframes
        % show the original frame
        f = rgb2gray(handles.mov(ii).cdata);
        subplot(handles.input_frame);
        image(f); axis off;
        set(handles.original_text, 'String', ...
            sprintf('Original Frame: %d of %d', ii, length(handles.mov)));
        % show the noisy frame
        fprime = imnoise(f, 'salt & pepper', how_much_noise);
        subplot(handles.noisy_frame);
        image(fprime); axis off;
        % have the DSP clean up the noisy video frame
        g = hpi_img_xfer(fprime, msg2target);
        % display denoised video frame
        subplot(handles.median_filtered_frame);
        image(g); axis off;
        drawnow
    end
catch
    errordlg(lasterr, 'video error');
end
```

There is not much more that can be said about the code at this point, since most of the custom pieces have already been discussed and the rest is basic MATLAB code where more detailed information can be obtained from the online help. Loading in an AVI file is achieved using `aviread`[4], which returns the individual frames in a movie data structure referenced in this code by `handles.mov`. The MATLAB movie data structure is an array of structures, in which each element in this array corresponds to a single frame of video. The `cdata` field, or `handles.mov(i).cdata`, is the RGB image data for the ith video frame.

Hence in `run_pushbutton_Callback`, we loop through each element in this array of frame data and extract a monochrome image using `rgb2gray`. Following this down-conversion, the current frame is corrupted with shot noise using `imnoise` (with the noise density coming from a user-interface edit box). The use of `imnoise` implies the presence of the Image Processing Toolbox, although in 4.5.3 a replacement M-file `add_shot_noise` is discussed and can very easily be used here in lieu of `imnoise`. The job of cleaning up the contaminated image frame falls on the DSP, who receives the frame from the MEX-file after the GUI invokes `hpi_img_xfer`. The GUI completely blocks until the EVM finishes processing the current frame, and the return value of `hpi_img_xfer`, the processed image matrix, is rendered to the screen.

A.4 IDEAS FOR FURTHER IMPROVEMENT

On a 2.8 GHz Pentium with 1 GB RAM and 533 MHz front side bus communicating with a C6701 EVM back-end, the total performance of this video demo is about four frames per second. While a portion of the performance lag is attributable to problems with the host HPI library (see the discussion in 6.2.3.2.2), there are certain optimizations that would make a large dent in the processing time. For starters, the target implementation should take advantage of DMA, which would drastically cut down on the amount of time taken to process each individual image frame. The effectiveness of this technique was demonstrated in Chapter 4, and in 4.5.4 a C6416 DSK median filter implementation that utilizes this very optimization is given.

Even though we are somewhat limited in the performance on the host side, due to the issues identified in the host HPI library, one modification that could help parallelize the rendering of video frames along with the processing of such video frames has already been mentioned. If a non-blocking form of `hpi_img_xfer` were used, then MATLAB could be off displaying image frames while pumping the EVM with data to process whenever it happened to be idle.

Finally, there are a few inefficiencies with regards to the whole column-major/row-major imbroglio. For the sake of clarity and simplicity, the image dimensions on both the host and target side are kept the same; however what one could do to avoid at least one column-major to row-major transformation is to deal with transposed data on the target side. That way, we can still transmit the contents of the MATLAB input data using a single HPI call, and we would need to swap X_SIZE and Y_SIZE in the target. This optimization would then entail a matrix transposition upon receipt of the image back from the EVM, so as to maintain the original frame dimensions.

REFERENCES

1. *MATLAB External Interfaces (Version 7)*.
2. The MathWorks, "Custom Building on Windows," Retrieved November 2004 from http://www.mathworks.com/access/helpdesk/help/techdoc/matlab_external/ch03cr18.html
3. *MATLAB External Interfaces Reference (Version 7)*.
4. *MATLAB Function Reference Volume 1: A-E (Version 7)*, aviread Function (refer to online documentation).

Appendix B

CODE OPTIMIZATION
Compiler Intrinsics and Optimization Techniques

Full blown code optimization of complicated algorithms is some of the most difficult work to do on a DSP; in fact it is some of the most difficult programming to do on *any* architecture. As soon as you find yourself in the situation where you are "cycle-counting", you may possibly be at the point where you have to resort to some fairly hard-core techniques. This may involve hand-tweaking assembly language code, a task definitely not for the faint of heart, especially for anything non-trivial on a VLIW architecture like the C6x. With VLIW architectures, the onus is on the compiler to generate optimal code, as the hardware provides no help in extracting any parallelism at the individual instruction level, in contrast to superscalar CISC CPUs like the Pentium. Thankfully, most developers never reach this point, especially if they have done their due diligence in selecting a DSP with sufficient horsepower and have performed the requisite algorithmic and high-level optimizations. The primary focus of this book has been implementing image processing algorithms on the C67x and C64x DSPs – along the way we have incorporated some high-level optimizations, mostly algorithmic modifications and the use of common-sense C coding practices. Due to advances in compiler technology, TI's compilers have progressed to the point where well-written C code using intrinsics and compiler directives can approach the performance of hand-coded assembly (I do not quote numbers here because that exercise is meaningless – there are too many factors to weigh in such as how well the C code is written, what compiler optimizations are enabled, the type of algorithm, and so on). Even with these advances, it still remains the case that writing efficient code on an embedded DSP requires deep knowledge of the algorithm, hardware architecture, and compiler. Knowing what we do about the performance of the TI compiler, a roadmap for developing efficient and optimized code on the C6x architecture, and for that matter any DSP architecture, might look like the following:

0. **A thorough understanding of the processing and overall algorithm**: It goes without saying that if the developer does not have an intimate understanding of the algorithm, any code optimization is doomed to be a fruitless endeavor. In this book we have extensively used MATLAB prototypes to this end, which are by their nature a more succinct representation of the types of data processing one frequently encounters in signal and image processing applications.

1. **First-cut, "out of the box" C/C++ code**: We typically took this route by developing and debugging test code in Visual Studio .NET 2003 – if one performs step 0, then this step generally entails MATLAB-to-C/C++ porting and produces ANSI C/C++. Since the goal is embedded deployment on the DSP, where for the most part one will be coding in the C language, a good strategy is to eschew "pushing-the-envelope" C++ and language features like templates, along with heavy-duty object oriented design patterns.

2. **Optimized C code**: This step may occur using a desktop IDE like Visual Studio or perhaps within CCStudio – what is done here is that first-order and relatively simple infrastructure and algorithm optimizations are implemented. This is typically the low-hanging fruit in the code optimization game, and some examples from the implementations in this book include reducing the memory footprint in the wavelet decomposition algorithm (see 6.2.3.3), moving frequently used data buffers and structures onto fast on-chip RAM (see Chapter 4), and if applicable, using DSPLIB, IMGLIB, or other canned routines.

3. **Further Optimized C code with intrinsics**: By this point we are entering the "low-level optimization regime" - we begin to apply some loop transformations (i.e., loop unrolling) and incorporate compiler intrinsics wherever appropriate.

4. **Linear Assembly**: And now we put our hard-hat on and really delve into an area where most software developers who do not deal with embedded systems rarely venture into. So-called linear assembly is TI's nomenclature for a simpler form of C6x assembly language where the developer can use C variable names and need not worry about specifying register allocation or which instructions are to be executed in parallel – such details are left to the assembly optimizer. There are some instances where the assembly optimizer may perform a better job of fully exploiting the C6x architecture compared to the C compiler, and there are some cases where the equivalent C code, heavily laden with compiler intrinsics, may in fact be more difficult to read than a version

implemented in linear assembly. Thus for code maintenance purposes, it may be advantageous to code such critical gating loops in linear assembly. Admittedly this scenario is rare.

5. **Hand-optimized Assembly**: The most difficult and arduous coding task, it is *extremely* challenging and very time consuming on an architecture as complex as the C6000. While steps 2-4 certainly require a fair amount of familiarity with the DSP architecture, hand-coded assembly requires an intimate familiarity with just about every aspect of the DSP architecture in order to improve upon what the compiler will come up with. Assembly language coding on a VLIW architecture is notoriously difficult, and TI recognizes this fact and stresses the need to try fully optimized C code first. For further information on this topic, the reader is referred to Chapter 8 of [3] and [4].

One should typically plan on always proceeding to at least step 2, and then care should be taken to not spend too much time up-front brooding over steps 3-5, unless it has been determined to be absolutely required (or you have a lot of time on your hands). With proper use of compiler directives (i.e. DATA_ALIGN and MUST_ITERATE) and "no-brainer" use of certain intrinsics like _nassert, the compiler will more often than not generate quite efficient code, provided that the developer understands the optimizing compiler options described in exhaustive detail in [5]. It is essential to profile the code at this point to garner whether or not the processing meets the stated time requirements (hopefully such requirements are available). After all, if the code is fast enough as is, there is not much to be gained from proceeding onwards, save for your own personal edification. Experimentation with the compiler and a full understanding of its capabilities must be stressed – it is far easier to play around or fiddle with the compiler optimizations than it is to code in assembly. The TI compiler has a "compiler directed feedback" feature that emits useful information on the utilization of the DSP's functional units, software pipelining, and other information that may point one towards which compiler optimizations may make sense in the current context (see Chapter 4 of [2] or [6] for additional information).

In this appendix we take a closer look at step 3. The information and code examples presented here are by no means an exhaustive look at the vast topic of DSP code optimization, but rather they serve to illustrate a few concrete examples and link them to some of the code discussed in this book. The references at the end of this appendix are a good place to start for a complete discussion of C6x code optimization.

B.1 INTRINSICS AND PACKED DATA PROCESSING

Intrinsics have been sprinkled throughout various code presented in this book. In 4.4.1, it was shown how the use of the _nassert intrinsic and the MUST_ITERATE pragma on the C6416 can be used to optimize the performance of the memclear memory initialization function. By providing hints to the compiler, it is then able to transform simple C code to use double word accesses. The _nassert intrinsic does not generate any code, and in this section we focus our attention on intrinsic functions that do generate code and show how they can be used to take advantage of a vital optimization technique in image processing applications – the "packed data processing" optimization. Along the way, we go back and revisit some of the image processing loops previously presented in the context of an encompassing algorithm, and explain the mechanics behind the use of intrinsics and how they apply to these particular loops.

Intrinsics may look like function calls, but in actuality they are something a bit different. Intrinsics provide the developer with access to the hardware while allowing one to remain in the friendly confines of the C programming language environment. Intrinsics are C extensions that translate into assembly instructions for processor-specific features that standard ANSI C cannot support, and they are inlined so there is no function call overhead. For example, the _abs intrinsic was used in Listings 6-7 (6.2.3.1) and 6-12 (6.2.3.3) to compute the absolute value of wavelet coefficients. The 6x DSP has an instruction that computes the absolute value of a register in a single clock cycle, which is going to be faster than a multi-cycle C ternary statement that includes a branching operation. Saturated arithmetic, whereby the result of an arithmetic operation of two pixels is clamped to the maximum or minimum value of a predefined range, is another basic operation one frequently encounters in image processing applications. For example, rather than code a multi-cycle series of C statements that implement the saturated add, one should instead use the _sadd intrinsic (see Example 2-6 in [2]). The use of appropriate intrinsics within loop kernels greatly increases the performance of such loops, and as we have seen throughout this book, image processing algorithms are dominated by loops operating over entire images or subsets of images.

Some C6x instructions are quite specialized, and many, though not all, C6x assembly instructions have intrinsic equivalents – a full list is enumerated in [2]. For example, knowing that division is enormously expensive, we found that using the reciprocal approximation in conjunction with Newton-Raphson iteration was a reasonable replacement for division on a fixed-point processor. However, it was a rather involved process (see Listing 4-19 in 4.6.5), entailing the construction of lookup tables for seeding the iterative procedure. The same situation holds on a floating-point processor, yet the C67x processor provides the RCPDP (double-precision) and RCPSP (single-precision) approximate reciprocal instructions that can be used to either seed a Newton-Raphson iterative procedure or perhaps as the reciprocal itself, as accuracy requirements warrant. The two aforementioned instructions can be accessed from C via _rcpdp and _rcpsp, but the focus here is not primarily on the use of intrinsics for specialized operations, but rather on using intrinsics within image processing loops for vectorizing code to use word-wide or double word-wide optimizations to operate on packed data.

B.1.1 Packed Data Processing

The DMA paging optimization described, implemented, and profiled in Chapter 4 is a recurring theme that pops up in all data-intensive operations running on the DSP. Equally salient to image processing and media processing in general is this idea of packed data processing, where the same instruction applies the identical operation on all elements in the data stream. In the computer architecture field, this general concept is known as SIMD (Single Instruction, Multiple Data), and has proven powerful because more often than not these arithmetic operations are independent of one another. This characteristic leads to code generation techniques that exploit this inherent parallelism – with the right optimization, code size is reduced and the throughput of the processor maximized, while placing a clamp on power consumption. As one might imagine, Texas Instruments is not the only chip developer to incorporate such functionality into their processor cores – Intel brought to market similar SIMD IA-32 instruction set extensions with MMX, SSE, SSE2, and SSE3, as did AMD with 3DNow!. In a nutshell, packed data processing boils down to storing multiple elements (mainly, in our case, pixels) in a single register and then using specialized processor instructions to operate on this data.

For example, consider the sum of products between two vectors, which is of vital importance in stream processing:

$$y[i] = \sum_{i=0}^{N-1} (a[i])(x[i])$$

This operation appears in various forms, most prominently in this book in the context of the weighted average, or FIR filter. Suppose a and x are 16-bit integer quantities, perhaps representing Q15 numbers. Then a high-level description of what the processor is doing within the loop kernel that performs this vector product sum would look like:

for each *i*
  1. load $a[i]$ from memory address $\&a+i$, and place in register 1
  2. load $x[i]$ from memory address $\&x+i$, and place in register 2
  3. multiply contents of register 1 and register 2, placing result in register 3
  4. add contents of register 3 to running sum stored in register 4
end

C6x registers are 32 bits wide, and by reading the data in 16-bit (half-word) chunks at a time, we are wasting half of the storage capacity of registers 1 and 2. By packing the data to store multiple elements of the stream within registers 1 and 2 we can reduce the load pressure on this loop:

for $i = 0$ ... N-1 in steps of 2
  1. load $a[i]$ and $a[i+1]$, starting from memory address $\&a+i$, placing $a[i]$ in the lower-half of register 1 and $a[i+1]$ in the upper-half of register 1
  2. load $x[i]$ and $x[i+1]$, starting from memory address $\&x+i$, placing $x[i]$ in the lower-half of register 2 and $x[i+1]$ in the upper-half of register 2
  3. multiply lower-half of register 1 by lower-half of register 2 and place 32-bit result in register 3
  4. multiply upper-half of register 1 by upper-half of register 2 and place 32-bit result in register 4
  5. add contents of register 3 to running sum stored in register 5
  6. add contents of register 4 to running sum stored in register 5
end

The above loop is actually an embodiment of two optimizations that go hand in hand: packed data and loop unrolling. We reduce the number of instructions and alleviate the load pressure in the loop kernel by using word-wide data access, i.e. replacing what would be LDH (Load Half-Word) instructions in the first loop with LDW (Load Word) instructions in the second loop, and then packing two 16-bit quantities into 32-bit registers. The same optimization holds (and is even more advantageous) if data elements in the stream are 8-bit quantities. Then, using the above example, each load would place four data elements in each register and operate on them accordingly. Such a strategy replaces four LDB (Load Byte) instructions with a single LDW instruction. Of course, specifying these sorts of loads and stores is not feasible in strict and portable ANSI C, and it is a risky proposition to completely rely on the compiler's optimization engine to generate code that takes full advantage of SIMD instructions operating on packed data. This is where compiler intrinsics come into play.

All C6x DSPs have instructions and corresponding intrinsics allowing for operations on 16-bit "fields" stored in the high and low parts of a 32-bit register, as illustrated in the second loop above. However, beyond this point one begins to see some divergence between the C62x, C67x, and C64x devices. The C62x line of DSPs is only capable of word-wide optimizations, however the newer C64x and C67x offer double word-wide access via the LDDW (Load Double Word) and STDW (Store Double Word) instructions and corresponding intrinsics. In this case, 64 bits worth of data are read into the register file, with elements packed into a pair of registers. Moreover, the C64x builds upon both the C62x and C67x packed data support by allowing for non-aligned accesses of memory via various instructions like LDNW/STNW (load/store non-aligned word) and LDNDW/STNDW (load/store non-aligned double word), which correspond to the _mem4 and _memd8 intrinsics, respectively. These non-aligned instructions are quite useful, especially in certain image and video processing scenarios when algorithms march along in 8-bit pixel (byte) boundaries; without these instructions one is locked out of the packed data optimization due to restrictions imposed by 32-bit or 64-bit alignment boundaries – for further information consult 6.2.8 of [2].

The aforementioned alignment restrictions shed some light on the need for _nassert and DATA_ALIGN, both of which are found in almost every program presented in this book. Except for the special non-aligned C64x instructions mentioned above, utilization of word-wide access to packed data on the C6x requires the stream (represented in the C language by an array) to be aligned on a 4-byte boundary. We assert this using DATA_ALIGN(buf,4) and _nassert((buf)%4==0), where buf is a C array containing the stream elements. Likewise, DATA_ALIGN(buf,8) and _nassert((buf)%8==0) are used as a hint to the compiler that double word-wide access is an option. Perhaps the simplest example of relying on the compiler optimizer to take advantage of packed data is the memclear function, which zeroes out an array. This function was explained in 4.4.1, and is repeated below:

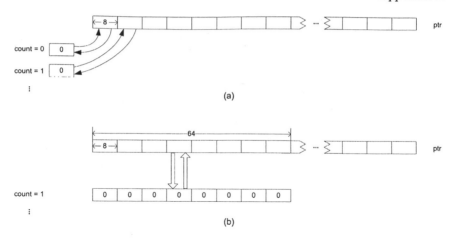

Figure B-1. Optimization of memory initialization function using packed data instructions. (a) Conservative implementation of memclear, where successive LDB/STB instructions are used for accessing and storing array elements. (b) One iteration of a "packed" loop showing how LDDW is used to load 64 bits worth of data into a register pair. The assembly code would use two MVK (Move Constant) instructions to zero out both registers each time through the loop, and then STDW to send the packed data back into the storage array pointed to by ptr.

```
void memclear( void * ptr, int count )
{
  long *lptr = ptr;
  _nassert((int)ptr%8==0);
  #pragma MUST_ITERATE (32);
  for (count>>=3; count>0; count--)
    *lptr++ = 0;
}
```

Figure B-1 depicts graphically how the packed data optimization helps in this context. In the next two sections we discuss the use of intrinsics to optimize more complicated and numerically-oriented operations.

B.1.2 Optimization of the Center of Mass Calculation on the C64x Using Intrinsics

In 5.2.5.1, a C6416 implementation of the isodata algorithm for automatic threshold computation was discussed. The isodata algorithm calls for computing the center-of-mass of two portions of the smoothed image histogram, where the image histogram is split by the current threshold. The isodata algorithm also entails repeatedly performing these center-of-mass computations, as the iterative procedure continues until a "good" threshold

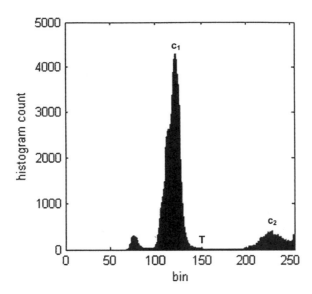

Figure B-2. The center-of-mass calculations in the isodata algorithm. The threshold $T=150$ bifurcates the histogram, and c_1 is the center-of-mass of the left portion of histogram, and c_2 is the center-of-mass to the right of T.

value is found. The center-of-mass calculation is illustrated in Figure B-2.

The optimized implementation of this numerical procedure is interesting to consider for two reasons. One is its broad applicability – the procedure in two dimensions returns the *centroid* of an area of an image, an important operation in machine-vision and pattern-recognition systems, and the techniques described here can be readily applied in the multi-dimensional case. Secondly, the nature of the algorithm is such that it illustrates the utility of the non-aligned double-word instructions present in the C64x line of DSPs.

A description of the process for computing the center-of-mass of a region of the histogram is simple enough. Referring to Figure B-2, computing the center-of-mass to the left of T requires multiplying each histogram count by the bin number, summing those results together, and then dividing by the sum of the counts from 0 to T. In mathematical terminology, that translates to

$$center\text{-}of\text{-}mass_{0 \rightarrow T} = \sum_{i=0}^{T} (i)(hist[i]) \bigg/ \sum_{i=0}^{T} hist[i]$$

Computation of the center-of-mass to the right of T is the same, except that the limits of summation change accordingly. We focus our attention on the numerator of the above expression, and note that we could state it in terms of a dot-product; in other words, the sum of products of two vectors, the actual histogram frequencies (y-axis) and the bins for that portion of the histogram (x-axis) we are summing over. So, instead of loops that look like the following:

```
/* "left" center-of-mass numerator */
for (ii=0; ii<T; ii++)
  sumofprod1 += ii*hist[ii];

/* "right" center-of-mass numerator */
for (ii=T+1; ii<MP; ii++) /* MP=255 for 8-bit images */
  sumofprod2 += ii*hist[ii];
```

we rewrite them, replacing ii in the loop kernels with an array indexing operation. In the function dotprod from Listing 5-10, this array is called pixval, and an updated version of the loops are then:

```
for (ii=0; ii<T; ii++) /* left */
  sumofprod1 += pixval[ii]*hist[ii];

for (ii=T+1; ii<MP; ii++) /* right */
  sumofprod2 += pixval[ii]*hist[ii];
```

We are now able to vectorize the operation because in this case we are willing to sacrifice memory usage (the pixval array for a 16-bit image would be quite large). As explained in 3.2.2, the IMGLIB function IMG_histogram returns the histogram of its input as an array of 16-bit short integers. If this histogram array, hist, and pixval are aligned on a double-word boundary (the default for global arrays when using the C64x compiler and simple to explicitly specify via the DATA_ALIGN pragma), we might consider replacing half-word accesses with double-word accesses, reading and writing four elements at a time and packing two elements per register. If we can make the assumption that the starting value of ii is divisible by four and $T-ii$ is also divisible by four (64 bits divided by 16-bit elements) – which is equivalent to saying the number of iterations through the loop is divisible by four – then the following code would suffice on both the C67x and c64x DSPs:

```
#pragma MUST_ITERATE(,,4)
_nassert((int)pixval%8 == 0)
_nassert((int)hist%8 == 0)
for (ii=0; ii<T; ii++) /* left */
  sumofprod1 += pixval[ii]*hist[ii];


#pragma MUST_ITERATE(,,4)
_nassert((int)pixval%8 == 0)
_nassert((int)hist%8 == 0)
for (ii=T+1; ii<MP; ii++) /* right */
  sumofprod2 += pixval[ii]*hist[ii];
```

The MUST_ITERATE pragma tells the compiler that the number of iterations is divisible by four, and this in conjunction with _nassert should be a sufficient trigger for the compiler to apply packed data optimizations using LDDW and STDW. Unfortunately, we can make no such assumption about the number of times through either loop and by extension can make no such claim about the starting value of the loop counter variable ii in the second loop. This fact precludes double word-wide accesses on the C67x, however with the C64x we can take advantage of the non-aligned double-word instructions LDNDW and STNDW to improve the loop performance. Listing B-1 shows the contents of the dotprod function, taken verbatim from Listing 5-10.

Listing B-1: Function used in the isodata implementation from 5.2.5.1.
```
unsigned long dotproduct(int lo, int hi)
{
  /* 0, 1, 2, ..., 255 */
  static const unsigned short pixval[]={0,1,2, /* 3,5,...,252 */ ,253,254,255};
  unsigned long sum1 = 0, sum2 = 0, sum3 = 0, sum4 = 0, sum;
  const int N = hi-lo;
  int ii=0, jj=lo, remaining;
  double h1_h2_h3_h4, b1_b2_b3_b4;
  unsigned int h1_h2, h3_h4, b1_b2, b3_b4;

  /* unrolled dot-product loop with non-aligned double word reads */
  for (; ii<N; ii+=4, jj+=4)
  {
    h1_h2_h3_h4 = _memd8_const(&smoothed_hist[ii]);
    h1_h2 = _lo(h1_h2_h3_h4);
    h3_h4 = _hi(h1_h2_h3_h4);

    b1_b2_b3_b4 = _memd8_const(&pixval[ii]);
    b1_b2 = _lo(b1_b2_b3_b4);
    b3_b4 = _hi(b1_b2_b3_b4);
```

```
    sum1 += _mpyu(h1_h2, b1_b2); /* (h1)(b1) */
    sum2 += _mpyhu(h1_h2, b1_b2); /* (h2)(b2) */
    sum3 += _mpyu(h3_h4, b3_b4); /* (h3)(b3) */
    sum4 += _mpyhu(h3_h4, b3_b4); /* (h4)(b4) */
  }
  sum = sum1 + sum2 + sum3 + sum4;
  /*
   * loop epilogue: if # iterations guaranteed to
   * be a multiple of 4, then this would not be required.
   */
  remaining = N - ii;
  jj = N - remaining;
  for (ii=jj; ii<N; ii++)
    sum += smoothed_hist[ii]*pixval[ii];

  return sum;
}
```

The use of double in the declaration for h1_h2_h3_h4 and b1_b2_b3_b4 may seem odd at first but the double type in this context is just used as a placeholder to signify 64 bits of storage. The loop has been unrolled by a factor of four, and we use the _memd8_const intrinsic, which generates code that uses LDNDW, to read 64 bits or two words worth of data into h1_h2_h3_h4 and b1_b2_b3_b4. Next, both of these 64-bit elements, which in reality are each stored in a register pair, are "split" into 32-bit upper and lower halves via the _lo and _hi intrinsics. At this point, we are left with a situation like that shown in Figure B-3, with four 32-bit elements.

Although we could split the contents of the four registers again, we need not resort to that course of action. There are assembly instructions that multiply the 16 LSBs (least significant bits) of one register by the 16 LSBs of another register. Similarly, there are assembly instructions for multiplying the 16 MSBs (most significant bits) of one register by the 16 MSBs of another register. There are actually four different variants of each of these instructions, pertaining to whether the data contained within the portion of the register are to be treated as signed or not, and whether the result should be truncated down to 16 bits or returned as a 32-bit entity. As the comments in Listing B-1 indicate, the intrinsics _mpyu (16 unsigned LSBs multiplied together and returned as 32-bit quantity) and _mpyhu (16 unsigned MSBs multiplied together and returned as 32-bit quantity) are used to perform the four packed-data multiplications. These quantities are then added to the accumulators, four of which are used to avoid loop carried dependencies that inhibit parallelism.

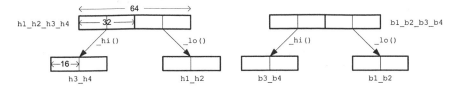

Figure B-3. Use of the _lo and _hi intrinsics to split 64-bit elements into 32-bit elements. 64-bit data elements are read into memory using the _memd8_const intrinsic.

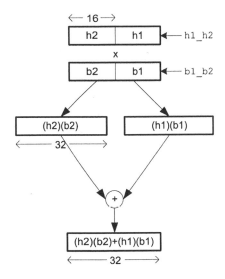

Figure B-4. Usage of the _dotp2 intrinsic in Listing B-2. The same operation is performed on h3_h4 and b3_b4.

A loop "epilogue" of sorts is required in order to clean up the computation. Because we make no claims about the loop count being divisible by four, we need to wrap up the dot product computation using a final "traditional" C loop which iterates at most three times. Alternatively, we could zero pad the arrays such that we are guaranteed to iterate a number of times that is divisible by four. We have actually succeeded in fully vectorizing the loop kernel, given the memory bandwidth of the architecture, but in fact have yet to reach the denouement of this story. A hallmark of DSP architectures is the fabled multiply-and-accumulate instruction, and it is not surprising that the C64x provides just the intrinsic we need here to cut down on the number of operations within this loop kernel. TI provides another set of intrinsics that map to so-called "macro" instructions, and one of these is DOTP2. This instruction, accessed in C with _dotp2, performs a

MAC-like operation on packed 16-bit data. That is, it returns the dot product between two pairs of packed 16-bit values, as shown in Figure B-4 for one of the _dotp2 "invocations" (remember, while they may look like function calls they are in reality inlined functions).

With this final modification, the number of add operations is further reduced, as _dotp2 subsumes two of the four adds present in Listing B-1. Listing B-2 contains the fully optimized dotproduct function that can be used within the isodata image segmentation implementation.

Listing B-2: A fully optimized version of dotprod.

```
unsigned long dotproduct(int lo, int hi)
{
  /* 0, 1, 2, ..., 255 */
  static const unsigned short
                    pixval[]={0,1,2, /* 3,5,...,252 */ ,253,254,255};
  unsigned long sum1 = 0, sum2 = 0, sum;
  const int N = hi-lo;
  int ii=0, jj=lo, remaining;
  double h1_h2_h3_h4, b1_b2_b3_b4;
  unsigned int h1_h2, h3_h4, b1_b2, b3_b4;
  /* unrolled dot-product loop with non-aligned double word reads */
  for (; ii<N; ii+=4, jj+=4)
  {
    h1_h2_h3_h4 = _memd8_const(&smoothed_hist[ii]);
    h1_h2 = _lo(h1_h2_h3_h4);
    h3_h4 = _hi(h1_h2_h3_h4);

    b1_b2_b3_b4 = _memd8_const(&pixval[ii]);
    b1_b2 = _lo(b1_b2_b3_b4);
    b3_b4 = _hi(b1_b2_b3_b4);

    sum1 += _dotp2(h1_h2, b1_b2); /* see Figure B-4 */
    sum2 += _dotp2(h3_h4, b3_b4);
  }
  sum = sum1 + sum2;
  /*
   * loop epilogue: if # iterations guaranteed to
   * be a multiple of 4, then this would not be required.
   */
  remaining = N - ii;
  jj = N - remaining;
  for (ii=jj; ii<N; ii++)
    sum += smoothed_hist[ii]*pixval[ii];

  return sum;
}
```

B.2 INTRINSICS AND THE UNDECIMATED WAVELET TRANSFORM

The increased memory bandwidth of the C64x and C67x DSPs allow us to apply certain packed data optimizations that processors without 64-bit support are excluded from. Section 6.2.3 features an implementation of a fixed-point multiscale edge detector using a simplified version of the "algorithme à trous", where at one point we need to compute the absolute value of the difference between two low-pass filtered versions of the input image. Image differencing is an important and common operation in image processing, and what makes this particular loop somewhat unique is that each "pixel" (they are actually wavelet coefficients) is a signed 32-bit quantity, the reason being that we need quite a few bits to achieve sufficient accuracy. The loop in question is shown in Listing B-3.

Listing B-3: The `calc_detail` function from Listing 6-12.

```
unsigned int calc_detail(int *p_approx1, int *p_approx2, int *p_detail)
{
  int *pa1 = p_approx1 + (Y_SIZE<<1) + 2,
      *pa2 = p_approx2 + (Y_SIZE<<1) + 2,
      *pd  = p_detail  + (Y_SIZE<<1) + 2;
  int irow, jcol;
  unsigned int sumlo = 0, sumhi = 0;

  for (irow=2; irow<(X_SIZE-2); ++irow) {
    for (jcol=2; jcol<(Y_SIZE-2); jcol+=2) {
      pd[0] = /* |pa2[0]-pa1[0]| >> 9 */
        _sshvr(_abs(_lo(_amemd8(pa2))-_lo(_amemd8(pa1))),9);
      sumlo += pd[0];
      pd[1] =
        _sshvr(_abs(_hi(_amemd8(pa2))-_hi(_amemd8(pa1))),9);
      sumhi += pd[1];
      pa1 += 2;
      pa2 += 2;
      pd += 2;
    }
    pa1 += 4;
    pa2 += 4;
    pd += 4;
  }
  return (sumlo+sumhi)>>LOG2_N_PIXELS;
}
```

With this loop, we are not shackled with the alignment issues that we were forced to deal with in the previous section, and consequently are free to use the LDDW intrinsic (_amemd8) to load 64 bits of data into a register pair from the appropriately aligned data buffers pa1 and pa2. Most of the interesting work occurs within the first two compound statements of the inner loop, where we are marching along the columns in the current row. The inner loop has been unrolled by a factor of two, and the double word-wide accesses load two 32-bit wavelet coefficients into the register file. A diagram that illustrates part of what is going on is shown in Figure B-5. The comments in the code should help parse how the intrinsics are being used. We use _lo and _hi to break apart the packed data, _abs to take the absolute value of the delta between two wavelet coefficients, and _sshvr to bit-shift to the right by 9 bits. Because this loop is reasonably simple, and the arrays are aligned correctly, with the correct combination of _nasserts and MUST_ITERATE pragmas, the compiler might be able to generate fully optimized assembly, although one would need to inspect the assembly listing

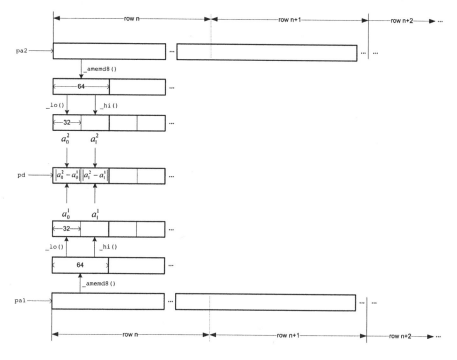

Figure B-5. Use of intrinsics in calc_detail. The buffers pointed to by pa1 (pointer to approximation image 1) and pa2 (approximation image 2) are flattened arrays and a combination of _amemd8, _lo and _hi is used to extract the wavelet coefficients from the two arrays. The absolute value of the delta between these two image matrices is then placed within the buffer pointed to by pd (pointer to detail image).

to verify this.

We then use intrinsics again in `wave_decomp_a_trous` to initialize memory a la `memclear`. The low-pass filtering that occurs in the undecimated wavelet transform uses a 5x5 convolution kernel, and the strategy used in the code for this book is to simply ignore the margins and set them to zero (see 4.1.4). While we could thoroughly ignore edge effects, this strategy would later present a problem because without inserting logic that would complicate and slow down the rest of the implementation, we would then end up including these invalid margin pixels in other downstream computations (i.e., the global mean).

Listing B-4 are the pertinent contents of the loop that constitutes the majority of `wave_deomp_a_trous`.

Listing B-5: Portions of `wave_decomp_a_trous` (see Listing 6-12 for the complete implementation).

```
for (ilevel=0; ilevel<nlevels; ++ilevel) {
    /*
     * zero out the margins of the current approximation coefficient
     * matrix - 1st and last couple of rows
     */
    p = p_approx2;
    for (kk = 0; kk < Y_SIZE; kk+=2)
        _amemd8(&p[kk]) =
        _amemd8(&p[kk+Y_SIZE]) =
        _amemd8(&p[kk+N_PIXELS-(Y_SIZE<<1)]) =
        _amemd8(&p[kk+N_PIXELS-Y_SIZE]) = 0;
    /*
     * first and last couple of columns (just the next 2 rows, the
     * remainder are done in the main convolution loop)
     */
    p += (Y_SIZE<<1);
    _amemd8(p) = _amemd8(p+Y_SIZE-2) = 0;
    p += Y_SIZE;
    _amemd8(p) = _amemd8(p+Y_SIZE-2) = 0;
    p += Y_SIZE;

    /* 5x5 convolution for approximation coeffs at level i */
    for (irow=2; irow<(X_SIZE-2); ++irow) {

        /* clearing out 1st & last two cols */
        _amemd8(p) = _amemd8(p+Y_SIZE-2) = 0;
        p += Y_SIZE;
```

```
    for (jcol=2; jcol<(Y_SIZE-2); ++jcol) {
        /* convolution occurs here (code omitted) */
    }
  }

    /* prep for next iteration (code omitted) */
  } /* end (for each decomposition level) */
```

Prior to the 5x5 convolution that occurs for each wavelet decomposition, this code sets the first, second, second-to-last, and last rows to zero. The approximation image buffers are aligned on a double-word boundary, and so we can use _amemd8 to initialize 64-bits, or two wavelet coefficients, at a time. While we could also call memclear to initialize the rows, we choose not to. As of version 2.21 of the TI compiler, certain loop optimizations − namely software pipelining − are disabled if the loop kernel calls *any* functions. Hence it makes sense to avoid any function calls in this particular loop. Continuing, once we are within the interior portion of the image we set the first two and last two columns of each row to zero − again, a perfect opportunity to use _amemd8 to cut down on the number of data accesses by a factor of two.

B.3 CONVOLUTION AND THE DWT

Sometimes a reordering of straight C code is enough to goad the TI compiler into generating more vectorized code. The following examples are indicative of the limitations that prevent the compiler from delving deep into the recesses of the programmer's mind and autonomously discern how best to optimize the code. At times the compiler needs a kick in the pants, and it is incumbent upon the programmer to do their due diligence. Sometimes this means taking a look at the generated assembly code.

In 6.3.2.1, a 2D D4 DWT and IDWT implementation on the C6416 was discussed. Performing the D4 DWT involves convolving a 4-tap filter with an input signal, which in the case of a 2D DWT is either a row or column of a matrix. The convolution essentially calls for flipping, or time-reversing, the filter coefficients and then "sliding" it across the input signal. At each convolution position, or "lag" as it is commonly referred to, the sum of vector products between the flipped filter array and a segment of the input signal is computed (see Figure 6-20). In Listing B-6, a portion of the wave_horz function from daub4.c (see Listing 6-16) is reproduced

below. This is but one of many convolution loops needed to implement the
2D D4 DWT and IDWT.

Listing B-6: Portions of `wave_horz` (see Listing 6-16 for the complete
code listing).
```
/* rest of the low-pass filter + decimation loop */
xptr = in_data + 1;
for (ii=0; ii<iters; ++ii, xptr+=2) {
  sum = Qr +
    xptr[0]*hLP[3] +
    xptr[1]*hLP[2] +
    xptr[2]*hLP[1] +
    xptr[3]*hLP[0];

    *out_data++ = sum>>15;
}
```

Listing B-7 is a portion of the generated assembly file (`daub4.asm`),
corresponding to the C code in Listing B-6, with the CCStudio (version
2.20) default release build options selected. To view generated assembly
code, use the –k option during compilation, which tells the TI C compiler to
retain all `.asm` files. To set this option within CCStudio, select
Project|Build Options, click on the "Compiler" tab, select the "Assembly"
category, and finally select the "Keep Generated .asm Files" checkbox.

Listing B-7: Assembly code for the convolution loop in Listing B-6, using
standard release build options of CCStudio (version 2.20).
```
   L5:    ; PIPED LOOP KERNEL
             LDNDW   .D1T1   *A8,A5:A4          ; |54| <0,0>
             LDNDW   .D1T1   *A3++(4),A7:A6   ; |54| <0,1>
             NOP             3
             MV      .D2X    A4,B6   ; |54| <0,1> Define a twin register

             MPYHL   .M2X    A7,B6,B6          ; |54| <0,6>
   ||        MPYHL   .M1     A5,A6,A5          ; |54| <0,6>

             MPYHL   .M1     A6,A5,A4          ; |54| <0,7>
             MPYHL   .M1     A4,A7,A4          ; |54| <0,8>
             ADD     .D1     A4,A5,A5          ; |54| <0,9>

             ADD     .D1     A4,A5,A4          ; |54| <0,10>
   ||  [B0]  BDEC    .S2     L5,B0             ; |55| <0,10>
```

```
NOP          1
ADD    .D2X     B6,A4,B6          ;  |54|  <0,12>  ^
ADD    .D2      B4,B6,B6          ;  |54|  <0,13>  ^
SHR    .S2      B6,15,B6          ;  |54|  <0,14>  ^
STH    .D2T2    B6,*B5++          ;  |54|  <0,15>  ^
```

Reading assembly is difficult on the eyes (even more so when reading parallelized C6x assembly), however the important point is that even though the compiler has generated code using LDNDW, it is still possible to improve this loop using techniques from the preceding section. In Listing B-8 is a rewritten loop utilizing intrinsics to take advantage of packed data processing via _dotp2. In order for this loop to work correctly, the hLP array has to be time-reversed in code, which then allows us to use the _dotp2 intrinsic. Rather than reverse the array at the beginning of each call to wave_horz, the code in Listing B-8 assumes that the caller passes in arrays containing flipped D4 filter coefficients.

Listing B-8: A vectorized version of the convolution loop shown in Listing B-6 that uses compiler intrinsics.

```
void wave_horz
(
  short *in_data, /* row of input pixels                         */
  const short *hLP, /* (flipped) low-pass 4-tap D4 decomposition filter */
  const short *hHP, /* (flipped) high-pass 4-tap decomposition filter   */
  short *out_data, /* row of output data                          */
  int    cols      /* length of input                             */
)
{
  int         ii, iters = (cols>>1) - 2;
  const short *xptr;
  int         sum;
  double x0_x1_x2_x3,
         hLP0_hLP1_hLP2_hLP3 = _memd8_const(hLP);
  int    hLP0_hLP1 = _lo(hLP0_hLP1_hLP2_hLP3),
         hLP2_hLP3 = _hi(hLP0_hLP1_hLP2_hLP3),
         x0_x1, x2_x3;

  /*  Convolve input with 4 tap D4 low-pass filter */

  /* (periodization omitted) */
```

```
for (ii=0; ii<iters; ++ii, xptr+=2) {
  x0_x1_x2_x3 = _memd8_const(xptr);
  x0_x1 = _lo(x0_x1_x2_x3);
  x2_x3 = _hi(x0_x1_x2_x3);
  sum = Qr +
        _dotp2(x0_x1, hLP0_hLP1) + _dotp2(x2_x3, hLP2_hLP3);
  *out_data++ = sum>>15;
}


/* rest omitted */
}
```

This loop does produce assembly code that uses DOTP2 and hence takes advantage of the C6416 instruction set. This is a perfectly valid solution, but it should be noted that due to this reordering of the filter coefficients in the hLP and hHP arrays, the compiler now has enough information to generate vectorized assembly on its own. In the chap6\daub4 directory, there is a C file daub4_optimized.c that performs the 2D D4 DWT and IDWT as described in 6.3.2.1, but rearranges the filter coefficients so that the compiler vectorizes the convolution loops in the fashion shown in Listing B-8. This can be seen in Listing B-9, which shows a portion of the wave_horz function from daub4_optimized.c file along with the generated assembly code. In contrast to the assembly code in Listing B-7, notice the presence of the DOTP2 instruction. The IMGLIB wavelet functions stipulate that the wavelet filter coefficients be flipped from how they would normally be specified, a la daub4_optimized.c, presumably for the same reason.

Listing B-9: Functionally the same convolution loop as shown in Listing B-6, but assumes the filter coefficients in hLP are time-reversed. This code comes from daub4_optimized.c, and the generated assembly from this loop kernel follows below.

```
/* rest of the low-pass filter + decimation loop */
xptr = in_data + 1;
for (ii=0; ii<iters; ++ii, xptr+=2) {
  sum = Qr +
    xptr[0]*hLP[0] +
    xptr[1]*hLP[1] +
    xptr[2]*hLP[2] +
    xptr[3]*hLP[3];
  *out_data++ = sum>>15;
}
```

```
L5:    ; PIPED LOOP KERNEL
          LDNDW    .D1T1   *A16,A5:A4           ; |54|<0,0>  ^
          LDNDW    .D1T1   *A9++(4),A7:A6       ; |54| <0,1> ^
          NOP              4
          DOTP2    .M1     A6,A4,A5             ; |54| <0,6> ^
          DOTP2    .M1     A7,A5,A4             ; |54| <0,7> ^
          NOP              1
  [ B0]   BDEC     .S2     L5,B0                ; |55| <0,9>
          NOP              1
          ADD      .D1     A4,A5,A4             ; |54| <0,11> ^
          ADD      .D1     A8,A4,A4             ; |54| <0,12> ^
          SHR      .S1     A4,15,A4             ; |54| <0,13> ^
          STH      .D1T1   A4,*A3++             ; |54| <0,14> ^
```

Something similar is done within the inverse horizontal wavelet function
invwave_horz, to the same effect. The synthesis filter arrays are *not*
flipped in daub4_optimized.c. In Listing B-10 the original "upsample
and convolution" loop from invwave_horz is shown. We can nudge the
compiler into using DOTP2 through the use of four two-element arrays
which serve as a reordering mechanism. This trick is shown in Listing B-11.
It turns out that with the flipping of the filter coefficient array for
wave_horz, and the appropriate array indexing modifications made within
wave_vert to account for this change, nothing more needs to be done to
optimize the vertical DWT. Optimization of invwave_vert is left as an
exercise for the reader.

Listing B-10: Portions of invwave_horz from daub4.c (see Listing 6-
16 for the complete code listing).

```
for (ii=0; ii<cols-4; ii+=2, ++pa, ++pd) {
  out1 = gLP[3]*pa[0]+gLP[1]*pa[1] + /* filtered approximation */
         gHP[3]*pd[0]+gHP[1]*pd[1]; /* filtered detail */
  *pout++ = out1>>15;
  out2 = gLP[2]*pa[1]+gLP[0]*pa[2] + /* filtered approximation */
         gHP[2]*pd[1]+gHP[0]*pd[2]; /* filtered detail */
  *pout++ = out2>>15;
}
```

Listing B-11: Demonstrating the use of a reordering mechanism so as to get
the compiler to use DOTP2. This code snippet is from the version of
invwave_horz in daub4_optimized.c.

```
short gLP3_gLP1[2], gLP2_gLP0[2],
```

```
        gHP3_gHP1[2], gHP2_gHP0[2];

  gLP3_gLP1[0]=gLP[3]; gLP3_gLP1[1]=gLP[1];
  gHP3_gHP1[0]=gHP[3]; gHP3_gHP1[1]=gHP[1];

  gLP2_gLP0[0]=gLP[2]; gLP2_gLP0[1]=gLP[0];
  gHP2_gHP0[0]=gHP[2]; gHP2_gHP0[1]=gHP[0];

  for (ii=0; ii<cols-4; ii+=2, ++pa, ++pd) {
    out1 = gLP3_gLP1[0]*pa[0]+gLP3_gLP1[1]*pa[1] + /* approx */
           gHP3_gHP1[0]*pd[0]+gHP3_gHP1[1]*pd[1];  /* detail */
   *pout++ = out1>>15;
    out2 = gLP2_gLP0[0]*pa[1]+gLP2_gLP0[1]*pa[2] + /* approx */
           gHP2_gHP0[0]*pd[1]+gHP2_gHP0[1]*pd[2];  /* detail */
   *pout++ = out2>>15;
  }
```

REFERENCES

1. Texas Instruments, *TMS320C64x Technical Overview* (SPRU395b).
2. Texas Instruments, *TMS320C6000 Programmer's Guide* (SPRU198g.pdf).
3. Chassaing, R., *DSP Applications Using C and the TMS320C6x DSK* (John Wiley and Sons, 2002).
4. Dahnoun, N., *Digital Signal Processing Implementation using the TMS320C000 DSP Platform* (Prentice Hall, 2000), Chapter 4, *Software Optimisation.*
5. Texas Instruments, *TMS320C6000 Optimizing Compiler User's Guide* (SPRU187k.pdf).
6. Brenner, J., Levy, M., "Code Efficiency & Compiler-Directed Feedback", *Dr. Dobb's Journal*, Dec. 2003; pp. 59-65.

Index